IONIZATION, CORRELATION, AND POLARIZATION IN ATOMIC COLLISIONS

Proceedings in the Series of

International Symposia on (e,2e), Double Photoionization, and Related Topics
and
International Symposia on Polarization and Correlation in Electronic and Atomic Collisions

Year		Held in	Publisher	ISBN
2005	13th	Buenos Aires, Argentina	AIP Conf. Proceedings Vol. 811	0-7354-0303-1
2003	12th	Königstein, Germany	AIP Conf. Proceedings Vol. 697	0-7354-0170-5
2001	11th	Rolla, Missouri, USA	AIP Conf. Proceedings Vol. 604	0-7354-0048-2

To learn more about these titles, or the AIP Conference Proceedings Series, please visit the webpage **http://proceedings.aip.org**

IONIZATION, CORRELATION, AND POLARIZATION IN ATOMIC COLLISIONS

Proceedings of the International Symposium on (e,2e), Double Photoionization, and Related Topics and the
Thirteenth International Symposium on Polarization and Correlation in Electronic and Atomic Collisions

Buenos Aires, Argentina   28 – 30 July 2005

EDITORS

Azzedine Lahmam-Bennani
Université Paris-Sud XI, France

Birgit Lohmann
Griffith University, Australia

All papers have been peer reviewed

SPONSORING ORGANIZATIONS
Consejo Nacional de Investigaciones Científicas y Técnicas
Agencia Nacional de Promoción Científica y Tecnológica
Instituto de Astronomía y Física del Espacio
Universidad de Buenos Aires
Centro Latino Americano de Física
Université Paris-Sud XI
Griffith University

Melville, New York, 2006
AIP CONFERENCE PROCEEDINGS ■ VOLUME 811

Editors

Azzedine Lahmam-Bennani
LCAM, Bâtiment 351
Université Paris-Sud XI
F-91405 Orsay cedex
France

Email: azzedine.l-bennani@lcam.u-psud.fr

Birgit Lohmann
School of Science
Griffith University
Nathan, Queensland 4111
Australia

E-mail: b.lohmann@griffith.edu.au

Authorization to photocopy items for internal or personal use, beyond the free copying permitted under the 1978 U.S. Copyright Law (see statement below), is granted by the American Institute of Physics for users registered with the Copyright Clearance Center (CCC) Transactional Reporting Service, provided that the base fee of $23.00 per copy is paid directly to CCC, 222 Rosewood Drive, Danvers, MA 01923, USA. For those organizations that have been granted a photocopy license by CCC, a separate system of payment has been arranged. The fee code for users of the Transactional Reporting Services is: 0-7354-0303-1/06/$23.00

© 2006 American Institute of Physics

Permission is granted to quote from the AIP Conference Proceedings with the customary acknowledgment of the source. Republication of an article or portions thereof (e.g., extensive excerpts, figures, tables, etc.) in original form or in translation, as well as other types of reuse (e.g., in course packs) require formal permission from AIP and may be subject to fees. As a courtesy, the author of the original proceedings article should be informed of any request for republication/reuse. Permission may be obtained online using Rightslink. Locate the article online at http://proceedings.aip.org, then simply click on the Rightslink icon/"Permission for Reuse" link found in the article abstract. You may also address requests to: AIP Office of Rights and Permissions, Suite 1NO1, 2 Huntington Quadrangle, Melville, NY 11747-4502, USA; Fax: 516-576-2450; Tel.: 516-576-2268; E-mail: rights@aip.org.

L.C. Catalog Card No. 2005937661
ISBN 0-7354-0303-1
ISSN 0094-243X

Printed in the United States of America

Contents

Preface ix
International Advisory Committees xi
Group Photo xiii

The Use of Correlated Wave Functions in Describing the Double
Ionization of Helium 1
    L. U. Ancarani, C. Dal Cappello, and T. Montagnese

Initial-state Correlation Effects in Low-energy Proton Impact
Ionization 7
    M. Foster, J. L. Peacher, A. Hasan, M. Schulz, and D. H. Madison

Fragmentation of Helium in Collisions With Slow Electrons: Three-
and Four-body Dynamics 12
    M. Dürr, B. Najjari, C. Dimopoulou, A. Dorn, and J. Ullrich

Nondipole Effects in Double Photoionization of He 18
    A. Y. Istomin, N. L. Manakov, A. V. Meremianin, and A. F. Starace

Measurements of Low-energy Electron Scattering from Atomic
Hydrogen 24
    J. G. Childers and M. A. Khakoo

Kinetic Correlation in Photo-Double-Ionization Processes: The
He-Isoelectronic Sequence 30
    S. Otranto and C. R. Garibotti

Theory of Electron Impact Ionization of Atoms 36
    A. T. Stelbovics, P. L. Bartlett, I. Bray, D. V. Fursa, and A. S. Kadyrov

Interference Effects in Single Ionization of H_2 Molecules by Fast
Electron Impact 42
    O. A. Fojón, C. R. Stia, and R. D. Rivarola

Double Photoionization of He and H_2 48
    J. Colgan, M. S. Pindzola, and F. Robicheaux

Photoionization of Xenon Clusters 54
    J. D. Bozek, B. S. Rude, and A. L. D. Kilcoyne

Ionization of Atoms with Spin Polarized Electrons 60
    J. Lower, S. Bellm, R. Panajotovic, E. Weigold, A. Prideaux,
    D. H. Madison, Z. Stegen, C. T. Whelan, and B. Lohmann

Study of Coherence in Electron-impact Excitation of Mercury ... 66
    F. Jüttemann, G. Außendorf, and G. F. Hanne

Theoretical Treatment of Electron-impact Ionization of Molecules ... 72
    J. Gao, J. L. Peacher, and D. H. Madison

New Results on Excitation, Ionization, and Ionization-Excitation by
Electron Impact 78
    A. Crowe

Calculation of Excitation-autoionization of Helium with Three Active
Electrons: Sharp Resonance Features in the SDCS ... 84
    C. W. McCurdy, D. A. Horner, and T. N. Rescigno

Interference of Exchange and Processes in the Case of Double
Ionisation of Argon..90
    F. Catoire, C. Dal Cappello, A. Lahmam-Bennani, and A. Duguet

(e,2e) and (e,3-1e) Studies on Double Processes of He Near the Bethe
Ridge..96
    N. Watanabe, Y. Khajuria, M. Takahashi, Y. Udagawa, P. S. Vinitsky,
    Y. V. Popov, O. Chuluunbaatar, and K. A. Kouzakov

(e,3-1e) Reactions at Large Momentum Transfer: The Plane-Wave
Second Born Approximation..102
    P. S. Vinitsky, Y. V. Popov, K. A. Kouzakov, N. Watanabe, and
    M. Takahashi

Angular Momentum Partitioning in the Dissociation of Diatomic
Molecules...108
    T. J. Gay, J. D. Bozek, J. E. Furst, G. A. Gallup, A. S. Green,
    A. L. D. Kilcoyne, J. R. Machacek, J. W. Maseberg, K. W. McLaughlin,
    and M. A. Rosenberry

Spin-Resolved Collisions of Electrons with Rubidium Atoms: A
Search for Relativistic Effects..114
    W. E. Guinea, G. F. Hanne, M. R. Went, M. L. Daniell, M. A. Stevenson,
    W. R. MacGillivray, and B. Lohmann

Multi-Auger Decay in Negative Ion Photodetachment.........................120
    R. C. Bilodeau, J. D. Bozek, G. D. Ackerman, N. D. Gibson, C. W. Walter,
    A. Aguilar, G. Turri, I. Dumitriu, and N. Berrah

Multiple Direct and Sequential Auger Effect in the Rare Gases.............126
    F. Penent, P. Lablanquie, J. Palaudoux, L. Andric, T. Aoto, K. Ito,
    Y. Hikosaka, R. Feifel, and J. H. D. Eland

First Principles Calculations of the Double Photoionization of Atoms
and Molecules Using B-Splines and Exterior Complex Scaling................132
    F. Martín, D. A. Horner, W. Vanroose, T. N. Rescigno, and C. W. McCurdy

Auger Electron-Photoelectron Coincidence Experiments in Ar................138
    P. Bolognesi, M. Coreno, A. De Fanis, V. Feyer, S. Turchini, N. Zema,
    T. Prosperi, and L. Avaldi

Analysis of the Resonant Auger Decay During Ultrafast
Fragmentation of CH_3F..144
    G. Prümper, V. Carravetta, Y. Muramatsu, X. J. Liu, K. Ueda, Y. Tamenori,
    M. Kitajima, M. Hoshino, and H. Tanaka

Correlation Effects in Electron-Diatomic Molecule Inelastic Collisions....150
    B. Joulakian, V. Serov, and N. Lahmidi

(γ, 2γ) Studies on Multiply Excited States Of H_2 and N_2 in the
Vacuum Ultraviolet Range..156
    T. Odagiri

Fragmentation Mechanism of Highly Excited C_{70} Cations in the
Extreme Ultraviolet...161
    K. Mitsuke, H. Katayanagi, J. Kou, T. Mori, and Y. Kubozono

Direct Measurement of Spectral Momentum Densities of Ordered and
Disordered Semiconductors by High Energy EMS..............................167
    C. Bowles, M. R. Went, A. S. Kheifets, and M. Vos

Towards Electron Momentum Spectroscopy Studies of Clusters:
A New Apparatus 173
    K. L. Nixon, G. Hewitt, B. Gilbert, A. Dunn, R. Northeast, M. Ellis,
    D. S. Slaughter, P. Euripides, W. D. Lawrance, and M. J. Brunger

Super-Elastic Scattering from Ca and Rb in a Magnetic Angle
Changing Spectrometer 179
    A. Murray, M. Hussey, W. MacGillivray, and G. King

Spin Up-Down Asymmetry in the Excitation of Kr $5p'$ $[3/2]_2$ by
Polarized Electrons 185
    D. H. Yu, D. Cvejanović, J. F. Williams, L. Pravica, R. Srivastava,
    A. Stauffer, P. A. Hayes, and S. Napier

Universal Scaling of Resonances in Vector Correlation
Photoionization Parameters 191
    A. N. Grum-Grzhimailo and M. Meyer

Correlation Spectroscopy of Condensed Matter Systems 197
    F. O. Schumann, J. Kirschner, K. A. Kouzakov, and J. Berakdar

A Hybrid DWBA−R-Matrix Approach for Charged-Particle Impact
Ionization of Atoms 203
    K. Bartschat

Conference Programs 209
Posters 215
List of Participants 217
Author Index 225

PREFACE

This volume contains papers presented at two atomic physics symposia, the *International Symposium on (e,2e), Double Photoionization and Related Topics* and the *13th International Symposium on Polarization and Correlation in Electronic and Atomic Collisions*, held jointly in Buenos Aires, Argentina from 28-30 July 2005. Both conferences were satellite meetings of the *XXIV International Conference on Photonic, Electronic and Atomic Collisions*, which was held in Rosario, Argentina from 20-26 July 2005. The manuscripts in this volume were peer-reviewed by selected delegates attending the two symposia.

The scientific program of these symposia emphasises current hot topics in the field. These include electron-electron correlation effects in excitation and in single and multiple ionization with electrons, ions and photons, as well as orientation, alignment and polarization effects in a range of atomic physics processes. Over 100 scientists from 17 countries around the world, including a large number of graduate students, attended the meetings in Buenos Aires to discuss recent progress in these areas.

Thirty stimulating talks were presented in each of the symposia. The program included a further five joint oral sessions, in which twelve talks of general interest to this community were presented. A joint poster session extended over two full days of the meeting, providing further opportunities for in-depth discussions among the participants.

The Editors would like to highlight the excellent local organisation of the symposia by Jorge Miraglia, Dario Mitnik and Claudia Montanari, the Local Organising Committee. The support of Consejo Nacional de Investigaciones Cientificas y Tecnicas, Agencia National de Promocion Cientifica y Tecnologica, Instituto de Astronomia y Fisica del Espacio, Universidad de Buenos Aires and Centro Latino Americano de Fisica is gratefully acknowledged. It was particularly pleasing to see that this support enabled a significant attendance by Argentinian delegates. Finally, the Editors would like to thank all the participants for their contributions, which resulted in two vibrant and thought-provoking meetings.

A. Lahmam-Bennani
B. Lohmann

INTERNATIONAL ADVISORY COMMITTEES

International Symposium on (e,2e), Double Photoionization and Related Topics

Lorenzo AVALDI (Italy)
Jamal BERAKDAR (Germany)
Claude DAL CAPPELLO (France)
Reinhard DÖRNER (Germany)
Nikolaï KABACHNIK (Russia)
George KING (UK)
Julian LOWER (Australia)
Don MADISON (USA)
Laurence MALEGAT (France)
William McCURDY (USA)

Andrew MURRAY (UK)
Bernard PIRAUX (Belgium)
Tim REDDISH (Canada)
Roberto RIVAROLA (Argentina)
Anthony STARACE (USA)
Giovanni STEFANI (Italy)
Masahiko TAKAHASHI (Japan)
Joachim ULLRICH (Germany)
Erich WEIGOLD (Australia)

13th International Symposium on Polarization and Correlation in Electronic and Atomic Collisions

Nils ANDERSEN (Denmark)
Klaus BARTSCHAT (USA)
Uwe BECKER (Germany)
Nora BERRAH (USA)
Igor BRAY (Australia)
Nikolay CHEREPKOV (Russia)
Albert CROWE (UK)
Danièle DOWEK (France)
Oscar FOJON (Argentina)
Tim GAY (USA)

Alexei GRUM-GRZHIMAILO (Russia)
Friedrich HANNE (Germany)
Rajesh SRIVASTAVA (India)
Al STAUFFER (Canada)
Peter TEUBNER (Australia)
Jim WILLIAMS (Australia)
Akira YAGISHITA (Japan)
Peter ZETNER (Canada)

The use of correlated wave functions in describing the double ionization of helium

L.U.Ancarani, C. Dal Cappello and T. Montagnese

L.P.M.C., Institut de Physique, Université de Metz, Technopôle 2000, 57078 Metz, France

Abstract. We study theoretically the (e,3e) process on helium at high incident energy (∼5.6 keV). We present model calculations performed with several correlated initial and final states in an attempt to further interpret the absolute coplanar measurements of Lahmam-Bennani et al. [1].

Keywords: double ionization; helium; correlation
PACS: 34.80.Dp

INTRODUCTION

Kinematically complete double ionization experiments by electron impact provide us with a very powerful tool to investigate the role of electron-electron correlation. The so-called (e,3e) experiments in which the three electrons (the two ejected and the scattered) are detected in coincidence allow one to check with a good accuracy the different theories (see the review paper [2]) and to test the quality of all wave functions involved.

In this contribution we compare calculated five-fold differential cross section (FDCS) for (e,3e) processes on helium with the absolute, coplanar, measurements of Lahmam-Bennani et al. [1]. These were performed at an incident energy of ∼5.6 keV, under a small projectile scattering angle of 0.45^0 (momentum transfer of 0.24 a.u.; dipolar regime), and the two ejected electrons detected with equal energy (10 eV). For these kinematics (high impact and scattered energies), the first Born approximation (FBA) - with respect to the interaction of the fast projectile with the target - should apply [3, 4]: consequently, the incoming and scattered electrons are described by plane waves [2]. The calculations presented here were done with several correlated initial and final wave functions.

Previous theoretical investigations have yielded contrasting results. The Green function expansion [5] with Slater-type helium wave function gives reasonably good agreement (shape and magnitude) in some but not all the experimental situations. The Convergent Close Coupling (CCC) model with a good Hylleraas-type helium wave function [6] yields agreement with the shapes of the FDCS but it is not able to reproduce the magnitude (factor of 3 too low). Calculations within the 3C model [7], and with hydrogenic and several Hylleraas-type wave functions, yield cross sections similar in shape and magnitude, but do not reproduce the experimental magnitude [3, 8]. This is true in both the first and second Born approximations [4]. Jones and Madison [3] made also a calculation with the 6C model, and found practically no difference with the 3C results. The same applies to the C4FS calculations of [1].

In the next two sections we shall discuss, separately, the role played by the initial Ψ_i and final Ψ_f^- states in the description of the cross sections for double ionization of helium, and try to see whether it is possible to state which one is responsible for the agreement/disagreement between experiments and theory.

In what follows, $Z = 2$ is the nucleus charge, r_1, r_2 are the distances between one of the helium electrons and its nucleus, $r_{12} = |\vec{r}_1 - \vec{r}_2|$ is the distance between the two bound electrons, and the momenta of the scattered, the first and second ejected electrons are denoted, respectively, by k_s, k_a and k_b.

INITIAL STATE

The non-relativistic Schrödinger equation (SE) for the $^1S_0$ ground state of helium reduces to the so-called Hylleraas equation (HE) for which there is no analytical solution today. On the other hand, many numerical wave functions have been proposed and optimized by application of the Rayleigh-Ritz variational method in order to get the minimal mean energy. Though obtaining a good mean energy is certainly important, it is not a sufficient criterion to state that a trial wave function is good [9].

To mathematically solve the HE, the Coulomb singularities (i.e. $r_1 = 0$, $r_2 = 0$ or $r_{12} = 0$) which appear in the potential terms of the hamiltonian must, somehow, be compensated by the kinetic terms. The study of the singularities has led Kato [10] to establish mathematical conditions that the bound state wave function must satisfy: the so-called "cusp conditions". However, it is known that imposing these conditions on a trial wave function has little or no influence on the calculation of the mean energy, and – seemingly – that it has little or no influence on calculated (e,3e) cross sections.

Various collision experiments involving helium allow for a direct or indirect test of the ground state wave function. The theoretical study of such collisions, necessitates the knowledge of a good wave function and not only good energy levels (the latter are enough for spectroscopical purposes). It is important to underline that no probing helium wave function (generally of the Hartree-Fock or Hylleraas type) – used up to now in calculations of, for example, single and double ionization cross sections – has the correct behavior both at small and large distance from the nucleus. The fact is that one has to compromise between building a wave function as close as possible to the formal solution on one hand, and practical (relatively simple) for numerical calculations on the other. Indeed, collision calculations involve multi-dimensional integrals which can be numerically quite demanding. One may question whether it is important to describe the wave function exactly close to the singularities and in the asymptotic regions. Indeed, these domains contribute little to the mean energy because of the small integration element close to the singularities and of the decreasing nature of the wave function at large distances. In other words, is it necessary to force a trial wave function to respect the cusp conditions or to have the correct asymptotic behavior? While this action does not necessarily improve the mean energy, it certainly makes the wave function closer to the formal solution.

Here, we investigate the role played by the ground state wave function by comparing calculated FBA (e,3e) cross sections to the absolute measurements of [1]. To do so, we fix the final state in the 3C model [7]. In [8] we have shown that calculations performed with several Hylleraas-type initial helium wave functions yield FDCSs similar in shape and magnitudes, the curves being bunched together. The magnitude is about a factor 1.7-2 larger and thus in disagreement with experimental data. As noted in [3], this is quite different from the factor 10 found in [1] with a three-parameter Hylleraas wave function.

Jones and Madison [3] have calculated 3C (e,3e) cross sections using the helium wave function introduced by Pluvinage [11] and have found overall good agreement (in 16 out of the 20 geometrical situations) with the experimental data of [1]. The authors conclude their paper stating that this agreement is related to the quality of Pluvinage wave function, in particular because it diagonalizes the Hamiltonian in all three Coulomb singularities and because the model deals with the initial and final states in a consistent way (Pluvinage wave function is the doubly bound analog of the 3C wave function). In [8] we have analyzed some properties of this and other initial state wave functions to check whether this is true. We have shown that: (i) the same kind of agreement can be obtained with another trial wave function which does not diagonalize the Hamiltonian; moreover, other wave functions which diagonalize the Hamiltonian do not reproduce the correct experimental cross section magnitude. Diagonalizing the Hamiltonian is therefore not the deciding factor in obtaining the agreement obtained by [3]; (ii) there is no reason to believe that the Pluvinage wave function is better than a good Hylleraas wave function. The agreement found by [3] has to be considered as a consequence of a lucky compensation of initial and final state deficiencies.

In [8] we have studied the HE in the case when the, non-singular, mixed partial derivatives term of the Hamiltonian is artificially ignored. This corresponds to replacing the 3-body system with three 2-body systems. The HE is then fully separable and the solution can be written as the product of three Coulomb-type functions (bound and continuum) and is analogous to the 3C final state (note that these separable solutions can never be exact solution of the 3-body problem since the full hamiltonian is not separable in the variables (r_1, r_2, r_{12})). Each of the two-body Coulomb interactions for the bound state is treated exactly and hence the Kato cusp conditions are automatically satisfied (making the Hamiltonian diagonal is one way of removing the Coulomb singularities). The best of such separable solutions is the Pluvinage wave function [11] given by

$$\Psi_i^{PLU} = \Psi_{11k} = 1s(r_1)1s(r_2)C_k(r_{12}) \\ = e^{-Zr_1}e^{-Zr_2}e^{-ikr_{12}}{}_1F_1(1-i\tfrac{1}{2k}, 2, 2ikr_{12}) \quad (1)$$

where $k = 0.41$ and $E = -2.878$ a.u.

In a subsequent paper Jones et al. [12] have attempted to improve this wave function by adding to it a symmetrized $1s2s$ term (see eq. (2) below). This new trial wave function includes more radial correlation, yields an improved mean energy ($E = -2.893$ a.u.), and improved ratios of single to total ionization by photoabsorption and for Compton scattering. However, it leads to strong disagreement with (e,3e) measurements. Their investigation led to the conclusion that Hylleraas wave functions are superior to that of Pluvinage, and they attribute the good agreement obtained with the Pluvinage wave function to the fact that the initial and final states are treated in a consistent manner. However, we have shown [8] that this is neither a sufficient nor a necessary criterion to achieve good agreement with the absolute (e,3e) measurements of [1].

Puzzled by these new results, we have pushed further the investigation by including more L=0 terms $(nl, n'l')$ in a Configuration Interaction approach, by multiplying each term by a different correlation function $C_k(r_{12})$ [13]

$$\begin{aligned}\Psi_i &= \sum_{n_i n_j k_l} a_{ijl} \Psi_{n_i n_j k_l} \\ &= 1s(r_1)1s(r_2)C_{k_1}(r_{12}) \\ &+ A[1s(r_1)2s(r_2) + 1s(r_2)2s(r_1)]C_{k_2}(r_{12}) \\ &+ B[1s(r_1)3s(r_2) + 1s(r_2)3s(r_1)]C_{k_3}(r_{12}) + \ldots \end{aligned} \quad (2)$$

Each term is individually diagonal, so that the trial wave functions automatically satisfy the Kato cusp conditions. The improved Pluvinage-type wave function of [12] corresponds to taking only the first two terms and $k_1 = k_2$ in eq. (2).

We have calculated, within the 3C/FBA, the FDCS with this kind of diagonal wave functions. In the Figure (panel A), only the FDCSs for the following two

$$\begin{aligned}\Psi_i^{III} &= \Psi_{11k_1} + A[\Psi_{12k_2} + \Psi_{21k_2}] & E = -2.894 \text{ a.u.} \\ \Psi_i^{VI} &= \Psi_{11k_1} + A[\Psi_{12k_2} + \Psi_{21k_2}] + B[\Psi_{13k_3} + \Psi_{31k_3}] & E = -2.897 \text{ a.u.} \end{aligned} \quad (3)$$

are compared to the Pluvinage result. For illustration purposes, only one of the 20 geometrical situations ($\theta_a = 319^0$; direction of the momentum transfer) is considered, and the FDCS is plotted as a function of the angle of the other ejected electron θ_b. It is clear from the Figure that, adding a $1s2s$ and/or $1s3s$ term leads to strong disagreement with the experimental data. We do not believe that this discrepancy is due to treating the initial and final state in a different manner [12], but rather that the FDCS is very sensitive to the description of the helium bound state at intermediate/large distances. Adding the $2s$ and $3s$ terms in the way described above leads to large increases of the cross sections. As more terms are included, one gets better mean energy, but the progression is very slow [13]. A similar convergence should be seen in the calculated FDCS.

FINAL STATE

Let us now turn to the role played by the final state $\Psi_f^-(\vec{r}_1, \vec{r}_2, \vec{k}_a, \vec{k}_b)$. In the 3C model the double continuum is described by the symmetrized product of three coulombic distortion factors (one for each 2-body Coulomb interaction) [7]. This asymptotically exact double continuum takes into account each two-body interaction (between each electron and the nucleus, and the electron-electron repulsion): it is characterized by three charges $Z_a = Z_b = -2$ and $Z_{ab} = 1$.

The 4 body-system of the final state is described by 6 Coulomb interactions, and the 6C model [3] would be more appropriate: it corresponds to a product of six Coulomb waves and takes into account the pairwise final state interactions between all four particles (the scattered electron, the two ejected electrons and the nucleus). Note that, this model implicitly includes some contributions from all higher terms of the Born series. One way to take into account the presence of the other particles is to subsume the 6 interactions to 3 interactions

$$-\frac{Z}{r_s} - \frac{Z}{r_a} - \frac{Z}{r_b} + \frac{1}{|\vec{r}_s - \vec{r}_a|} + \frac{1}{|\vec{r}_s - \vec{r}_b|} + \frac{1}{|\vec{r}_b - \vec{r}_a|} = \frac{Z_a}{r_a} + \frac{Z_b}{r_b} + \frac{Z_{ab}}{|\vec{r}_b - \vec{r}_a|} \quad (4)$$

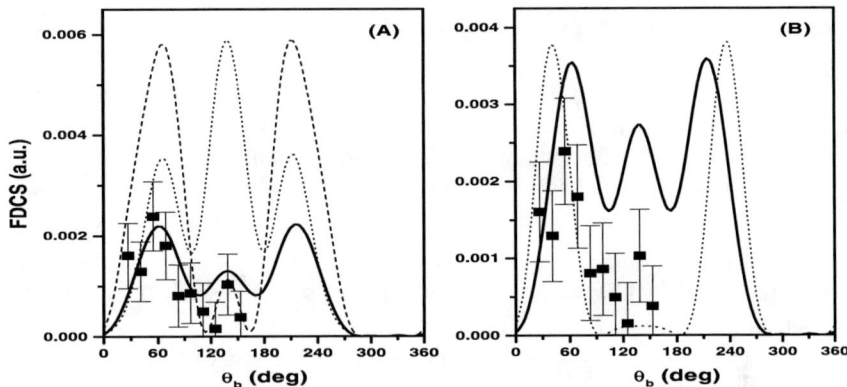

FIGURE 1. FDCS (a.u.) for (e,3e) ionization of the helium ground state, as a function of the angle of one of the ejected electrons θ_b, and the other ejected electron is detected at $\theta_a = 319^0$. The absolute experimental data [1]: full squares. On panel A, the FDCSs are obtained with the 3C model but with different initial wave functions for the helium ground state: Pluvinage wave function Ψ_i^{PLU} (solid line), Ψ_i^{III} (dotted line), Ψ_i^{VI} (dashed line). On panel B, the FDCSs are obtained with the same initial helium wave function (Bonham and Kohl - number 14 [17]) but with different final states: 3C model (solid line), C4FS model (dotted line) with the effective charges given by eq. (5).

where Z_a, Z_b and Z_{ab} are effective charges which depend on all relative positions. The 3C model with effective charges is called C4FS [2]. While there is an infinite number of solutions to eq. (4), the choice is guided by the requirement that the 3-body wave function should match some known solution of the Schrödinger equation in some limiting cases [14]. Moreover, since position dependent charges are difficult to treat, one makes the transformation $r \sim k$ to obtain momentum dependent effective charges.

The simplest solution to (4) is given by the 3C model: $Z_a = Z_b = -2$ and $Z_{ab} = 1$. Another choice is given in [1]; however, in the experimental kinematics under scrutiny here, the scattered electron is fast ($k_s \gg k_a, k_b$) so that its interaction with the nucleus and the ejected electrons can be neglected, and one then essentially recovers the 3C model.

We have considered Berakdar effective charges [14] (used with success in symmetric and asymmetric (e,2e) DS3C calculations [15, 16]) and applied them to (e,3e) calculations. Since in the experiments we are considering, $E_a = E_b$, we have simply

$$Z_a = Z_b = -(Z - \frac{1}{4}\sin\theta)$$
$$Z_{ab} = 1 - \sin^2\theta \tag{5}$$

where $\cos(2\theta) = \hat{k}_a \cdot \hat{k}_b$. Let us take for the initial state a good Hylleraas-type wave function: number 14 of Bonham and Kohl ($E = -2.903$ a.u.) [17]. On panel B of the Figure, we illustrate the results obtained with the 3C model and with the set of effective charges given by eq. (5). One clearly sees that both the shape and the magnitude are strongly affected when using set of charges (5). We are presently looking for a more appropriate set of charges which yield a less dramatic modification.

CONCLUDING REMARKS

We have analyzed theoretically the absolute (e,3e) differential cross sections at high impact energy [1]. The calculated cross sections are very sensitive to the choice of both the helium initial bound state and the final double continuum.

For fixed final state (3C model) we have compared the FDCSs obtained with several correlated initial state helium wave functions :

- the overall good agreement obtained with the Pluvinage wave function can be equally obtained with a non-diagonal wave function [8]. Somehow, the deficiency of the initial and final states compensate each other.
- more elaborate diagonal wave functions are worse because they describe badly the intermediate/large distance behavior.
- good helium wave functions reproduce the FDCS shapes but with a magnitude 1.7-2 too high [3, 4, 8].

For a good initial state (number 14 of Bonham and Kohl [17]) we have compared the FDCSs obtained with several correlated final states (3C and C4FS models):

- the 3C model does not work (factor 1.7-2 too high). Note that the C4FS model with Lahmam-Bennani et al [1] effective charges give practically the same results.
- the use of the effective charges given by eq. (5) [15] has a very strong effect on both the shape and the magnitude of the FDCS.

Since the 3C model does not work at ejected energies of 10 eV, it would be interesting to test its validity by comparing the FDCS to new absolute measurements at the same high incident energy but at slightly higher ejected energies, for example 25 eV.

REFERENCES

1. A. Lahmam-Bennani *et al.*, *Phys. Rev. A*, **59**, 3548 (1999).
2. J. Berakdar, A. Lahmam-Bennani and C. Dal Cappello, *Phys. Rep.*, **374**, 91 (2003).
3. S. Jones and D. H. Madison, *Phys. Rev. Lett.*, **91**, 73201 (2003).
4. L.U. Ancarani, T. Montagnese and C. Dal Cappello, in *Electron and Photon Impact Ionization and Related Topics*, edited by B. Piraux, IOP Conference Series 183, Bristol and Philadelphia, 2005, p21.
5. J. Berakdar, *Phys. Rev. Lett.*, **85**, 4036 (2000).
6. A.S. Kheifets *et al.*, *J. Phys. B*, **32**, 5047 (1999).
7. M. Brauner, J. Briggs and H. Klar, *J. Phys. B*, **22**, 2265 (1989); B. Joulakian, C. Dal Cappello and M. Brauner, *J. Phys. B*, **25**, 2863 (1992).
8. L.U. Ancarani, T. Montagnese and C. Dal Cappello, *Phys. Rev. A*, **70**, 12711 (2004).
9. L.U. Ancarani and Yu. V. Popov, in *Correlations, polarization, and ionization in atomic systems*, edited by D.H. Madison and M. Schulz, AIP Conference Proceedings, New York, 2002, p115.
10. T. Kato, *Comm. Pure Appl. Math.*, **10**, 151 (1957).
11. P. Pluvinage, *Ann. Physique*, **5**, 145 (1950); *J. Phys. Radium*, **12**, 789 (1951).
12. S. Jones, J.H. Macek and D.H. Madison, *Phys. Rev. A*, **70**, 12712 (2004)
13. L.U. Ancarani *et al.*, *in preparation*.
14. J. Berakdar, *Phys. Rev. A*, **53** 2314 (1996).
15. J. Berakdar and J.S. Briggs, *J. Phys. B*, **27** 4271-80 (1994).
16. S. Zhang, *J. Phys. B*, **33** 3545-53 (2000).
17. R. A. Bonham and D. A. Kohl, *J. Chem. Phys.*, **45**, 2471 (1966).

INITIAL-STATE CORRELATION EFFECTS IN LOW-ENERGY PROTON IMPACT IONIZATION

M. Foster, J. L. Peacher, A. Hasan, M. Schulz, and D. H. Madison

Laboratory for Atomic, Molecular and Optical Research, Physics Department, University of Missouri-Rolla, Rolla, Missouri 65409-0640

Abstract. In this paper, we will report on fully differential cross sections (FDCS) for single ionization of helium by 75 keV proton impact for fixed ejected electron energies and different momentum transfers. These measurements show major discrepancies in the absolute magnitude between experiment and the theoretical, 3DW (three-distorted-wave) model. The 3DW model treats the collision as a three-body process (projectile, ion, ejected electron), and for the scattering plane it has accurately predicted the FDCS for higher energy C^{6+} impact ionization of helium. The lack of agreement between the 3DW model and experiment for low energy collisions suggests that a three-body model may not be appropriate for lower collision energies. We will present a four-body model that includes full initial-state correlation.

Keywords: ground states; wave functions; ion-atom collisions, ionization, helium neutral atoms
PACS: 34.10.+x, 34.85.+x, 03.65.Nk, 34.50.Fa

INTRODUCTION

Recent fully differential cross section (FDCS) measurements, using the COLTRIMS technique, have been reported for kinematical conditions previously unstudied for low energy (75 keV) proton impact ionization of helium [1-2]. Initially, it was thought that at large projectile energies theoretical models like the three-distorted-wave (3DW) model or even the less sophisticated first-Born-approximation-Hartree-Fock (FBA-HF) should produce an accurate FDCS for single ionization of helium by proton impact in the scattering plane. The FBA-HF approximation varies from the standard FBA model in the choice of the final state wavefunction for the ejected electron. The FBA-HF approximation uses an ejected electron wavefunction that is calculated as an eigenfunction of the Schrödinger equation using a Hartree-Fock potential for the ion. Thus, the effective charge seen by the ejected electron varies from two close to the nucleus to unity asymptotically. The FBA-HF model treats the projectile as a plane wave in both the initial and final state. The use of the Hartree-Fock wavefunction for the ionized electron has been shown to provide much better agreement with the magnitude for the absolute FDCS for 100 MeV/u C^{6+} ionization of helium [3]. Both the FBA-HF and the 3DW models employ Hartree-Fock initial state wavefunctions for the helium atom. For the cases of 100 MeV/u and 2 MeV/u C^{6+} ionization of helium, both the FBA-HF and 3DW models were able to reproduce the overall magnitude of the experimental data accurately. The 3DW approach is an improved fully quantum mechanical version of the standard CDW (continuum-distorted-wave) approximation [4-8] that has been used successfully for decades to study single and double differential cross sections for heavy ion collisions.

A rough measure of the accuracy of using the first term in a perturbation theory expansion is the ratio between the projectile's charge, Z_p, to the incoming projectile velocity, v_a. The charge-to-velocity ratios for the 100 MeV/u and 2 MeV/u C^{6+} are 0.1 and 0.7, respectively. For 75 keV protons, the charge-to-velocity ratio is 0.6. By this measure, one would expect that the 3DW models should yield satisfactory results for 75 keV protons similar to the 2 MeV/u C^{6+} results.

Figure 1 compares the absolute experimental data of Hasan et al. [2] and Maydanyuk et al. [1] with the 3DW and FBA-HF calculations. In figure 1, the ejected electron

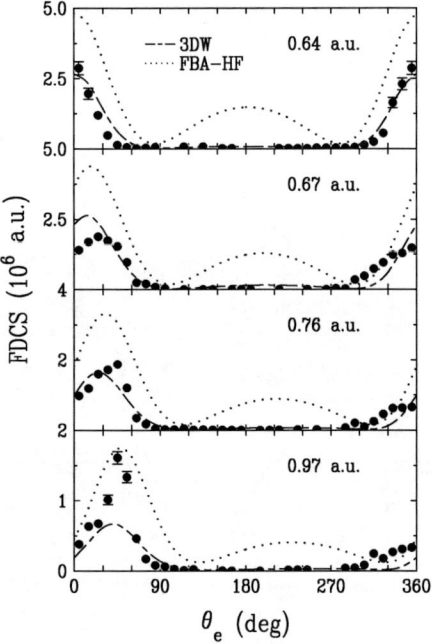

FIGURE 1. Fully differential cross sections for 75 keV p^+ impact ionization of helium in the scattering plane. All of the experimental data are absolute values in the centre of mass frame. The ejected electron energy E_e is 5.5 eV and the magnitude of the momentum transfer, |q|, is indicated in each part of the figure. The emission angle θ_e of the ejected electron in the scattering plane is measured clockwise from the beam direction. The solid circles are the absolute measurements and the theoretical curves: dotted line FBA-HF, and long dash short dashed line 3DW model multiplied by a factor of 0.25 line.

is emitted into the scattering plane (i.e. the plane that contains the initial and final projectile momentum, k_i and k_f) with an energy E_e of 5.5 eV and momentum transfers of 0.64 a.u., 0.67 a.u., 0.76 a.u., and 0.97 a.u. respectively. The peak between 0° and 90° is the binary peak and the peak between 180° and 270° is the recoil peak. While the qualitative agreement between the 3DW and the experiment for the shape of the

FDCS is good, the absolute magnitudes are in very poor agreement (factor of 4 greater than experiment). The FBA-HF model is actually in better agreement with the magnitude of the measurements than the 3DW results (factor of 1.6 greater than experiment). On the other hand, it is also seen that the experimental and 3DW results only have a binary peak while the FBA-HF approximation predicts both a binary and recoil peak. Thus, both models do not provide an adequate description of the data! We have therefore developed a new model which accounts for the four-body dynamics, i.e. the passive electron is treated as a separate particle.

ANALYSIS

The T-Matrix for single ionization of the helium atom is given by

$$T_{fi} = \langle \chi_f^-(r_1, r_2, r_3) | V_i | \psi_i(r_1, r_2, r_3) \rangle \tag{1}$$

Here V_i is the initial channel interaction potential between the projectile and helium atom,

$$V_i = 2/r_1 - 1/r_{12} - 1/r_{13} \tag{2}$$

The initial-state wavefunction ψ_i is a product of a plane wave for the projectile and a correlated initial-state wavefunction for the helium atom. Thus

$$\psi_i = (2\pi)^{-3/2} \exp(i\mathbf{k}_i \cdot \mathbf{r}_1) \phi(\mathbf{r}_2, \mathbf{r}_3) \tag{3}$$

where $\phi(\mathbf{r}_2, \mathbf{r}_3)$ is the correlated ground state wavefunction for the helium atom (correlation refers to the electron-electron interaction). Calculations have been performed using three types of correlated initial-state wavefunctions: a 20-parameter Hylleraas wavefunction [9], the Le Sech wavefunction [10] and the Pluvinage wavefunction [11]. The 20-parameter Hylleraas wavefunction is considered the benchmark wavefunction for the helium atom because of the precision to which the ground-state energy of helium can be calculated (equal to the exact ground-state energy to six significant digits - see Hart and Herzberg [12] for the specific values of the parameters). However, the Hylleraas wavefunction does not satisfy the Kato cusp condition [13]. In order for the cusp condition to be met, the local energy must be a constant as $r_{23} \to 0$. For the Hylleraas wavefunction, the local energy is infinite as $r_{23} \to 0$. The second correlated initial-state wavefunction tested was the Le Sech [10] wavefunction. The Le Sech wavefunction is a three parameter analytic wavefunction that does meet the cusp conditions requirements and yields the helium ground-state energy to within three significant digits. The final correlated initial-state wavefunction was the Pluvinage wavefunction [11]. The Pluvinage wavefunction is also satisfies the Kato cusp conditions, but is the simplest wavefunction, and as a result, the ground state energy of helium is not as accurate as the previous two wavefunctions (~1% off the exact value). One of the attractive features of the Pluvinage wavefunction is the bound state equivalent to 3DW final state wavefunction. There is growing evidence that this is important treatment of the T-Matrix [13]. In a previous study of double ionization of helium, the Pluvinage wavefunction in conjunction with a final-state equivalent wavefunction yielded better agreement with experiment than calculations using a more accurate Hylleraas wavefunction [13-14].

The final state wavefunction that we have used for this study is a plane wave for the projectile, a Hartree-Fock distorted wave for the ejected electron and a bound state for the passive electron.

$$\chi^-_{final} = (2\pi)^{-3} \exp(i\mathbf{k}_f \cdot \mathbf{r}_1) \, \phi^-_e(\mathbf{k}_2,\mathbf{r}_2) \, \psi_{1s}(r_3) \qquad (4)$$

Here $\psi_{1s}(r_3)$ is the bound state wavefunction for the passive electron which is modeled as a hydrogenic wave function with the full nuclear charge of two. The Hartree-Fock distorted wave [15] ϕ^-_e for the ejected-electron-helium-ion subsystem is a numerical solution of the Schrödinger equation

$$\left(-\frac{1}{2}\nabla^2_{r_2} - U_{ion}(r_2) + \frac{k^2_2}{2}\right) \phi^-_e(\mathbf{k}_2,\mathbf{r}_2) = 0 \qquad (5)$$

where U_{ion} is the static Hartree-Fock potential for the helium ion.

To investigate the importance of initial state correlation effects between the ionized electron and an atomic passive electron for single ionization of helium by the impact of a 75 keV proton, figure 2 compares three different FBA-HF FDCS calculations using the various initial-state wavefunctions: FBA-HY (long dashed line), FBA-LS (dotted line), and FBA-PL (short dashed line) with the absolute experimental data (solid dots) (Maydanyuk et al. 2005, Hasan et al. 2004). For figure 2, the electron is ejected into the scattering plane with an energy, E_e, equal to 5.5 eV and four different momentum transfer values ($|\mathbf{q}| = 0.64$ a.u., 0.67 a.u., 0.76 a.u., and 0.97 a.u.). The most distinctive feature of all three of the theoretical curves is the large

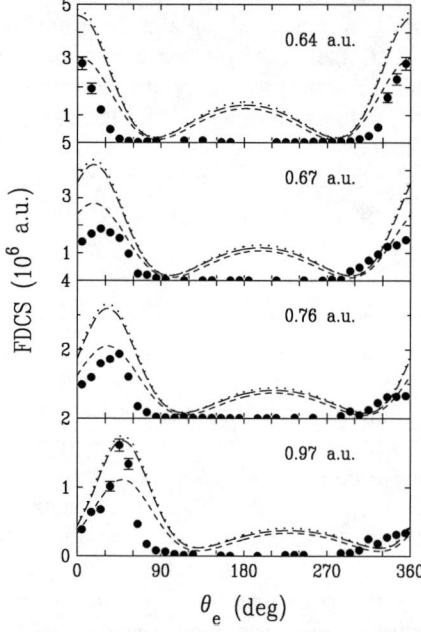

FIGURE 2. Same kinematical conditions as figure 1, the solid circles are the absolute measurements and the theoretical curves: dotted line FBA-LS, long dashed line FBA-HY model, and short dashed line FBA-PL.

recoil peak in the backward direction. In the FBA model, the recoil peak is always at about 180^0 from the binary peak and is understood as a double scattering event – the projectile interacts with the active electron and then the active electron backscatters from the ion. Since the final state distorted wave for the ejected electron is an elastic scattering wavefunction for the ejected electron in the field of the ion, the FBA-HF approach contains the physics necessary for a recoil peak and all the FBA calculations predict a recoil peak whereas no recoil peak is seen in the experimental data. Consequently, some additional physical effects not in the FBA-HF must suppress the recoil peak. For the binary peak, both, the FBA-HY and the FBA-LS results are virtually identical in both shape and scale. This observation suggests that satisfying the Kato cusp condition is not important for the initial state since the Hylleraas wavefunction does not satisfy the cusp condition whereas the Le Sech wavefunction does. Interestingly, the FBA-PL results are nearly a factor of 1.5 lower in absolute magnitude and in closer agreement with the absolute measurements. For the case of double ionization of helium, it was suggested that the Pluvinage wavefunction gave better agreement with experiment due to the fact that the initial and final states were then treated symmetrically. However, that is not the case here with a HF final state. On the other hand, the final state contains no direct correlation at all. To better understand the results of figure 2, a similar study of the effects of correlation on the final state is needed and we are currently in the process of performing such a study. In conclusion, these results indicate that the passive electron may play a more important role than previously assumed. However, a detailed study of final state correlation effects is needed before a definite conclusion can be drawn.

ACKNOWLEDGMENTS

This work was supported by the NSF under Grant. No. PHY-0070872.

REFERENCES

1. N. V. Maydanyuk, A. Hasan, M. Foster, B. Tooke, E. Nanni, D. H. Madison and M. Schulz, *Phys. Rev. Lett.* **94**, 24320 (2005).
2. A. Hasan, N. V. Maydanyuk, Bernard Fendler, A. Voitkiv, B. Najjari and M. Schulz, *J. Phys. B.* **37**, 1923-1930 (2004).
3. D. H. Madison, M. Schulz, S. Jones, M. Foster, R. Moshammer and J. Ullrich, *J. Phys. B.* **35**, 3297 (2002).
4. A. Salin, *J. Phys. B* **2**, 631-639 (1969).
5. P. J. Redmond, (unpublished) as discussed in L. Rosenberg, *Phys. Rev. D* **8**, 1833 (1973).
6. M. Brauner, J. S. Briggs and H. Klar, *J. Phys. B.* **22**, 2265 (1989).
7. L. Gulyás, P. D. Fainstein and A. Salin, *J. Phys. B.* **28**, 245 (1995).
8. S. Jones and D. H. Madison, *Phys. Rev. A.* **65**, 052727 (2002).
9. E. A. Hylleraas, *Z. Phys.* **54**, 347 (1929).
10. C. Le Sech, *J. Phys. B.* **30**, L47-L50 (1997).
11. P. Pluvinage, *Ann. Phys. (N. Y.)* **5**, 145 (1950).
12. J. F. Hart and G. Herzberg, *Phys. Rev.* **106**, 79-82 (1957).
13. S. Jones, Joseph H. Macek and D. H. Madison, *Phys. Rev. A* **70**, 012712 (2004).
14. S. Jones and D. H. Madison, *Phys. Rev. Lett.* **91**, 073201 (2003).
15. S. Jones and D. H. Madison, *J. Phys. B.* **27**, 1423 (1993).

FRAGMENTATION OF HELIUM IN COLLISIONS WITH SLOW ELECTRONS : THREE- AND FOUR-BODY DYNAMICS

M. Dürr*, B. Najjari*, C. Dimopoulou*, A. Dorn* and J. Ullrich*

*Max-Planck-Institute for Nuclear Physics, Saupfercheckweg 1, 69117 Heidelberg, Germany

Abstract. With a multi-coincidence imaging technique (reaction microscope) the fragmentation of helium atoms by electron impact was explored at 106 eV incident energy. Kinematically complete information for both single- and double-ionization is obtained. The large acceptance achieved for low-energy electrons (E < 15 eV) covering almost the entire solid-angle allows to gain three-dimensional images of the electron emission for single-ionization with asymmetric kinematics. Strong deviations in the fully differential cross section between experimental data and a 3C-calculation appear in the non-coplanar geometry. Furthermore, for the first time differential experimental data on double-ionization of helium at the same impact energy (27 eV above the threshold) are presented. Preferred configurations of electron emission after the collision can be identified. The angular correlation reveals, that the dynamics is dominated by the mutual repulsion of the electrons.

Keywords: Electron-impact ionization of atoms; (e,2e); (e,3e); Wannier threshold-law
PACS: 34.80.Dp, 34.80.Pa

INTRODUCTION

In the field of atomic collisions single- and double-ionization of helium through electron-impact represents one of the most fundamental few-body problems, involving four interacting particles. Recent theoretical models (e.g. [1], [2]) have been very successful in reproducing fully differential data of (e,2e) experiments on helium even at very low impact energies. The wide majority of these experiments is restricted to geometries, where the momentum-vectors of the participating particles are all in one plane - the scattering plane. In this case three- or four-body effects, higher order interactions, etc. may be masked by occuring binary collisions so that these effects might only become visible outside the scattering plane. In fact dedicated experiments at low impact energies exploring the non-coplanar geometry have been reported [3, 4], which are also well described by the mentioned approaches [5, 6], nevertheless so far no three-dimensional image of the process has been published. In this contribution a 'complete' image of the single-ionization of helium by 106 eV electrons is presented, which can provide a further benchmark for theoretical calculations of the few-body problem. In addition, more complex systems like the full fragmentation of helium, the so called (e,3e) process in particular at impact energies close to threshold have not been explored experimentally. So far, experimental studies were restricted to primary beam energies beyond 500 eV, where the projectile-target interaction can be treated pertubatively. Near the ionization threshold the projectile has to be included in a non-pertubative manner, which turns out to be difficult to include in theoretical descriptions, since in total four continuum

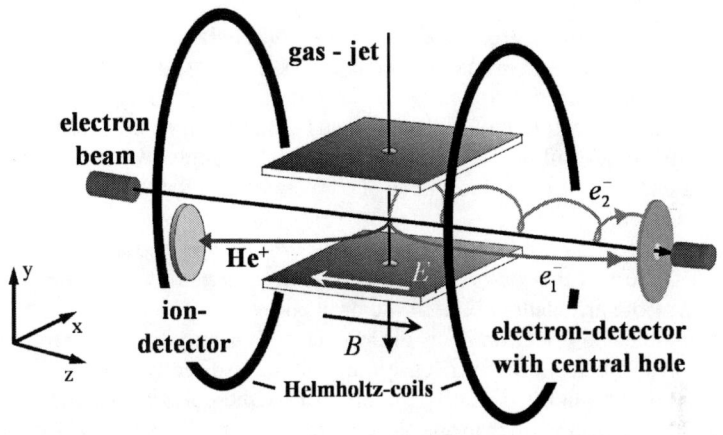

FIGURE 1. Schematic view of the 'reaction microscope' for studying electron-atom collisions. The projectile beam is crossed with a internally cold He beam inducing fragmentation of He. A homgeneous electric field generated by two parallel arrays of electrodes extracts the electrons and ions from the interaction-point to the detectors.

particles are involved. We have performed a kinematical complete experiment for double ionization of helium at $E_0 = 106$ eV impact energy, which is only 27 eV above the threshold for the full fragmentation. This provides detailed insight into the strongly correlated three-electron continuum and the dynamics of a four-body system.

EXPERIMENTAL METHOD

The apparatus - the 'reaction microscope' (RM) - is applied to study fragmentation processes induced by ion-impact, electron-impact and photons (an overview is given by [7]). The method is based on the coincident detection of the target fragments after crossing the projectile beam with an internally cold atomic beam and determining the momentum components. This is achieved by mapping the trajectories of the ion and the electrons onto two position-sensitive detectors with electric and magnetic fields yielding all three momentum components, so complete kinematical information of the reaction is obtained.

As discussed in [7] the particular difficulty of studying electron collisions with a RM is given by the fact that the trajectory of the primary projectile beam is strongly affected by the electric and magnetic fields used for the imaging of the secondary particles. In the present arrangement (Fig. 1) the primary beam is aligned collinearly with the both fields resulting in a straight trajectory which is essentially independent of the strength of the imaging fields. Thus the beam is pointed directly to the 80 mm diameter electron detector with a 6 mm diameter central bore and a subsequent Faraday cup where the non-scattered projectiles finally are dumped. This modified detector design is essential to prevent the intense projectile beam ($\approx 10^8$ electrons/second) from hitting and degrading the sensitive multichannelplates and to avoid backscattering of part of the projectiles which would give rise to a strong electron background signal rate. Since the primary electron beam is transported over a rather large distance of more than 0.6 m without any electrostatic refocusing elements, in particular for low primary beam energies, the magnetic field is also required to guide and focus the projectiles into the target and into the Farady cup. As a result, the present set up allows to apply low energetic projectile beams. So far energies of E_0 = 30 eV have been realized and for the future the sub 10 eV regime is envisaged.

An essential step forward of this spectrometer with respect to all previous reaction microscopes is the fact that projectiles inducing a fragmentation or ionization reaction and which therefore are scattered out of the projectile beam impinge on the electron detector and are detected (electron e_1 in Fig. 1). Thus, not only the slow target electrons and ions but also the scattered projectile momentum is directly measured and the recoil ion momentum is not necessarily required to fix the collision kinematics. Besides critical tests of the spectrometer properties this redundant information allows also kinematically complete experiments with improved resolution for the momentum transfer and with heavy or warm target species where the recoil ion momentum resolution is poor. Furthermore, if molecular fragmentation processes are studied the fragment ion detection can be used to obtain information on the molecular orientation in space during the collision and the collision kinematics can be obtained even if occurring neutral fragments are not detected.

EXPERIMENTAL RESULTS

Single Ionization of Helium

The fully differential cross section (FDCS) for the (e,2e) process is evaluated in the offline analysis considering events only if a single-charged ion and two electrons were detected in coincidence. The momentum-transfer $\vec{q} = \vec{p}_0 - \vec{p}_1$ is determined from the momentum-vector of the scattered projectile $\vec{p}_1$ and the given initial momentum $\vec{p}_0 = (0,0,2.79)$ a.u. By calculating the energy-sum of the two outgoing electrons information on the total energy-loss of the projectile is obtained (the Q-value), and single ionization leaving the residual ion in the $n = 1$, $n = 2$ and $n \geq 3$ state can be separated from each other. Identifying the 'fast' electron as the scattered projectile, the scattering angle of recorded events, which is the angle between the projectile axis and the momentum vector $\vec{p}_1$, ranges between 8 and 30 degrees. To evaluate the FDCS we chose events where the scattering angle of the 'fast' electron Θ_1 and the energy of the second ejected electron E_2 is fixed (and therefore the momentum-transfer $\vec{q}$), as it is usually practiced in conventional (e,2e) experiments. The emission angle of the second electron remains as the free kinematical parameter. With the RM, emission angles covering almost the full solid angle can be detected, giving a complete image of the emission process. An example can be seen in Fig. 2 showing the (relative) FDCS in a three-dimensional polar plot for $\Theta_1 = 22°$ and $E_2 = 10$ eV. The electron emission shows two lobes, one pointing roughly in direction of the momentum transfer $\vec{q}$ which is known as the binary peak resulting from binary collision between the projectile and the ionized target-electron. The lobe in backward direction - the recoil peak - is often interpreted as

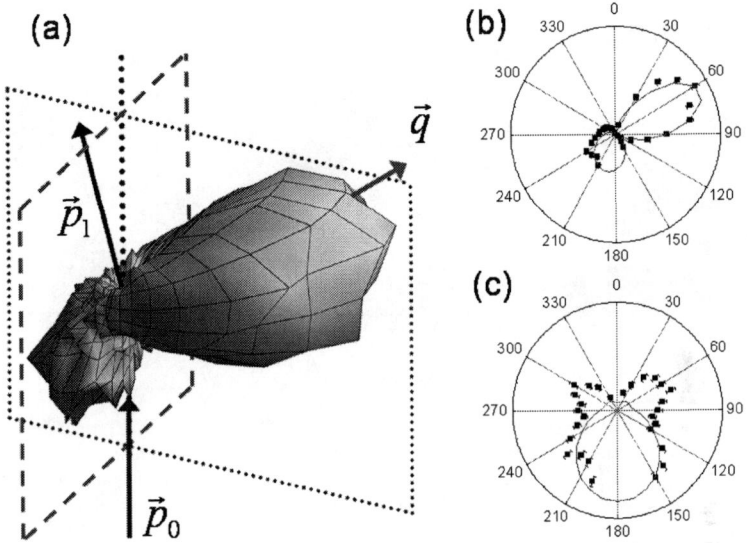

FIGURE 2. (a) Three-dimensional polar-plot of the electron-emission for an electron with 10 eV after single-ionization of helium by electron impact with 106 eV energy. The scattering plane and the plane perpendicular to it are indicated by dotted lines and dashed lines respectively. The incoming and scattered projectile-direction ($\vec{p}_0$ and $\vec{p}_1$) as well as the direction of the momentum transfer $\vec{q}$ is shown. For both planes cuts are compared with a 3C-calculation: in the scattering plane (b) and the plane perpendicular containing the projectile-axis (c).

a second rescattering of the outgoing electron on the ionic potential. Both lobes are tilted in backward direction, which is an effect due to the post-collision interaction with the scattered projectile. For comparison with theory two cuts were applied: the first cut in the scattering plane which is spanned by the momentum vectors of the incoming and scattered projectile (dotted lines in Fig. 2 (a)). The second cut was done in the plane perpendicular to the scattering plane containing the vector of the incoming projectile $\vec{p}_0$ (dashed lines in Fig. 2 (a)). In the scattering plane a 3C-Calculation in the length form (according to the approach in [2]) shows good agreement, whereas in the cut perpendicular to the scattering plane, structures appear, which are absent in the calculation (Fig. 2 (c)). Similar features were observed by Schulz et. al. [8] studying single-ionization of helium by ion impact in the pertubative regime. The authors propose an additional scattering of the projectile with the residual ion, transferring additional momentum, which could also explain the structures observed in the case presented here. This example illustrates how theoretical descriptions of few-body-processes can be tested with increased sensitivity in the out-of-plane geometry.

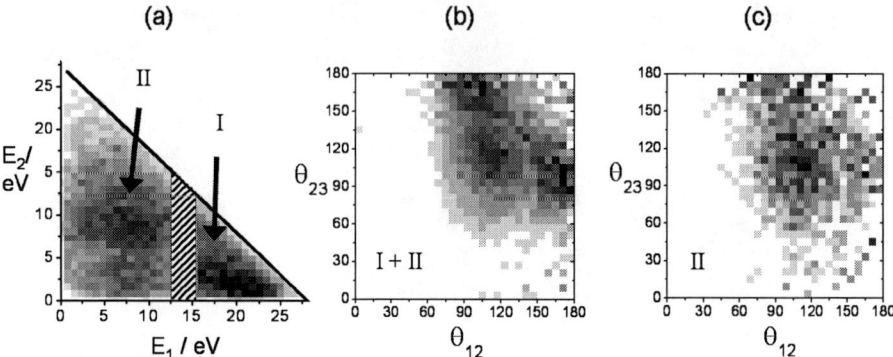

FIGURE 3. Integrated cross-sections of the (e,3e) reaction with 106 eV initial projectile-energy. The energy-sharing between the electrons (left plot) shows two regions with considerably high cross-section (In the blanked area there is no full acceptance for electron e_1). One with unequal energy-sharing (I) and a second where all three electrons have similar energy (II). The evaluation of mutual angles between the three electrons reveals two favoured configurations: one with $180°/90°$ and one with $120°/120°$ (b). In the case of symmetric energy-sharing mainly mutual angles of $120°/120°$ appear (c).

Double Ionization of Helium

With the same apparatus double-ionization of helium by electron impact was recorded. Kinematical complete information of the four continuum-particles is achieved by coincident detection of two electrons and the double-charged recoil-ion. The momentum of the third undetected particle is obtained by momentum conservation. The excess energy of 27 eV = 106 eV - IP is shared among the three electrons, since due to its large mass the recoiling ion carries negligible kinetic energy. The energy-sharing of the excess energy among the three continuum-electrons can be seen in Fig. 3 (a), showing the measured cross-section depending on the energy E_1 and E_2 of the two detected electrons. The energy of the third electron is given by energy conservation, $E_3 = 27eV - E_2 - E_3$. The two detected electrons are labeled by their arrival time on the detector, such that the first electron always carries a bigger momentum in longitudinal direction than the second. Area (I) in Fig. 3 (a) involves electrons with asymmetric energy-sharing, the electron with index 1 carrying the majority of excess energy, leaving the other two with small kinetic energy. In the second case (II) all three electrons carry approximately equal energies $E_1 \approx E_2 \approx E_3 \approx 9\ eV$. The angular correlation between the three electrons was analyzed and mainly two configurations appear: one where the electrons escape preferably with relative angles of $120°$, and a second configuration with $90°/180°$ angles (Fig. 3 (b)). The first case is predominant in the symmetric energy-sharing case, which can be seen in Fig. 3 (c), where the angular correlation is shown only for the case of symmetric energy-sharing. The configuration with mutual angles of $120°$ is predicted by Wannier's theory [9] extended to three-electrons by Klar and Schlecht [10] describing the motion of three electrons in an attractive potential close to threshold. The second configuration can be seen as a process where the high-energy electron is identified as the scattered projectile quickly leaving the fragments behind, where the two low-energy target electrons escape the ionic potential with opposite momenta (back to back), similar to the Wannier-configuration describing the motion of two electrons with small energies. Note that this is merely a rough overview and that the integrated cross-sections presented do not cover all scattering angles for the first and second electron. A further analysis is underway evaluating angular correlations with respect to the projectile axis and fully differential cross sections.

SUMMARY AND OUTLOOK

With a newly developed reaction microscope that allows for low impact energies, we have performed (e,2e) and (e,3e) experiments on helium at 106 eV projectile-energy. For single ionization experimental fully differential cross sections covering almost the entire solid angle for ejected electrons with low-energy are presented, giving a full image of the process. In the plane perpendicular to the scattering plane big discrepancies between experimental data and a 3C-calculation appear. To gain a better understanding of the origin for this discrepancies, an (e,2e) experiment studying the single ionization of helium at 1000 eV projectile-energy has been performed and will be reported elsewhere. In this regime the projectile acts as small pertubation and higher-order processes should be less important. Furthermore, it can be compared with the ion-impact experiment reported in [8] with the identical pertubation parameter $Z_p/v_p \approx 0.1$, where similar deviations have been found, and whose origins are so far completely unknown.

Secondly, first experimental data for double-ionization of helium with the same projectile-energy are presented. Apparently two configurations are favored in the reaction. In the first, one electron carries the majority of the excess energy. The remaining electrons have only small kinetic energy and the angular correlation between the electrons can be interpreted such that they escape the ion with relative angles of 180°, i.e. back-to-back. The second configuration, in which the electrons share the excess energy rather equally, favours relative angles of 120°, where the mutual repulsion between all three electrons is minimized. However, to fully elucidate this system, the evaluation of higher differential cross sections is necessary.

REFERENCES

1. I. Bray, and D. V. Fursa, *Phys. Rev. Lett.*, **76**, 2674–2677 (1996).
2. M. Brauner, J. S. Briggs, and H. Klar, *J. Phys. B*, **22**, 2265–2287 (1989).
3. A. J. Murray, M. B. Woolf, and F. H. Read, *J. Phys. B*, **25**, 3021–3036 (1992).
4. T. Rösel, J. Röder, L. Frost, K. Jung, and H. Ehrhardt, *Phys. Rev. A*, **46**, 2539–2552 (1992).
5. J. Berakdar, and J. S. Briggs, *Phys. Rev. Lett.*, **72**, 3799–3802 (1994).
6. A. T. Stelbovics, I. Bray, D. V. Fursa, and K. Bartschat, *Phys. Rev. A*, **17**, 1592–1599 (2005).
7. J. Ullrich, R. Moshammer, A. Dorn, L. P. Schmidt, and H. Schmidt-Böcking, *Rep. Prog. Phys.*, **66**, 1463–1545 (2003).
8. M. Schulz, R. Moshammer, D. Fischer, H. Kollmus, D. H. Madison, S. Jones, and J. Ullrich, *Nature*, **422**, 48–50 (2003).
9. G. H. Wannier, *Phys. Rev.*, **90**, 817–825 (1953).
10. H. Klar, and W. Schlecht, *J. Phys. B*, **9**, 1699–1711 (1976).

Nondipole Effects in Double Photoionization of He

A. Y. Istomin*, N. L. Manakov†, A. V. Meremianin† and A. F. Starace*

*Department of Physics and Astronomy, The University of Nebraska, Lincoln, NE 68588-0111
†Physics Department, Voronezh State University, Voronezh 394006, Russia

Abstract. Lowest-order nondipole effects are studied in double photoionization (DPI) of the He atom. *Ab initio* parametrizations of the quadrupole transition amplitude for DPI from the $^1S_0$-state are presented in terms of the exact two-electron reduced matrix elements. Parametrizations for the dipole-quadrupole triply differential cross section (TDCS) and doubly differential cross section (DDCS) are presented in terms of polarization-independent amplitudes for the case of an elliptically polarized photon. Expressions for the DDCS in terms of the reduced two-electron matrix elements are also given. A general analysis of retardation-induced asymmetries of the TDCS including the circular dichroism effect at equal energy sharing is presented. Our numerical results exhibit a nondipole forward-backward asymmetry in the TDCS for DPI of He at an excess energy of 450 eV that is in qualitative agreement with existing experimental data.

Keywords: double ionization, nondipole effects, circular dichroism, helium, quadrupole amplitude
PACS: 32.80.Fb

GENERAL RESULTS FOR THE DPI QUADRUPOLE TRANSITION AMPLITUDE

We consider double photoionization (DPI) from the $^1S_0$ state $|0\rangle$ to the final two-electron singlet state $|\mathbf{p}_1\mathbf{p}_2\rangle$, with asymptotic electron momenta $\mathbf{p}_1$ and $\mathbf{p}_2$, in the nonrelativistic domain of photon energies taking into account lowest-order retardation corrections. Because neither orbital nor spin-dependent parts of the magnetic dipole interaction contribute to the transition amplitude A for nonrelativistic photon energies [1], the spin dependence of the two-electron wave functions is suppressed in our analysis. The dipole-quadrupole TDCS for DPI is:

$$\frac{d^3\sigma}{d\varepsilon_1 d\Omega_1 d\Omega_2} \equiv \sigma = \mathscr{A}|A|^2, \quad (1)$$

where $\mathscr{A} = 4\pi^2 \alpha p_1 p_2/\omega$ is a normalization factor, and $\alpha = 1/137.036$. Atomic units are used throughout this paper. The amplitude A involving E1 and E2 components has the form (where $\mathbf{e}$ is the photon polarization vector and $\mathbf{k}$ is the photon wavevector),

$$A = A_d + A_q = \langle \mathbf{p}_1\mathbf{p}_2|(\mathbf{e}\cdot\mathbf{D}) + (\{\hat{\mathbf{k}}\otimes\mathbf{e}\}_2 \cdot Q_2)|0\rangle, \quad (2)$$

for both velocity (V) and length (L) gauges of the electron-photon interaction. $\{\mathbf{a}\otimes\mathbf{b}\}_{2m}$ is the irreducible tensor product of rank 2. In the V-gauge, $\mathbf{D} \equiv \mathbf{D}^{(V)} = -i(\nabla_1 + \nabla_2)$ and $Q_{2m} \equiv Q_{2m}^{(V)} = \alpha\omega(\{\mathbf{r}_1\otimes\nabla_1\}_{2m} + \{\mathbf{r}_2\otimes\nabla_2\}_{2m})$. In the L-gauge, $\mathbf{D} \equiv \mathbf{D}^{(L)} = i\omega(\mathbf{r}_1 + \mathbf{r}_2)$ and $Q_{2m} \equiv Q_{2m}^{(L)} = -(1/2)\alpha\omega^2(\{\mathbf{r}_1\otimes\mathbf{r}_1\}_{2m} + \{\mathbf{r}_2\otimes\mathbf{r}_2\}_{2m})$. (See Ref. [2] for details.)

In the electric dipole approximation (EDA), the parametrization of the EDA amplitude A_d in terms of scalar products of the vectors $\mathbf{e}$, $\hat{\mathbf{p}}_1$, and $\hat{\mathbf{p}}_2$ is well-known [3, 4]:

$$A_d = f_1(\mathbf{e} \cdot \hat{\mathbf{p}}_1) + f_2(\mathbf{e} \cdot \hat{\mathbf{p}}_2). \tag{3}$$

In this equation, $f_1 \equiv f(p_1, p_2, \cos\theta)$ and $f_2 \equiv f(p_2, p_1, \cos\theta)$ [where $\theta_{12} \equiv \theta$ is the mutual ejection angle, $\cos\theta = (\hat{\mathbf{p}}_1 \cdot \hat{\mathbf{p}}_2)$] are defined by a single function,

$$f(p, p', \cos\theta) = \sum_{l=1}^{\infty} (-1)^{l+1} \left[\sum_{l'=l\pm 1} D_{ll'}(p, p') \right] P_l^{(1)}(\cos\theta), \tag{4}$$

where $P_l^{(n)}(x)$ is the n-th derivative of the Legendre polynomial $P_l(x)$, $P_l^{(n)}(x) = (d^n/dx^n) P_l(x)$. The energy-dependent coefficient $D_{ll'}(p, p')$ is given by $D_{ll'}(p, p') = [(2l+1)(2l'+1) \max(l,l')]^{-1/2} d_{ll'}(p,p')$, where $d_{ll'}(p,p') \equiv \langle pp'; (ll')1||\mathbf{D}||0\rangle$ is the reduced matrix element of the operator $\mathbf{D}$ between the $^1S_0$-state and the P-wave component of the final state $|\mathbf{pp'}\rangle$, with photoelectron angular momenta l and $l' = l \pm 1$. We use an expansion of $|\mathbf{pp'}\rangle$ in terms of modified bipolar harmonics $\mathscr{C}_{LM}^{ll'}(\hat{\mathbf{p}}, \hat{\mathbf{p}}')$,

$$|\mathbf{pp'}\rangle = \sum_{ll'LM} \mathscr{C}_{LM}^{ll'\,*}(\hat{\mathbf{p}}, \hat{\mathbf{p}}') |pp'; (ll')LM\rangle. \tag{5}$$

Similarly to the derivations of Eqs. (3) and (4) in Ref. [3], to derive a parametrization of the quadrupole amplitude A_q we use the expansion (5) and the reduction formulae [3] for the rank-2 bipolar harmonics. Thus, A_q may be presented as follows [2, 5]:

$$A_q = g_1(\mathbf{e}\cdot\hat{\mathbf{p}}_1)(\hat{\mathbf{p}}_1\cdot\hat{\mathbf{k}}) + g_2(\mathbf{e}\cdot\hat{\mathbf{p}}_2)(\hat{\mathbf{p}}_2\cdot\hat{\mathbf{k}}) + g_s\left[(\mathbf{e}\cdot\hat{\mathbf{p}}_1)(\hat{\mathbf{p}}_2\cdot\hat{\mathbf{k}}) + (\mathbf{e}\cdot\hat{\mathbf{p}}_2)(\hat{\mathbf{p}}_1\cdot\hat{\mathbf{k}})\right], \tag{6}$$

where the generally complex parameters $g_{1,2}$ and g_s depend on p_1, p_2, and θ. The parameter g_s is symmetric in p_1 and p_2, $g_s \equiv g_s(p_1, p_2, \cos\theta) = g_s(p_2, p_1, \cos\theta)$, while g_1 and g_2 are expressed in terms of a single function, $g(p, p', \cos\theta)$, with $g_1 \equiv g(p_1, p_2, \cos\theta)$ and $g_2 \equiv g(p_2, p_1, \cos\theta)$. The explicit forms of the functions $g(p, p', \cos\theta)$ and $g_s(p, p', \cos\theta)$ are as follows:

$$g_s(p, p', \cos\theta) = \sum_{l=1}^{\infty} (-1)^{l+1} \left[\sum_{l'=l\pm 2} Q_{ll'}(p,p') P_{\frac{l+l'}{2}}^{(2)}(\cos\theta) \right.$$
$$\left. + \sqrt{6} Q_{ll}(p,p') \left(P_{l+1}^{(2)}(\cos\theta) - \frac{2l+3}{2} P_l^{(1)}(\cos\theta) \right) \right],$$

$$g(p, p', \cos\theta) = \sum_{l=2}^{\infty} (-1)^l \left[\sum_{l'=l\pm 2} Q_{ll'}(p, p') + \sqrt{6} Q_{ll}(p, p') \right] P_l^{(2)}(\cos\theta), \tag{7}$$

where $Q_{ll'}(p, p') = [4(l+l'-2)!/(l+l'+3)!]^{1/2} q_{ll'}(p,p')$ and $q_{ll'}(p,p') \equiv \langle pp'; (ll')2||Q_2||0\rangle$ is the reduced matrix element of the operator Q_{2m} between the initial $^1S_0$-state, $|0\rangle$, and the D-wave component of the two-electron continuum state $|\mathbf{pp'}\rangle$ with photoelectron angular momenta l and $l' = l, l \pm 2$.

An alternative parametrization of the DPI amplitude A in terms of orthogonal vectors $\mathbf{p}_+ = (\hat{\mathbf{p}}_1 + \hat{\mathbf{p}}_2)/2$ and $\mathbf{p}_- = (\hat{\mathbf{p}}_1 - \hat{\mathbf{p}}_2)/2$ and the symmetrized ($f^{(g)} = f_1 + f_2$ and $g_\pm^{(g)} = g_1 + g_2 \pm 2g_s$) and antisymmerized ($f^{(u)} = f_1 - f_2$ and $g^{(u)} = g_1 - g_2$) amplitudes is rather obvious and can be easily obtained from Eqs. (3), (4), (6), and (7) [2]. The results in Eqs. (3), (4), (6) and (7) are general and do not depend upon the dynamical model used for calculation of the reduced matrix elements $d_{ll'}(p,p')$ and $q_{ll'}(p,p')$.

PARAMETRIZATIONS AND NUMERICAL ESTIMATES FOR THE DIPOLE-QUADRUPOLE TDCS

For the most general case of DPI by elliptically polarized light (described by the complex polarization vector $\mathbf{e}$ [$(\mathbf{e} \cdot \mathbf{e}^*) = 1$]), the TDCS in Eq. (1) (neglecting the small terms $\sim |A_q|^2$) has a model-independent parametrization similar to that for dipole DPI [3],

$$\sigma = \mathscr{A}\{c_1|\mathbf{e}\cdot\hat{\mathbf{p}}_1|^2 + c_2|\mathbf{e}\cdot\hat{\mathbf{p}}_2|^2 + \operatorname{Re} c_3\left[(1-\ell)\left((\hat{\mathbf{p}}_1\cdot\hat{\mathbf{p}}_2) - (\hat{\mathbf{k}}\cdot\hat{\mathbf{p}}_1)(\hat{\mathbf{k}}\cdot\hat{\mathbf{p}}_2)\right) \right. $$
$$\left. + 2\ell(\hat{\boldsymbol{\varepsilon}}\cdot\hat{\mathbf{p}}_1)(\hat{\boldsymbol{\varepsilon}}\cdot\hat{\mathbf{p}}_2)\right] + \xi\operatorname{Im} c_3\left(\hat{\mathbf{k}}\cdot[\hat{\mathbf{p}}_1\times\hat{\mathbf{p}}_2]\right)\}, \qquad (8)$$

where, in contrast to the dipole case, the coefficients c_i depend upon $\mathbf{k}$:

$$c_1 = |f_1|^2 + 2\operatorname{Re}\left[f_1 g_1^*(\hat{\mathbf{k}}\cdot\hat{\mathbf{p}}_1) + f_1 g_s^*(\hat{\mathbf{k}}\cdot\hat{\mathbf{p}}_2)\right],$$
$$c_2 = |f_2|^2 + 2\operatorname{Re}\left[f_2 g_2^*(\hat{\mathbf{k}}\cdot\hat{\mathbf{p}}_2) + f_2 g_s^*(\hat{\mathbf{k}}\cdot\hat{\mathbf{p}}_1)\right],$$
$$c_3 = f_1 f_2^* + (f_1 g_s^* + f_2^* g_1)(\hat{\mathbf{k}}\cdot\hat{\mathbf{p}}_1) + (f_2^* g_s + f_1 g_2^*)(\hat{\mathbf{k}}\cdot\hat{\mathbf{p}}_2). \qquad (9)$$

The parameters ℓ and ξ in Eq. (8) are the degrees of linear and circular polarization of an elliptically polarized photon; $\ell = \mathbf{e}^2 = \sqrt{1-\xi^2}$, $\xi \equiv i(\hat{\mathbf{k}}\cdot[\mathbf{e}\times\mathbf{e}^*])$, and the unit vector $\hat{\boldsymbol{\varepsilon}}$ is directed along the major axis of the polarization ellipse. The photon polarization dependence of the dipole-quadrupole TDCS is thus determined by *four* real $\hat{\mathbf{k}}$-dependent parameters, c_1, c_2, $\operatorname{Re} c_3$, and $\operatorname{Im} c_3$. These parameters may be determined from four measurements with different polarizations of the photon beam, e.g., two experiments with linearly polarized photons and two experiments with circularly polarized photons. For circularly polarized photons ($\ell = 0$, $\xi = \pm 1$) Eq. (8) simplifies,

$$\sigma = \frac{\mathscr{A}}{2}\left(c_1[\hat{\mathbf{k}}\times\hat{\mathbf{p}}_1]^2 + c_2[\hat{\mathbf{k}}\times\hat{\mathbf{p}}_2]^2 + \operatorname{Re} c_3\left([\hat{\mathbf{k}}\times\hat{\mathbf{p}}_1]\cdot[\hat{\mathbf{k}}\times\hat{\mathbf{p}}_2]\right)\right.$$
$$\left. + 2\xi\operatorname{Im} c_3\left(\hat{\mathbf{k}}\cdot[\hat{\mathbf{p}}_1\times\hat{\mathbf{p}}_2]\right)\right). \qquad (10)$$

The TDCSs in Eqs. (8) and (10) both contain a term that is proportional to ξ, which is responsible for the circular dichroism (CD) effect. This effect is usually characterized by the absolute CD parameter, $\Delta_{cd} \equiv \sigma(\xi = +1) - \sigma(\xi = -1)$:

$$\Delta_{cd} = 2\mathscr{A}\operatorname{Im} c_3\left(\hat{\mathbf{k}}\cdot[\hat{\mathbf{p}}_1\times\hat{\mathbf{p}}_2]\right). \qquad (11)$$

The term Δ_{cd} involves both the dipole-dipole and dipole-quadrupole contributions,

$$\Delta_{cd} = \Delta_{cd}^{(dip)} + \Delta_{cd}^{(quadr)}, \quad \text{where } \Delta_{cd}^{(dip)} = 2\mathscr{A}\operatorname{Im}(f_1 f_2^*)\left(\hat{\mathbf{k}}\cdot[\hat{\mathbf{p}}_1\times\hat{\mathbf{p}}_2]\right), \text{ and}$$
$$\Delta_{cd}^{(quadr)} = 2\mathscr{A}\operatorname{Im}[(f_1 g_s^* + f_2^* g_1)(\hat{\mathbf{k}}\cdot\hat{\mathbf{p}}_1) - (f_2 g_s^* + f_1^* g_2)(\hat{\mathbf{k}}\cdot\hat{\mathbf{p}}_2)]\left(\hat{\mathbf{k}}\cdot[\hat{\mathbf{p}}_1\times\hat{\mathbf{p}}_2]\right). \qquad (12)$$

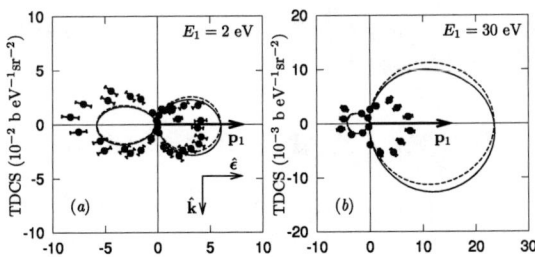

FIGURE 1. Comparison of our LOPT results for the TDCS at an excess energy of 450 eV for the case of linear polarization with the normalized experimental data of Ref. [10]. The directions of the photon wavevector $\hat{\mathbf{k}}$ and polarization $\hat{\boldsymbol{\varepsilon}}$ are as shown in (a); the electron having energy E_1 is ejected along $\boldsymbol{\varepsilon}$. Full curves: dipole-quadrupole results; dashed curves: EDA results.

Within the EDA, the CD effect is described by the term $\Delta_{cd}^{(dip)}$ [3, 6], which vanishes at equal energy sharing (because for $p_1 = p_2$ one has $f_1 = f_2$). However, the quadrupole term, $\Delta_{cd}^{(quadr)}$, produces a non-zero CD effect even at equal energy sharing. The dichroic term $\mathrm{Im}\, c_3(\hat{\mathbf{k}} \cdot [\hat{\mathbf{p}}_1 \times \hat{\mathbf{p}}_2])$ in Eqs. (8) and (10) results also in an unusual feature of the TDCS, a different symmetry of the TDCS with respect to two transformations: (i) $\mathbf{k} \to -\mathbf{k}$ (i.e., the inversion of the photon beam direction) and (ii) $(\theta_1, \theta_2) \to (\pi - \theta_1, \pi - \theta_2)$ (i.e., the reflection of the photoelectron pair in the polarization plane) [2].

The expression for the TDCS in terms of the symmetrized amplitudes has a form identical to those in Eqs. (8)-(10) provided the following substitutions are made: $\{\hat{\mathbf{p}}_1, \hat{\mathbf{p}}_2, f_1, f_2, g_1, g_2, g_s\} \to \{\mathbf{p}_+, \mathbf{p}_-, f^{(g)}, f^{(u)}, g_+^{(g)}, g_-^{(g)}, g^{(u)}\}$. This parametrization leads to a simpler form of the TDCS for equal energy sharing [7]:

$$\sigma^{(eq)} = \mathscr{A}\left\{ \left[|f^{(g)}|^2 + 2\mathrm{Re}\{f^{(g)*}g_+^{(g)}\}(\hat{\mathbf{k}} \cdot \mathbf{p}_+)\right] |\mathbf{e} \cdot \mathbf{p}_+|^2 + \mathrm{Re}\{f^{(g)*}g_-^{(g)}\}(\hat{\mathbf{k}} \cdot \mathbf{p}_-) \right.$$
$$\left. \times \left[2l(\hat{\boldsymbol{\varepsilon}} \cdot \mathbf{p}_+)(\hat{\boldsymbol{\varepsilon}} \cdot \mathbf{p}_-) + (l-1)(\hat{\mathbf{k}} \cdot \mathbf{p}_+)(\hat{\mathbf{k}} \cdot \mathbf{p}_-)\right] \right\} + (\xi/2)\Delta_{cd}^{(eq)},$$
$$\Delta_{cd}^{(eq)} \equiv \Delta_{cd}^{(quadr)}|_{p_1=p_2} = 2\mathscr{A}\, \mathrm{Im}\{f^{(g)*}g_-^{(g)}\}(\hat{\mathbf{k}} \cdot [\mathbf{p}_- \times \mathbf{p}_+])(\hat{\mathbf{k}} \cdot \mathbf{p}_-). \quad (13)$$

Numerical estimates of nondipole effects in TDCSs calculated within lowest-order perturbation theory (LOPT) in the interelectron interaction [8, 9] have been reported in Refs. [2, 5, 7] for excess energies ranging from tens to hundreds of eV. In Fig. 1 we present an example of such predictions for an excess energy of 450 eV that are compared to the experimental data of Ref. [10]. Despite the deviation of our LOPT predictions within EDA from the experimental data, one sees that both the experimental data and our nondipole results exhibit a noticeable forward-backward asymmetry as compared to our EDA results: the angular distributions of the fast electron are shifted along the direction of the vector $\mathbf{k}$, especially in the angular ranges $0 < \theta_2 < \pi/2$ and $3\pi/2 < \theta_2 < 2\pi$. The relative magnitudes of such an asymmetry at particular angles θ_2, measured by the quantity $R = [\sigma(2\pi - \theta_2) - \sigma(\theta_2)]/\sigma(2\pi - \theta_2)$, appear to be in reasonable agreement with the experimental data. Indeed, for $\theta_1 = 0.174$ (in radians), one has the theoretical value $R^{th} = 0.051$ vs. the experimentally measured value $R^{exp} = 0.049$; for $\theta_1 = 0.516$,

$R^{th} = 0.143$ vs. $R^{exp} = 0.117$; for $\theta_1 = 0.868$, $R^{th} = 0.212$ vs. $R^{exp} = 0.181$; and for $\theta_1 = 1.216$, $R^{th} = 0.229$ vs. $R^{exp} = 0.398$.

DIPOLE-QUADRUPOLE DDCS

The DDCS for DPI from an initial ^1S state can be written in a form that is identical to that of the angle-differential cross section for single photoionization,

$$\frac{d^2\sigma}{d\Omega_{\mathbf{p}_1}dE_1} = \frac{\sigma_0}{4\pi}\left\{1 + \beta P_2(|\mathbf{e}\cdot\hat{\mathbf{p}}_1|) + [\delta + \gamma|\mathbf{e}\cdot\hat{\mathbf{p}}_1|^2](\hat{\mathbf{k}}\cdot\hat{\mathbf{p}}_1)\right\}, \quad (14)$$

where, however, the singly-differential cross section (SDCS) σ_0, the dipole asymmetry parameter β, and the nondipole asymmetry parameters γ and δ depend upon both the photon frequency ω *and* the energy sharing (i.e., upon the energy of one of the photoelectrons). For linear polarization ($\mathbf{e} = \mathbf{e}^* \equiv \hat{\varepsilon}$), $P_2(|\mathbf{e}\cdot\hat{\mathbf{p}}|) = P_2(\cos\alpha)$ is a Legendre polynomial, where α is the angle between the vectors $\hat{\mathbf{p}}$ and $\hat{\varepsilon}$; for circular polarization, $P_2(|\mathbf{e}\cdot\hat{\mathbf{p}}|) = (-1/2)P_2(\cos\phi)$, where ϕ is the angle between the vectors $\hat{\mathbf{p}}$ and $\hat{\mathbf{k}}$.

We have derived *ab initio* representations for σ_0, β, γ, and δ at two different levels of detail. The first one is in terms of integrals of the polarization-invariant amplitudes,

$$\sigma_0 = \mathscr{A}\frac{8\pi^2}{3}\int_{-1}^{1}\left[|f_1|^2 + |f_2|^2 + 2\mathrm{Re}(f_1 f_2^*)x\right]dx,$$

$$\beta = \mathscr{A}\frac{(4\pi)^2}{3\sigma_0}\int_{-1}^{1}\left[|f_1|^2 + |f_2|^2 P_2(x) + 2\mathrm{Re}(f_1 f_2^*)x\right]dx = 2 - \mathscr{A}\frac{8\pi^2}{\sigma_0}\int_{-1}^{1}|f_2|^2(1-x^2)dx,$$

$$\gamma = \mathscr{A}\frac{(4\pi)^2}{\sigma_0}\int_{-1}^{1}\mathrm{Re}\left\{f_1^*[g_1 + g_2 P_2(x) + 2g_s x] + f_2^*[g_1 x + g_2 P_3(x) + 2g_s P_2(x)]\right\}dx,$$

$$\delta = \mathscr{A}\frac{8\pi^2}{\sigma_0}\int_{-1}^{1}\mathrm{Re}\left[f_2^*(g_s + g_2 x)\right](1-x^2)dx, \quad (15)$$

where the amplitudes $f_{1,2}(x)$, $g_{1,2}(x)$, and $g_s(x)$ depend upon $x \equiv \cos\theta$. The second parametrization is in terms of an infinite sum over the reduced matrix elements,

$$\sigma_0 = (4\pi)^2 \frac{\mathscr{A}}{3} \sum_{l_2=0}^{\infty} \sum_{l_1=l_2\pm1} \frac{|d_{l_1 l_2}|^2}{(2l_1+1)(2l_2+1)},$$

$$\beta = \frac{(4\pi)^2}{\sigma_0}\mathscr{A}\sqrt{\frac{2}{3}} \sum_{l_2=0}^{\infty} \sum_{l_1=l_2\pm1} \sum_{l_1'=l_2+1} \frac{(-1)^{l_2}}{2l_2+1} C_{l_1 0 l_1' 0}^{20} \left\{\begin{array}{ccc} 1 & l_1' & l_2 \\ l_1 & 1 & 2 \end{array}\right\} d_{l_1 l_2} d_{l_1' l_2}^*,$$

$$\gamma = -\frac{(4\pi)^2}{\sigma_0}\sqrt{10}\mathscr{A} \sum_{l_2=2}^{\infty} \sum_{l_1=l_2\pm1} \sum_{l_1'=l_2,l_2\pm2} \frac{(-1)^{l_2}}{2l_2+1} C_{l_1 0 l_1' 0}^{30} \left\{\begin{array}{ccc} 2 & l_1' & l_2 \\ l_1 & 1 & 3 \end{array}\right\} \mathrm{Re}(d_{l_1 l_2} q_{l_1' l_2}^*),$$

$$\delta = \frac{(4\pi)^2}{\sigma_0} \frac{\mathscr{A}}{\sqrt{5}} \sum_{l_2=2}^{\infty} \sum_{l_1=l_2\pm 1} \sum_{l'_1=l_2,l_2\pm 2} \frac{(-1)^{l_2}}{2l_2+1} \operatorname{Re}(d_{l_1 l_2} q^*_{l'_1 l_2})$$
$$\times \left(\sqrt{3} C^{10}_{l_1 0 l'_1 0} \begin{Bmatrix} 2 & l'_1 & l_2 \\ l_1 & 1 & 1 \end{Bmatrix} + \sqrt{2} C^{30}_{l_1 0 l'_1 0} \begin{Bmatrix} 2 & l'_1 & l_2 \\ l_1 & 1 & 3 \end{Bmatrix} \right). \tag{16}$$

The dipole-quadrupole terms, which are of the order of ω/c, do not appear in the SDCS given by the parameter σ_0 in Eqs. (15) and (16), i.e., the lowest-order nondipole corrections that contribute to the SDCS are the quadrupole-quadrupole and dipole-octupole terms, which are of the order $(\omega/c)^2$ and are thus not accounted for here. Also, the parameter δ is generally non-zero. This is in contrast to SPI, for which δ^{SPI} vanishes for ionization from atomic s subshells [11].

We have calculated the SDCS, σ_0, and the asymmetry parameters β, γ, and δ by the LOPT approach for an excess energy of 450 eV for two energy sharings used in the experiment of Ref. [10]. For the angular distribution of the electron having energy $E_1 = 448$ eV, we find that $\sigma_0 = 3.06$ b/eV, $\beta = 1.93$, $\gamma = 0.51$, and $\delta = 0.0043$, as compared to the CCC results $\sigma_0 = 2.54$ b/eV, $\beta = 1.92$ [10]. For the angular distribution of the electron having energy $E_1 = 420$ eV, we find that $\sigma_0 = 1.19$ b/eV, $\beta = 1.83$, $\gamma = 0.46$, and $\delta = 0.011$, as compared to the CCC results, $\sigma_0 = 0.73$ b/eV, $\beta = 1.78$ [10].

Comparisons of the LOPT predictions for dipole-quadrupole TDCSs, as well as for the parameters γ and δ, with those calculated by a more elaborate approach would be of great interest and are now in progress [12].

ACKNOWLEDGMENTS

This work was supported in part by the U.S. Department of Energy, Office of Science, Division of Chemical Sciences, Geosciences, and Biosciences, under grant DE-FG03-96ER14646, by RFBR Grant 04-02-16350, and by the joint Grant VZ-010-0 of the CRDF and the RF Ministry of Education and Sciences (NLM and AVM).

REFERENCES

1. A.I. Mikhailov, I.A. Mikhailov, A.N. Moskalev, A.V. Nefiodov, G. Plunien, and G. Soff, *Phys. Rev. A* **69**, 032703 (2004).
2. A.Y. Istomin, N.L. Manakov, A.V. Meremianin, and A.F. Starace, *Phys. Rev. A* **71**, 052702 (2005).
3. N.L. Manakov, S.I. Marmo, and A.V. Meremianin, *J. Phys. B* **29**, 2711 (1996).
4. J.S. Briggs and V. Schmidt, *J. Phys. B* **33**, R1 (2000).
5. A.Y. Istomin, N.L. Manakov, A.V. Meremianin, and A.F. Starace, *Phys. Rev. Lett.* **92**, 063002 (2004).
6. J. Berakdar and H. Klar, *Phys. Rev. Lett.* **69**, 1175 (1992).
7. A.Y. Istomin, N.L. Manakov, A.V. Meremianin, and A.F. Starace, *Phys. Rev. A* **70**, 010702(R) (2004).
8. A.Y. Istomin, N.L. Manakov, and A.F. Starace, *J. Phys. B* **35**, L543 (2002).
9. A.Y. Istomin, N.L. Manakov, and A.F. Starace, *Phys. Rev. A* **69**, 032713 (2004).
10. A. Knapp, A. Kheifets, I. Bray, Th. Weber, A.L. Landers *et al.*, *Phys. Rev. Lett.* **89**, 033004 (2002).
11. J.W. Cooper, *Phys. Rev. A* **47**, 1841 (1993).
12. A.Y. Istomin, A.F. Starace, N.L. Manakov, A.V. Meremianin, A.S. Kheifets, and I. Bray, papers are in preparation.

Measurements of Low Energy Electron Scattering from Atomic Hydrogen

J. G. Childers* and M. A. Khakoo*

Department of Physics, California State University, Fullerton, California 92834, USA

Abstract. Normalized doubly-differential cross-sections and differential cross-sections for the electron impact ionization, excitation to the $n = 3$ and 4 levels, and elastic scattering from atomic hydrogen have been measured at incident electron energies ranging from 14.6 eV to 40 eV for scattering angles of 10° to 130°. The measurements are normalized to the accurate differential cross-section for the electron-impact excitation of the H (n=2) level. These measurements were made possible by the use of a new moveable target source which enables the collection of hydrogen energy loss spectra that are free of all backgrounds. Our measurements are in very good agreement with the results of recent theoretical models—the Convergent Close-Coupling model and the Exterior Complex Scaling model.

Keywords: atomic, hydrogen, ionization, excitation, elastic scattering, cross-sections
PACS: 34.80.Bm, 34.80.Dp

INTRODUCTION

Low-energy electron scattering from atomic hydrogen attracts great interest in collision physics. The interaction of an electron with a ground-state hydrogen atom is the simplest example of the three-body Coulomb problem incorporating fermions and therefore has significant theoretical interest. This problem has led to the development of two recent *ab initio* theoretical models: the Convergent Close-Coupling (CCC) model of Bray and Stelbovics [1] and Bray and Fursa [2], and the more recent Exterior Complex Scaling (ECS) model of Rescigno et al. [3]. In this paper, we compare select results of a number of recent measurements to previous measurements and to the theoretical calculations. The full results of the measurements are available [4, 5].

Ionization. Because of significant differences between the low-energy doubly-differential cross-section (DDCS) measurements of Shyn [6] and the CCC and ECS, and questions concerning the normalization of the triply-differential cross-section measurements of Röder et al. [7, 8], new DDCS measurements were needed.

Excitation of $n = 3$ and 4 levels. Recent measurements of the differential cross-section (DCS) for the excitation of the $n = 3$ and 4 levels by Sweeney et al. [9] show good agreement with the CCC, but these measurements were normalized using the elastic scattering DCS measurements of Shyn and Cho [10] and Shyn and Grafe [11]. At large scattering angles, these elastic DCSs differ from both the CCC and the earlier measurements of Williams [12], which are in good agreement with each other. Therefore, the DCS measurements of the $n = 3$ and 4 levels bear repeating.

Elastic Scattering. As mentioned above, there are significant disagreements between the elastic DCS measurements of Shyn and Cho [10], Shyn and Grafe [11], and Williams

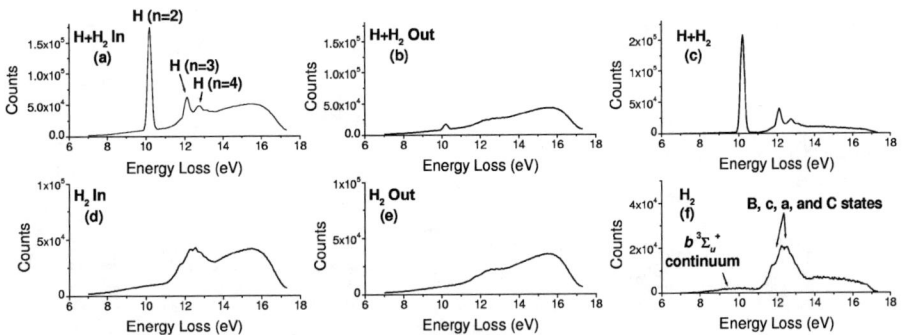

FIGURE 1. (a) Electron energy loss spectrum taken at $E_0 = 17.6$ eV and $\theta = 20°$ with the discharge on and gas beam needle aligned with the electron beam. (b) Same as (a) but with gas beam needle displaced away from the electron beam. (c) Spectrum in (b) subtracted from that in (a) without scaling. (d) Electron energy loss spectrum taken at $E_0 = 17.6$ eV and $\theta = 20°$ with the discharge off and gas beam needle aligned with the electron beam. (e) Same as (d) but with gas beam needle displaced away from the electron beam. (f) Spectrum in (e) subtracted from that in (d) without scaling.

[12]. This discrepancy needs to be resolved to complete the picture.

EXPERIMENTAL METHOD

Our apparatus has been discussed previously [13], so only a brief summary follows. The atomic beam is generated by an outside-silvered glass capillary needle of 0.5 mm internal diameter, and is made to cross a monochromatic beam of electrons from an electrostatic electron spectrometer in a conventional beam-beam configuration. Scattered electrons are detected by an electrostatic analyzer as a function of energy loss (E_L) and scattering angle (θ). The analyzer has a restricted depth-of-field so that it observes electrons only from a small volume of the collision region close to the capillary needle (about a 5 – 6 mm region). A four-element zoom lens enables it to transmit electrons over a wide range of kinetic energies with good efficiency. The efficiency of the analyzer is determined by measuring the spectrum of He at 31.7 eV incident electron energy (E_0) and 90° scattering angle. At this energy, the He ionization continuum is flat within 10% as observed by Keenan et al. [14]. The production of secondary electrons from surfaces in the experiment was reduced by using large electron-optic filling factors (> 0.5), tightly collimating the incident electron beam, and maintaining open, heavily sooted, and grounded areas around the collision region. The spectrometer performs with a typical incident electron current of $\sim 50 - 100$ nA with an energy resolution of about 120 – 150 meV (FWHM). This spectrometer is stable over long periods ($\sim$ 1 year). The unit is baked at ≈ 140 °C to maintain stability against oil contamination. The spectrometer is enclosed in a double μ-metal shield to reduce the Earth's magnetic field to less than 5 mG. Our atomic H source is detailed in Paolini and Khakoo [15]. Teflon tubing is used to conduct the atoms from the H-discharge tube to the outside-silvered glass capillary. This source delivers an intense, stable H beam with a dissociation fraction

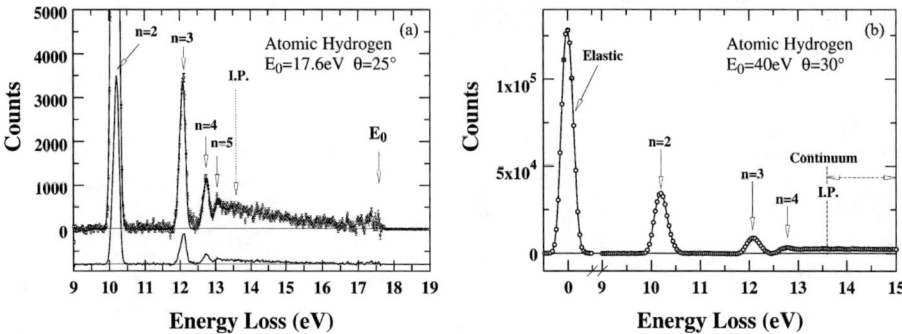

FIGURE 2. Electron energy loss spectra of H obtained after the application of the background subtraction procedures described in the text. (a) A spectrum of the $n = 2, 3$ and partially-resolved $n = 4$ features. The line is the unfolding fit to the spectrum. (b) A spectrum of the elastic scattering feature and the $n = 2$, 3 and partially-resolved $n = 4$ features.

of approximately 82 – 85%. Our measurements are comprised of electron energy loss spectra covering the elastic electron scattering peak and the energy loss range of 6.5 eV to $E_0 + 1$ eV. This covers the molecular hydrogen $b^3\Sigma_u^+$ continuum plus the full range of H_2 excited states including the ionization continuum of H_2 starting at 15.94 eV [16]. A major difficulty in these experiments is the isolation of the atomic hydrogen related scattering signal from background signals. To accomplish this, we have developed a new moveable source technique [17]. Spectra are collected as the capillary needle is rotated "In" and "Out" of the collision region by a "Hobby-Shack" servo-motor, as shown in Figure 1. A simple subtraction of the "Out" spectrum from the corresponding "In" spectrum leaves a spectrum containing only the contribution from gas related scattering. Excellent background determination free from additional electrons is observed for energies up to threshold. In the final step, the background-free discharge-off H_2 spectrum (Fig. 1f) is subtracted from the corresponding background-free discharge-on $H+H_2$ spectrum (Fig. 1c) after applying a scaling factor and allowing for small adjustments (< 60 meV) in the energy loss scale. This scaled subtraction was determined (within 6% on average) by viewing the resultant spectrum and ensuring that there was no residual background in the energy loss region between the $n = 2, 3$, and 4 energy loss features (Fig. 2). The resulting pure spectrum of H consists of discrete states resolved up to $n = 3$, the partially resolved $n = 4$ state, and the continuum. A typical resultant H spectrum following all corrections is shown in Fig. 2a, and a similar spectrum that includes elastic scattering is shown in Fig. 2b. The well-established CCC H ($n = 2$) DCS [1, 18] is used as the normalization standard to place our measurements of the continuum on an absolute scale. By fitting the continuum to a polynomial in energy loss of order ≤ 4, we obtained the continuum DDCSs. Error bars include statistical uncertainties propagated by all subtractions, uncertainties in determining the subtraction parameters, uncertainties in transmission of the analyzer, and uncertainties in the polynomial fitting to the continuum. We do not assume any errors in the DCS for the H ($n = 2$) feature from the CCC.

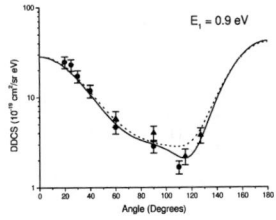

FIGURE 3. Doubly-differential cross-section for the electron impact ionization of H at $E_0 = 14.6$ eV obtained from the present experiments (before ● and after ▲ modification of the spectrometer to increase the angular range) compared to the ECS [19] (———) and the CCC [20] (···).

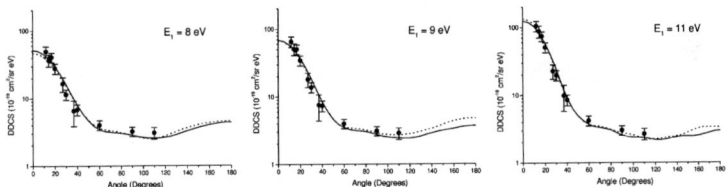

FIGURE 4. Doubly-differential cross-sections for the electron impact ionization of H at $E_0 = 25$ eV obtained from the present experiments (●) compared to the ECS [19] (———) and the CCC [20] (···) shown for different E_1 values.

RESULTS AND DISCUSSION

Doubly-differential ionization cross-sections. At $E_0 = 14.6$ eV and residual energy $E_1 = 0.9$ eV, shown in Fig. 3, our measurements agree very well with the calculations. This agreement is remarkable since this lowest E_0 presents the highest difficulty for both the calculations and the experiment. Figure 4 shows measurements taken at the intermediate energy $E_0 = 25$ eV which are also in excellent agreement with the CCC and ECS. Measurements taken at the highest energy used in these experiments, $E_0 = 40$ eV, are shown in Fig. 5. Again, we see excellent agreement with the calculations with the exception of low residual energies at forward scattering angles. Our measurements there show a forward scattering peak that is absent in both of the calculations. To verify

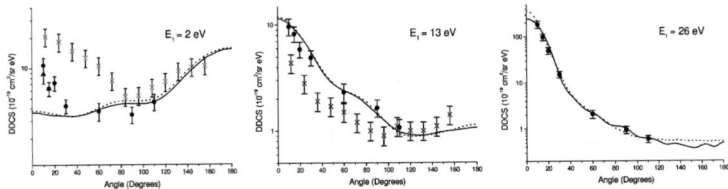

FIGURE 5. Same as Fig. 4, but for $E_0 = 40$ eV. Also shown are later measurements (▲) performed to verify the disagreement with the calculations at low residual energies, and the measurements of Shyn [6] (×) where available.

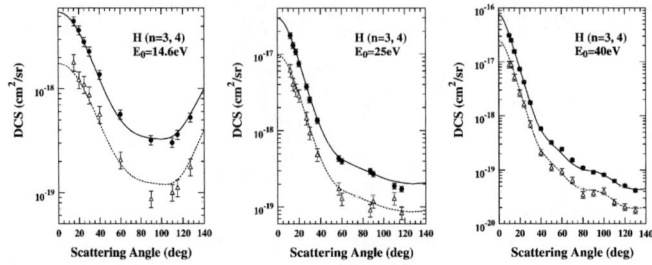

FIGURE 6. DCSs for the electron impact excitation of the $n = 3$ and $n = 4$ levels of H at $E_0 = 14.6$ eV, 25 eV, and 40 eV. Legend: • present work ($n = 3$), △ present work ($n = 4$), ——— ($n = 3$) and – – – ($n = 4$) CCC [1, 18]. Error bars constitute one standard deviation of uncertainty.

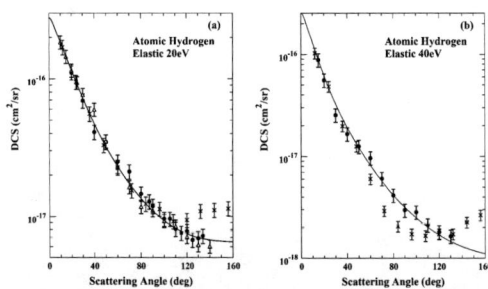

FIGURE 7. Relative elastic DCSs for electron scattering from H at (a) $E_0 = 20$ eV and (b) $E_0 = 40$ eV. Legend: • present work, △ Williams [12], × Shyn and Cho [10] (20 eV) and Shyn and Grafe [11] (40 eV), ——— CCC [1, 18]. The present results are normalized to the CCC DCS about $\theta = 20°$. Error bars constitute one standard deviation of uncertainty.

this disagreement with the calculations, we performed additional measurements at 10° and 15° scattering angles, shown as triangles in Fig. 5. However, we presently believe this to be a problem with the moveable source method and not with the calculations. Additional tests of the moveable source method at high E_0, low E_1, and small θ are under consideration by our group. At higher residual energies, especially at the highest $E_1 = 26$ eV, the agreement with the calculations is excellent.

Inelastic scattering differential cross-sections. Figure 6 shows our H $n = 3$ and 4 DCSs using the accurate CCC H ($n = 2$) DCS as a calibration standard. Typical uncertainties for $n = 3$ are in the region of 7% to 10%, and for $n = 4$ in the region of 14% to 19%. The agreement between our results and the CCC is excellent at all incident energies and scattering angles.

Elastic scattering differential cross-sections. Figure 7 shows our relative elastic DCSs compared with the CCC [1, 18] and with the experimental DCSs of Williams [12] and Shyn and Cho [10] at 20 eV (Fig. 7a) and with Shyn and Grafe [11] at 40 eV (Fig. 7b). Our relative DCSs are normalized to the absolute DCSs about $\theta = 20°$. We note that in Fig. 7a our results are in very good agreement with those of the CCC and Williams for all θ of this work. At large θ, our DCSs remain below those of Shyn and Cho [10]. At

40 eV (Fig. 7b) we see that our results are again in good agreement with the CCC but are in disagreement with the DCSs of Shyn and Grafe [11] around $\theta = 60°-100°$. The present results clearly demonstrate that the CCC obtains excellent values for the DCSs for elastic scattering from H.

CONCLUSIONS

Using a new moveable source method, we have measured accurate DDCSs for the electron-impact ionization of atomic H and DCSs for electron scattering from H for both elastic and inelastic processes at near-threshold to intermediate energies. We are able to obtain, after a relatively simple and direct data analysis, an energy loss spectrum of background-free H. The results show very good agreement with the recent CCC and ECS calculations over a wide range of incident electron energies and scattering angles.

ACKNOWLEDGMENTS

This project was funded by a grant from the National Science Foundation under grant NSF-RUI-PHY-0096808. We acknowledge the expert help of technical staff Mssrs. Jorge Meyer (glass blowing shop), David Parsons (machine shop) and Hugo Fabris (electronics shop), as well as undergraduate students Mark Hughes and Kenneth James.

REFERENCES

1. I. Bray and A. Stelbovics, *Phys. Rev. A*, **46**, 6995 (1992).
2. I. Bray and D. V. Fursa, *Phys. Rev. A*, **54**, 2991 (1996).
3. T. N. Rescigno, M. Baertschy, W. A. Isaacs, and C. W. McCurdy, *Science*, **286**, 2474 (1999).
4. J. G. Childers, K. E. James, Jr., I. Bray, M. Baertschy, and M. A. Khakoo, *Phys. Rev. A*, **69**, 022709 (2004).
5. K. E. James, Jr., J. G. Childers, and M. A. Khakoo, *Phys. Rev. A*, **69**, 022710 (2004).
6. T. W. Shyn, *Phys. Rev. A*, **45**, 2951 (1992).
7. J. Röder, J. Rasch, K. Jung, C. T. Whelan, H. Ehrhardt, R. J. Allan, and H. R. J. Walters, *Phys. Rev. A*, **53**, 225 (1996).
8. J. Röder, H. Ehrhardt, C. Pan, A. F. Starace, I. Bray, and D. V. Fursa, *Phys. Rev. Lett.*, **79**, 1666 (1997).
9. C. J. Sweeney, A. Grafe, and T. W. Shyn, *Phys. Rev. A*, **64**, 032704 (2001).
10. T. W. Shyn and S. Y. Cho, *Phys. Rev. A*, **40**, 1315 (1989).
11. T. W. Shyn and A. Grafe, *Phys. Rev. A*, **46**, 2949 (1992).
12. J. F. Williams, *J. Phys. B*, **8**, 2191 (1975).
13. M. A. Khakoo, M. Larsen, B. Paolini, X. Guo, I. Bray, A. Stelbovics, I. Kanik, S. Trajmar, and G. K. James, *Phys. Rev. A*, **61**, 012701 (1999).
14. G. A. Keenan, I. C. Walker, and D. F. Dance, *J. Phys. B*, **15**, 2509 (1982).
15. B. P. Paolini and M. A. Khakoo, *Rev. Sci. Instrum.*, **69**, 3132 (1998).
16. G. Herzberg, *Spectra of Diatomic Molecules*, D. Van Nostrand Company, New York, 1950.
17. M. Hughes, K. E. James, Jr., J. G. Childers, and M. A. Khakoo, *Meas. Sci. Technol.*, **14**, 841 (2003).
18. I. Bray, Private Communication (2003).
19. M. Baertschy, T. N. Rescigno, and C. W. McCurdy, *Phys. Rev. A*, **64**, 022709 (2001).
20. I. Bray, *Phys. Rev. Lett.*, **89**, 273201 (2002).

Kinetic Correlation In Photo-Double-Ionization Processes: The He-Isoelectronic Sequence.

S. Otranto[1†] and C. R. Garibotti[‡]

†Dto. de Física, Universidad Nacional del Sur, 8000 Bahía Blanca, Argentina
Physics Department, University of Missouri-Rolla, Rolla MO 65401, USA
‡CONICET and Centro Atómico Bariloche, S. C. de Bariloche 8400, Argentina.

Abstract. Analytical models proposed to represent the two-electron continuum are revisited. Main results obtained with these models are summarized. Recent studies of the photo-double-ionization (PDI) of the He-isoelectronic sequence by means of the recently introduced SC3 model are shown and compared with the results predicted by classical and semi-classical Wannier approaches. By fitting the triply differential cross sections (TDCSs) with the usual dipolar Gaussian form we find that the width has a power dependence on excess energy with exponent 0.25 in the near threshold region and departs from this law with increasing energy.

Keywords: Photo double ionization. Three body Coulomb problem.
PACS: 32.80.Fb

INTRODUCTION

During the last few years, the three body Coulomb problem has been considered in several articles in order to improve the theoretical description of atomic collisions processes [1-7]. The main idea underlying in these articles is to generalize an existing analytical model denoted as C3 [8-9] which fulfills the Redmond asymptotic conditions in the Ω_0 region, where the three particles are far from each other. These generalizations were built requiring a satisfactory description of other possible asymptotic configurations. For many atomic ionization processes, these models lead to a good qualitative description of the cross sections at high and intermediate collision energies, and allow for a partially analytic evaluation of transition amplitudes.

Nowadays, powerful numerical methods have gone beyond the existing analytical models leading to very accurate results for the cross sections for different processes [10-13]. Meanwhile, the description of an approximate continuum wave function for a three Coulomb interacting particles system remains as a challenging theoretical problem.

In a recent paper, we have presented a new analytical model for the two electron continuum, which we have denoted SC3 [14]. This model was formulated as to improve the C3 results for the photo double ionization (PDI) of He in the threshold region. From an approximated analysis in this region we have shown that the scaling

---
[1] Electronic address: sotranto@uns.edu.ar

of the interelectronic distance with a multiplicative energy dependent factor leads to a better description of the evaluated cross sections.

In this report, we summarize and discuss the above mentioned generalizations of the C3 model and we show recent results obtained with the SC3 model for the He-isoelectronic sequence. By fitting the TDCS with the usual dipolar Gaussian form [15], we obtain the energy dependence of the widths and compare our results with classical and semi-classical Wannier approaches [16-21].

Atomic units are used unless where explicitly stated.

THE THREE BODY COULOMB PROBLEM

The Schrödinger equation for the two-electron atom is a non-separable differential equation that couples the three possible pairs of particles through the usually denominated non-diagonal kinetic energy W_1 given by

$$W_1 = -\nabla_1 \cdot \nabla_{12} + \nabla_2 \cdot \nabla_{12} \quad (1)$$

The analytical C3 model [8-9] represents an approximated solution of the Schrödinger equation where only the diagonal terms of the kinetic energy are considered (ie. $W_1 = 0$). Since 1993, several models were developed in order to include the information contained in Eq. (2). C3-type models with coordinates or momenta-dependent charges have been developed by asking the satisfaction of several physical desirable asymptotic limits [1-3,7]. Models based on several variables hypergeometric functions have also been developed with the same aim [4-5]. Most of these models were initially designed and tested in the (e,2e) context. However, they have not been pushed further in the PDI field and up to our knowledge, only a few works have been devoted to their implementation [22-23]. For the screened models, the small number of asymptotic conditions imposed in order to determine the effective charges, leaves considerable freedom to potential models in the reaction region where the particles are close to each other and the information contained in Eq. (2) is supposed to be of vital importance and should be properly accounted.

Very recently, we have introduced another C3-type wave function which boosts the interelectronic distance through the introduction of an energy dependent multiplicative factor [14]. This model, hereafter denoted as SC3 model, was designed in order to correct the well known exponential decreasing behavior given by the C3 model in the threshold region. The latter is well known to be due to an overestimation of the electronic repulsion at low excess energies [24]. In the following section we focus our attention in the PDI process for the He-isoelectronic sequence and we present the results obtained with the SC3 model.

THE PDI PROCESS

The PDI process has proven to be one of the more stringent tests for any theory posed to describe the two-electron continuum. Due to the mono-electronic character of the operators conforming the dipole Hamiltonian, this process only can be achieved through electronic correlation and in general, results turn to be extremely sensitive on the modeling performed in both initial and final states [25]. Maulbetsch and Briggs studied the PDI of He using the C3 model to represent the two-electron continuum. The C3 model, even showing nice angular distributions given its simplicity, evidenced the necessity of new models mainly at low energies at which experimental data started to be available.

As has been shown by Lucey *et al.* [22] the use of approximations for both the initial and final states which do not correspond to the same Hamiltonian, leads to gauge discrepancies due to the lack of satisfaction of the Heisenberg equation of motion. In their work, a C3 model with coordinate and momentum-dependent charges proposed by Berakdar in 1996 [2] was used and disagreement between gauges was obtained for a whole set of initial state wave functions of different accuracy.

However, for the equal energy sharing regime where both electrons leave the reaction region with the same velocity, angular shapes obtained in velocity and length gauge are in good agreement with each other even when magnitude differences are obtained. Huetz *et al.* [15] pointed out that in this regime the TDCS can be rewritten as the product of a correlation factor times an angular factor:

$$\frac{d\sigma}{d\Omega_1 d\Omega_2 dE_1} = C(\theta_{12}, E_f, Z)(\cos\theta_1 + \cos\theta_2)^2. \quad (2)$$

During the last few years, the Gaussian shape for the correlation factor has become of standard use with its full width at half maximum considered as an empirical parameter over a wide range of excess energies [26-27]:

$$C(\theta_{12}, E_f, Z) = A(E_f, Z)\exp\left[-\frac{4\ln 2(\theta_{12} - 180°)^2}{\Gamma(E_f, Z)^2}\right]. \quad (3)$$

Semi-classical Wannier approaches have generally assumed the following energy dependence for the width Γ, even when they obtain somewhat different values for Γ_0.

$$\Gamma(E_f, Z) = \Gamma_0(Z) E_f^{1/4}. \quad (4)$$

In order to show the contrast between models we can point out that Altick model leads to the Z-independent value $\Gamma_0 = 66.7$ eV [20] while Rau predicts a constant value for Γ for $Z \geq 3$, indicating the complete absence of correlation footprints [16].

The PDI of the He-isoelectronic sequence was initially considered by Kornberg and Miraglia who proposed scalings for the TDCS, singly differential cross sections (SDCS) and total cross sections (TCS) [28]. The validity of the TCS scaling was also confirmed with the CCC method [10]. The validity of the TDCS scaling was recently shown by Otranto and Garibotti [14] with the SC3 model. According to Kornberg and Miraglia, the TDCS should verify the following scaling,

$$\frac{d\sigma}{d\Omega_1 d\Omega_2 dE_1}(E_f, Z) = \frac{1}{Z^6} \frac{d\sigma}{d\Omega_1 d\Omega_2 d(E_1/Z^2)}(E_f/Z^2, 1) \quad (5)$$

In Fig. 1 we present the TDCS in the SC3 model [14] for different ions and $E_f/Z^2=5eV$. For increasing nuclear charge, the angular distributions tend to the uncorrelated distribution $(1+\cos\theta_2)^2$. The maximum of the two lobes goes to $\theta_1=0$ and the distributions approach the envelope. It can be seen that convergence predicted by the scaling is achieved in the infinite nuclear charge limit.

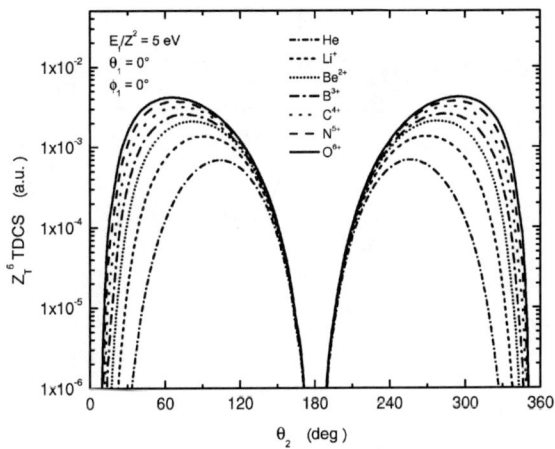

FIGURE 1. Scaled TDCS for $E_f/Z^2=5eV$ and $Z=2,..,8$.

In Fig. 2 we present the ratio of the theoretical TDCS to the angular factor. That is the SC3 correlation factor. Four different ions are considered at the excess energy of 20eV. As can be inferred from the figure, Gaussian fits properly reproduce the angular behavior of the correlation factor [14].

In Fig. 3 a) and b) we represent the $\Gamma(E_f,Z)$ as a function of E_f and $E_f^{1/4}$ respectively. It can be seen that for low excess energies, the widths behave like predicted by Eq. (5) and that departure from this law is obtained as energy increases. However this departure is charge dependent. As the nuclear charge increases, a wide window of the Wannier region is obtained. From Fig. 3a) we can see that the SC3 tends to agree with the predictions of the Z-independent model of Altick [20]. By the other side, from Figs. 2 and 3 we see that for charge values higher than 3 a clear dependence of the correlation factor with the interelectronic angle is obtained in contrast with the flat prediction of Rau [16]. A mean linear fitting of the present results leads $\Gamma(E_f,Z)=63.6\ E_f^{1/4}$ which is valid up to 40eV excess energy for the larger charges here considered.

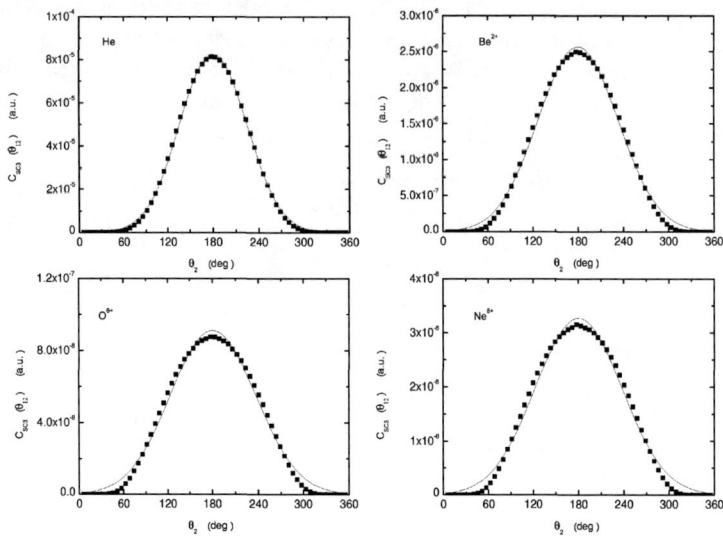

FIGURE 2. SC3 Correlation coefficients for ions with charge Z=2,4,8,10. The excess energy is 20eV. The Gaussian fit (solid line) has been included for comparison.

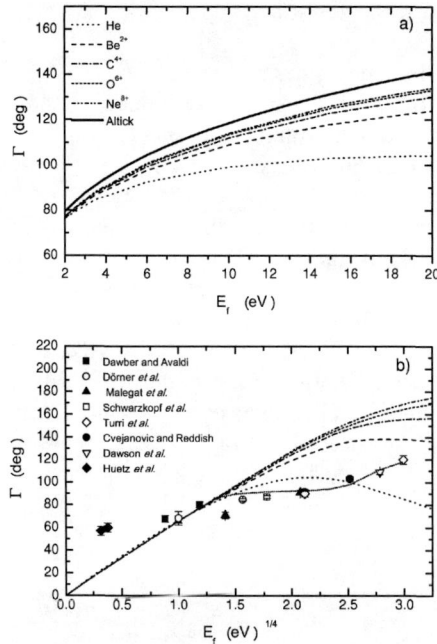

FIGURE 3. $\Gamma(E_f, Z)$ as a function of E_f a) and $E_f^{1/4}$ b). The short-dotted-line corresponds to the CCC theory. The experimental points included for the He target are those shown in Ref. [29].

CONCLUSIONS

We presented results given by the SC3 method for the PDI of the He-isoelectronic sequence ions and in the equal energy sharing regime. We showed that the SC3 results are well described by a Gaussian parametrization up to relatively high energies and nuclear charges. The resulting widths converge to the prediction of Altick's model, indicating that the electron-electron correlation keeps relevant for increasing nuclear charges. This contradicts other Wannier-type models like those developed by Rau [16] and Gailitis and Peterkop [17]. Even when the SC3 model is a perturbative model, the width behavior it displays is in concordance with the Wannier theory at low energies. Furthermore, we find that the threshold region extends up to higher energies as the nuclear charge increases, indicating an energy widening of the Wannier region.

ACKNOWLEDGMENTS

This work was partially supported by the ANPCYT under Grant No. PICTR 03/00437 and UNS under Grant No. PGI 24/F027 (Argentina). NSF and DOE are also acknowledged for partial support of SO travel expenses.

REFERENCES

1. E. O. Alt and A. M. Mukhamezhanov, *Phys. Rev. A* **47**, 2004 (1993).
2. J. Berakdar, *Phys. Rev. A* **53**, 2314 (1996).
3. Z. Chen, S. Zhang, Q. Shi, J. Chen and K. Xu, *J. Phys. B: At. Mol. Opt. Phys.* **30**, 4963 (1997).
4. G. Gasaneo, F. D. Colavecchia, C. R. Garibotti, J. E. Miraglia and P. Macri, *Phys. Rev. A* **55**, 2809, 1997.
5. J. E. Miraglia, M. G. Bustamante and P. A. Macri, *Phys. Rev. A* **60**, 4532 (1999).
6. G. Gasaneo and F. D. Colavecchia, *Nucl. Instr. and Meth. B* **192**, 150 (2002).
7. J. R. Götz, M. Walter and J. S. Briggs, *J. Phys. B: At. Mol. Opt. Phys.* **38**, 1569 (2005).
8. C. R. Garibotti and J. E. Miraglia, *Phys. Rev. A* **21** 572 (1980).
9. M. Brauner, J. S. Briggs and H. Klar, *J. Phys. B: At. Mol. Opt. Phys.* **22**, 2265 (1989).
10. A. S. Kheifets and I. Bray, *Phys. Rev. A* **62**, 065402 (2000).
11. J. Colgan, M. S. Pindzola M and F. Robicheaux, *J. Phys. B: At. Mol. Opt. Phys.* **34**, L457 (2001).
12. L. Malegat, P. Selles and A. K. Kazansky, *Phys. Rev. Lett.* **85**, 4450 (2000).
13. T.M. Rescigno, M. Baertschy, W.A. Isaacs, C.W. McCurdy, *Science* **286**, 2474 (1999).
14. S. Otranto and C. R. Garibotti, *Eur. Phys. J. D* **21**, 285 (2002).
15. A. Huetz, P. Selles, D. Waymel and J. Mazeau, *J. Phys. B: At. Mol. Opt. Phys.* **24**, 191 (1991).
16. A.R.P. Rau, *J. Phys. B: At. Mol. Opt. Phys.* **9**, L283 (1976).
17. M. Gailitis and R. Peterkop, *J.Phys.B: At. Mol. Opt. Phys.* **22**, 1231 (1989).
18. R. Peterkop, *J.Phys.B: At. Mol. Opt. Phys.* **4**, 513 (1971).
19. J.M. Feagin, *J. Phys. B: At. Mol. Opt. Phys.* **17**, 2433 (1984).
20. P. L. Altick, *J. Phys. B: At. Mol. Opt. Phys.* **18**, 1841 (1985).
21. A.K. Kazansky and V.N. Ostrovsky, *J. Phys. B: At. Mol. Opt. Phys.* **26**, 2231 (1993).
22. S. P. Lucey, J. Rasch, C. T. Whelan and H. R. J. Walters, *J. Phys. B: At. Mol. Opt. Phys.* **31**, 1237 (1998).
23. S. Otranto and C. R. Garibotti, *Eur. Phys. J. D* **27**, 215 (2003).
24. J. S. Briggs and V. Schmidt, *J. Phys. B: At. Mol. Opt. Phys.* **33**, R1 (2000).
25. F. Maulbescht and J. S. Briggs, *J. Phys. B: At. Mol. Opt. Phys.* **26**, L647 (1993); **27**, 4095 (1994).
26. S. Cvejanovic and T.J. Reddish, *J. Phys. B: At. Mol. Opt. Phys.* **33**, 4691 (2000).
27. P. Bolognesi et al., *J. Phys. B: At. Mol. Opt. Phys.* **34**, 3139 (2001).
28. M. Kornberg and J. E. Miraglia J E, *Phys. Rev. A* **48**, 3714 (1993); *Phys. Rev. A* **49**, 5120 (1994).
29. A. Huetz and J. Mazeau, *Phys. Rev. Lett.* **85**, 530 (2000).

Theory of Electron Impact Ionization of Atoms

A. T. Stelbovics, P. L. Bartlett, I. Bray, D. V. Fursa and A. S. Kadyrov

Centre for atomic, Molecular and Surface Physics, Division of Science and Engineering, Murdoch University, Perth 6150, Australia

Abstract. The theory of electron impact ionization of one- and two-electron atoms has advanced significantly in the past two years. This paper will summarize the progress that the members of our research center have contributed to.

Keywords: electron-impact ionization
PACS: 34.80.Dp

INTRODUCTION

We wish to report on our recent progress in several aspects of the electron-impact ionization from atoms. First we summarize the considerable developments in the formal aspects of the theory of ionization. Then we turn to a discussion of recent near-threshold numerical studies of electron impact from an atomic hydrogen target. This is followed by some examples of our recent work on electron impact ionization from the two-electron atoms, including helium and calcium.

FORMAL THEORY

The theoretical development we wish to report concerns the three body scattering wave function. We were able to derive the asymptotic form of the three-body scattering wave function [1, 2] and use it to establish integral representations for the scattering amplitude that are free of divergence problems characteristic of previous formulations.

It is well known that the ionization amplitude can be represented in terms of a trial integral which has a structure well suited for practical calculations. Here we show that it can be written in similar form without recourse to external trial quantities which is the requirement of a formally complete scattering theory. First we note that the total scattering wave Φ_i^+ developed from the initial two-fragment state $\Phi^{(i)}$ satisfies (in notation of Ref. [2])

$$(E-H)\Phi_i^{(\text{sc})+}(r_1, r_2) = \overline{V}_i \Phi^{(i)}(r_1, r_2), \qquad (1)$$

where $\overline{V}_i$ is the projectile-target interaction potential, and we separated Φ_i^+ according to $\Phi_i^+ = \Phi^{(i)} + \Phi_i^{(\text{sc})+}$. Therefore, for the ionization amplitude in prior form we can write that

$$T^{(\text{prior})}(k_1, k_2) \equiv \langle \Psi_f^- | \overline{V}_i | \Phi^{(i)} \rangle$$

$$\begin{aligned}
&= \langle \Psi_f^- | E - \vec{H} | \Phi_i^{(\text{sc})+} \rangle \\
&= \langle \Psi_f^- | \overleftarrow{H} - E | \Phi_i^{(\text{sc})+} \rangle + \langle \Psi_f^- | E - \vec{H} | \Phi_i^{(\text{sc})+} \rangle \\
&= \langle \Psi_f^- | \overleftarrow{H}_0 - \vec{H}_0 | \Phi_i^{(\text{sc})+} \rangle \equiv T^{(a)}(\boldsymbol{k}_1, \boldsymbol{k}_2),
\end{aligned} \qquad (2)$$

where a left (right) arrow on the differential Hamiltonian operator indicates that it acts on the *bra* (*ket*) state. This is a new surface-integral form for the ionization amplitude.

Next we note that the total scattering wave Ψ_f^+ developed from the final three-fragment state $\Psi^{(f)-}$ satisfies

$$(E - H_0)\Psi_f^-(\boldsymbol{r}_1, \boldsymbol{r}_2) = V\Psi_f^-(\boldsymbol{r}_1, \boldsymbol{r}_2). \qquad (3)$$

In addition, we note that

$$(E - H_0)\Phi^{(i)}(\boldsymbol{r}_1, \boldsymbol{r}_2) = V_i\Phi^{(i)}(\boldsymbol{r}_1, \boldsymbol{r}_2). \qquad (4)$$

In the light of Eqs. (3) and (4) we get

$$\begin{aligned}
T^{(\text{prior})}(\boldsymbol{k}_1, \boldsymbol{k}_2) &\equiv \langle \Psi_f^- | V - V_i | \Phi^{(i)} \rangle \\
&= \langle \Psi_f^- | E - \overleftarrow{H}_0 - (E - \vec{H}_0) | \Phi^{(i)} \rangle \\
&= -\langle \Psi_f^- | \overleftarrow{H}_0 - \vec{H}_0 | \Phi^{(i)} \rangle \equiv T^{(b)}(\boldsymbol{k}_1, \boldsymbol{k}_2).
\end{aligned} \qquad (5)$$

If we separate the unscattered and scattered parts of wave function Ψ_f^- according to $\Psi_f^- = \Psi^{(f)-} + \Psi_f^{(\text{sc})-}$ then we can also introduce

$$T^{(c)}(\boldsymbol{k}_1, \boldsymbol{k}_2) = \langle \Psi^{(f)-} | \overleftarrow{H}_0 - \vec{H}_0 | \Phi_i^+ \rangle, \qquad (6)$$

$$T^{(d)}(\boldsymbol{k}_1, \boldsymbol{k}_2) = -\langle \Psi_f^{(\text{sc})-} | \overleftarrow{H}_0 - \vec{H}_0 | \Phi_i^+ \rangle. \qquad (7)$$

Forms $T^{(a)}$, $T^{(b)}$, $T^{(c)}$ and $T^{(d)}$ are convenient for numerical calculations as the result depends only on the asymptotic behaviour of the scattered wave functions. The asymptotic forms of Φ_i^+ and Ψ_f^- have been given in [2, 3].

We emphasize the importance of the new form of the ionization amplitude $T^{(c)}$ given by Eq. (6), from the point of view of the general scattering theory. Eq. (6) leads to a well-defined conventional volume-integral form of the ionization amplitude in terms of the total three-body scattering wave function Φ_i^+, being developed from the initial two-fragment channel $\Phi^{(i)}$. In the stationary-state scattering theory the *post* form of the breakup amplitude is defined by

$$T^{(\text{post})}(\boldsymbol{k}_1, \boldsymbol{k}_2) = \langle \boldsymbol{k}_1, \boldsymbol{k}_2 | V | \Phi_i^+ \rangle, \qquad (8)$$

where $\langle \boldsymbol{r}_1, \boldsymbol{r}_2 | \boldsymbol{k}_1, \boldsymbol{k}_2 \rangle = e^{i\boldsymbol{k}_1 \cdot \boldsymbol{r}_1 + i\boldsymbol{k}_2 \cdot \boldsymbol{r}_2}$ is the undistorted three-body plane wave. However, this form is valid only when interaction between particles is short-ranged. The commonly accepted stationary theory of scattering fails to define the same for long-range interactions. At the same time from the c-form of the ionization amplitude we get

$$T^{(c)}(\boldsymbol{k}_1,\boldsymbol{k}_2) = \langle \Psi^{(f)-}|\overleftarrow{H}_0+V-E-\overrightarrow{H}_0-V+E|\Phi_i^+\rangle \quad (9)$$
$$= \langle \Psi^{(f)-}|\overleftarrow{H}_0+V-E|\Phi_i^+\rangle. \quad (10)$$

This allows us to introduce

$$T^{(\text{post})}(\boldsymbol{k}_1,\boldsymbol{k}_2) = \langle \Psi^{(f)-}|\overleftarrow{H}-E|\Phi_i^+\rangle. \quad (11)$$

Eq. (11) takes the form of Eq. (8) when the full interaction is short-ranged. Thus, Eq. (11) extends the definition of the *post*-form of the breakup amplitude to long-range potentials including the Coulomb interaction.

e–H NUMERICAL SOLUTION

We have recently made important progress in the direct numerical solution of electron-hydrogen ionizing collisions. Our method, propagating exterior complex scaling (PECS), combines exterior complex scaling [4] with a highly efficient numerical technique to obtain solutions to the full time-independent Schrödinger equation for *e*–H collisions in coordinate space [5]. This method was used to calculate accurate total and differential ionization cross sections at energies to within 0.01 a.u. of threshold [6], which provided convincing fully-quantal *ab initio* support for the Wannier [7] and related ionization threshold laws.

Our TICS results, shown in Fig. 1a, are scaled by $E^{1.127}$ to give emphasis to the low-energy results and highlight the improvement in precision over older convergent close-coupling (CCC) and R-matrix with pseudo states (RMPS) calculations. The estimated

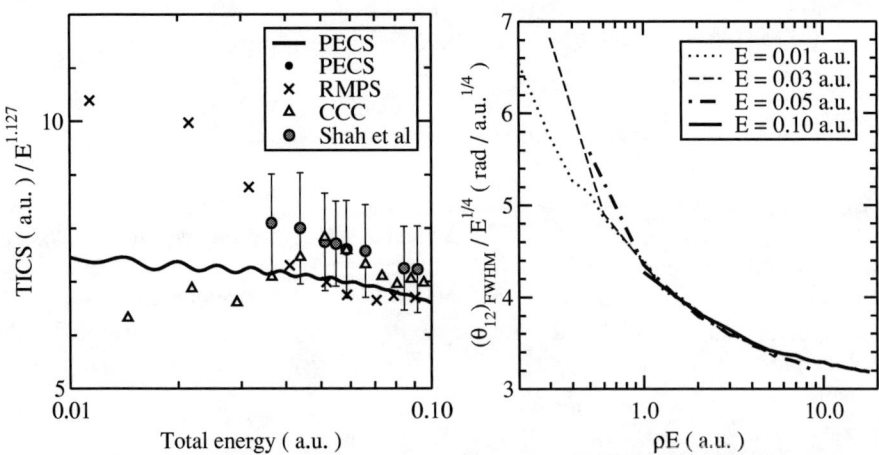

FIGURE 1. PECS calculations for *e*–H ionizing collisions near threshold. (a) Total ionization cross section (TICS) scaled by $E^{1.127}$, and (b) Full-width-half-maximum (FWHM) of the angle between ejected electrons scaled by $E^{1/4}$. CCC and RMPS calculations from [8] and measurements from [9].

standard errors of our calculations range from ±3% at 0.01 a.u. to ±1% as 0.10 a.u.. We applied non-linear curve-fitting techniques to these results and calculated the threshold power law to be $\sigma \propto E^{1.122\pm 0.015}$, and the L=0 triplet power law to be $\sigma \propto E^{3.36\pm 0.02}$, in accord with Wannier theory. Analysis of the single-differential cross sections showed a deviation of approximately 4% from energy-independence of the outgoing electrons, and we estimated the spin-asymmetry to approach the limiting value $A_s = 0.54 \pm 0.01$, as threshold is approached.

Wannier theory predicts that the full-width-half-maximum of the angular separation of the outgoing electrons is given by $(\theta_{12})_{\text{FWHM}} = \alpha E^{1/4}$, but there is disagreement on the value of α (see [6] and references therein). An important outcome from our near-threshold calculations was that we were able to establish that $\alpha = 3.0 \pm 0.2$ (in atomic units), which is in the middle of the range of values predicted by semi-classical methods (2.66 to 3.55). Figure 1b shows our results for $(\theta_{12})_{\text{FWHM}}$ as a function of ρE, where ρ is the hyperradius where the cross sections are calculated, and scaled by $E^{1/4}$. The results have not fully converged with respect to increasing ρE, but give convincing evidence of a $E^{1/4}$ relationship, and an estimate for α was made from an asymptotic extrapolation. For the energies considered here, we estimate that full convergence of α would require a calculation grid extending to $\rho \approx 2000\ a_0$, well beyond the 180 a_0 used for these calculations, and well beyond our present computational resources.

TWO-ELECTRON TARGETS

Two electron-atoms present a more challenging vista for theorists. As we have indicated, the atomic three-body problem is largely solved numerically. The treatment of the four-body problem requires further simplifying assumptions to reduce it to a form suitable for practical solution. In this section we illustrate how the CCC method can be applied to the problem of electron impact ionization of two-electron atoms. We will consider the case of equal-energy sharing for the ionizing collisions as the CCC theory for this case is has received detailed analysis and is now well understood. We will contrast recent studies for Helium where there are numerous investigations with Calcium which has been far less studied.

Helium

Helium is the ideal target for the experimentalists with by far the most data available than for any other target. What is particularly helpful is that much of these data are on the absolute scale allowing for quantitative comparison with theory. From the theoretical perspective while helium is not as ideal as atomic hydrogen, it turns out that by far the most dominant transitions involve the excitation of only one electron. This allows the treatment of the e-He problem as predominantly a three-body problem. The details of the CCC theory for e-He ionization have been given some time ago [10, 11]. The case of calculating electron-helium ionization with equal-energy outgoing electrons has been studied in great detail very recently [12]. Here, in Fig. 2, we show a representative ex-

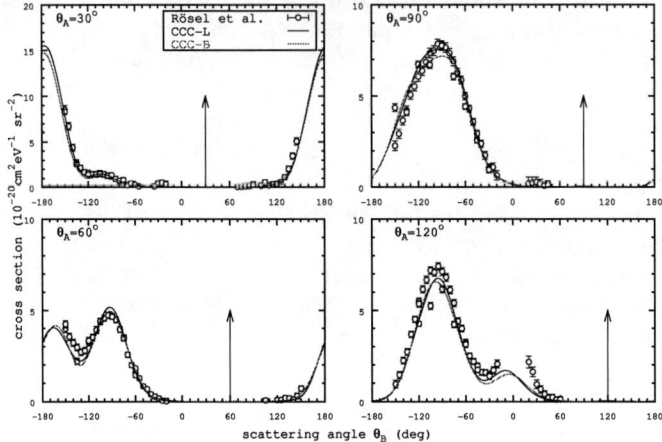

FIGURE 2. 26.6 eV electron-impact ionization of He with two 1 eV outgoing electrons. The theory and experiment are from Refs. [12] and [13], respectively.

ample of the excellent quantitative agreement between the CCC theory and experiment.

Calcium

We treat the calcium target as a two-electron atom with an inert Hartree-Fock core [15]. This way the theory developed for ionization of helium is equally applicable to

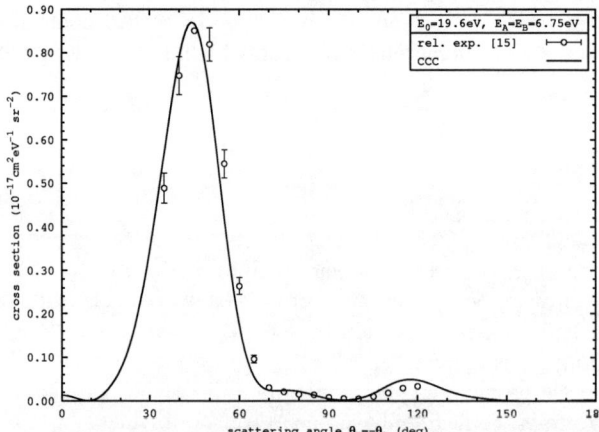

FIGURE 3. 19.6 eV e-Ca ionisation with two 6.75 eV outgoing electrons. The relative measurements of Murray and Cvejanovic [14] have been normalised to the present CCC calculation.

calcium. In Fig. 3 we give the results of a previously unpublished CCC calculation and compare with existing measurements [14]. While agreement looks excellent for the case considered, as the threshold is approached substantial discrepancies arise. This is currently under investigation.

CONCLUDING REMARKS

The review has presented snapshots of our most recent work. The progress in our understanding of the theory that has led to the closing of some formal long-standing problems is satisfying. Our next work will be to fully implement the new approach as a viable calculational framework. Similarly for electron-hydrogen scattering, following on from the earlier successes of CCC and ECS methods the PECS calculations have been able to reveal details of the near threshold ionization behaviour and to confirm the predictions of the semiclassical models. Our continuing focus will be now to model the two-electron systems in detail. While complete solution of the full Schrödinger equation for a three-electron system is still not possible, it should be within the next few years. One line of our research will be to extend the PECS method to this system. From a practical point of view the CCC formulation of ionization is well advanced and it now remains to investigate more fully application to quasi two-electron targets.

ACKNOWLEDGMENTS

The work was supported by the Australian Research Council.

REFERENCES

1. A. S. Kadyrov, A. M. Mukhamedzhanov, A. T. Stelbovics, and I. Bray, *Phys. Rev. Lett.*, **91**, 253202 (2003).
2. A. S. Kadyrov, A. M. Mukhamedzhanov, A. T. Stelbovics, and I. Bray, *Phys. Rev. A*, **70**, 062703 (2004).
3. A. S. Kadyrov, A. M. Mukhamedzhanov, A. T. Stelbovics, I. Bray, and F. Pirlepesov, *Phys. Rev. A*, **68**, 022703 (2003).
4. T. N. Rescigno, M. Baertschy, W. A. Isaacs, and C. W. McCurdy, *Science*, **286**, 2474–2479 (1999).
5. P. L. Bartlett, A. T. Stelbovics, and I. Bray, *J. Phys. B*, **37**, L69 (2004).
6. P. L. Bartlett, and A. T. Stelbovics, *Phys. Rev. Lett.*, **93**, 233201 (2004).
7. G. H. Wannier, *Phys. Rev.*, **90**, 817–825 (1953).
8. K. Bartschat, and I. Bray, *J. Phys. B*, **29**, L577–L583 (1996).
9. M. B. Shah, D. S. Elliot, and H. B. Gilbody, *J. Phys. B*, **20**, 3501–3514 (1987).
10. D. V. Fursa, and I. Bray, *Phys. Rev. A*, **52**, 1279–1298 (1995).
11. I. Bray, and D. V. Fursa, *Phys. Rev. A*, **54**, 2991–3004 (1996).
12. A. T. Stelbovics, I. Bray, D. V. Fursa, and K. Bartschat, *Phys. Rev. A*, **71**, 052716(13) (2005).
13. T. Rösel, J. Röder, L. Frost, K. Jung, H. Ehrhardt, S. Jones, and D. H. Madison, *Phys. Rev. A*, **46**, 2539–2552 (1992).
14. A. J. Murray, and D. Cvejanovic, *J. Phys. B*, **36**, 4875–4888 (2003).
15. D. V. Fursa, and I. Bray, *J. Phys. B*, **30**, 5895–5913 (1997).

Interference effects in single ionization of H_2 molecules by fast electron impact

O.A. Fojón[*,†], C.R. Stia[*] and R.D. Rivarola[*,†]

[*]*Instituto de Física Rosario, CONICET-UNR, Av. Pellegrini 250, 2000 Rosario, Argentina*
[†]*Escuela de Ciencias Exactas y Naturales, FCEIA, UNR, Rosario, Argentina*

Abstract. Interference effects in the ejected electron spectra of (e,2e) reactions with hydrogen molecules is studied. A three continuum molecular model is employed to obtain multiple differential cross sections. Theoretical predictions are compared with available data for hydrogen and deuterium molecules. In particular, it is shown that constructive interferences are produced at the binary region.

Keywords: ionization, hydrogen molecule, interference
PACS: 34.80.Gs,03.75.-b,31.15.-p,34.10.+x

INTRODUCTION

Interference phenomena have been of crucial importance in the foundation of Quantum Mechanics. Cohen and Fano [1] predicted by using a simple model (hereafter CF) that the spectrum of photoelectrons emerging from diatomic molecules is modulated by interferences. Consequently, interference patterns coming from the coherent emission from both molecular centres should be observed at high photon impact energies. However, these patterns are difficult to detect as the cross section decreases quickly as the impact energy increases. Only recently, measurements of the electron emission spectra in collisions of fast ions impacting on H_2 provided experimental evidence for these interference patterns, renewing the interest on these subjects [2]. The validity of the basic assumptions of the CF formula was studied in the case of photoionization of H_2 molecules. In particular, it was found that at high impact energies there is a good qualitative agreement with the oscillations predicted by the CF formula. However, at lower energies discrepancies with ab-initio results and experiments were observed. It was concluded that screening and correlation effects not present in the simple CF formula are responsible of these failures [3].

In this work we focus on the interference effects in the high energy part of the ejected electron spectra of (e,2e) reactions with H_2 molecules. We present theoretical multiply differential cross sections obtained by using a molecular three-continuum approximation taking into account the Coulomb interactions present in the final channel of the reaction [4]. Atomic units are used except where otherwise stated.

THEORY

We consider the single ionization process of hydrogen molecules following the impact of fast electrons, i.e.,

$$e^- + H_2(^1\Sigma_g^+) \to 2e^- + H_2^+(^2\Sigma_g^+). \tag{1}$$

We are interested in this work in highly asymmetric geometries. Therefore, exchange effects are neglected. Moreover, the fixed nuclei approximation is used as only high impact energies are considered, i.e., the molecule has no time to rotate nor vibrate during the fast ionization process. In consequence, the ionization reaction may be approximated by vertical electronic transitions from the electronic ground state of H_2 to the electronic ground state of H_2^+ at the fixed equilibrium distance of H_2. Therefore, the five-fold differential cross section (5DCS) reads,

$$\sigma^{(5)} = \frac{d^5\sigma}{d\Omega_\rho d\Omega_e d\Omega_s d(k_e^2/2)} \simeq 2(2\pi)^4 \frac{k_e k_s}{k_i} |t_{fi}^e(\rho_0)|^2 \tag{2}$$

where the emitted electron is ejected with momentum $\mathbf{k}_e$ into the solid angle Ω_e with respect to the incidence direction defined by the momentum $\mathbf{k}_i$ of the incident electron, and the projectile is scattered with momentum $\mathbf{k}_s$ into the solid angle Ω_s. ρ (ρ_0) is the (equilibrium) internuclear distance of H_2. In order to take into account the two electrons of the H_2 molecule, an extra factor 2 is included in the expression (2). The *prior* version of the electronic transition matrix element reads,

$$t_{fi}^e(\rho_0) = \left\langle \Psi_f^-(\rho_0, \mathbf{R}, \mathbf{r}_1, \mathbf{r}_2) | V_i | \Psi_i(\rho_0, \mathbf{R}, \mathbf{r}_1, \mathbf{r}_2) \right\rangle \tag{3}$$

where Ψ_i is the non-perturbed electronic wave function in the initial channel and Ψ_f^- is the final exact electronic wave function with correct asymptotic conditions. In equation (3), V_i represents the perturbation in the initial channel given by,

$$V_i = \frac{1}{r_{1p}} + \frac{1}{r_{2p}} - \frac{Z_T}{R_a} - \frac{Z_T}{R_b} \tag{4}$$

with $Z_T = 1$, the charge of the molecular nuclei. The position vector of each electron in the initial bound state with respect to the centre of mass of the molecule is denoted by $\mathbf{r}_1$ and $\mathbf{r}_2$. The position vectors of the incident electron with respect to nucleus a, to nucleus b and to the centre of mass of H_2 are given by $\mathbf{R}_a$, $\mathbf{R}_b$ and $\mathbf{R}$, respectively. The position vectors of the incident electron with respect to each bound electron are given by $\mathbf{r}_{1p}$ and $\mathbf{r}_{2p}$.

The initial wave function Ψ_i is represented as

$$\Psi_i = \frac{e^{i\mathbf{k}_i \cdot \mathbf{R}}}{(2\pi)^{3/2}} \Phi_i(\rho, \mathbf{r}_1, \mathbf{r}_2) \tag{5}$$

where the initial molecular bound state Φ_i is described here by a Heitler-London type wave function,

$$\Phi_i(\rho_0, \mathbf{r}_1, \mathbf{r}_2) = N_{HL}(\rho_0) \left\{ e^{-\alpha^* r_{1a}} e^{-\alpha^* r_{2b}} + e^{-\alpha^* r_{1b}} e^{-\alpha^* r_{2a}} \right\} \tag{6}$$

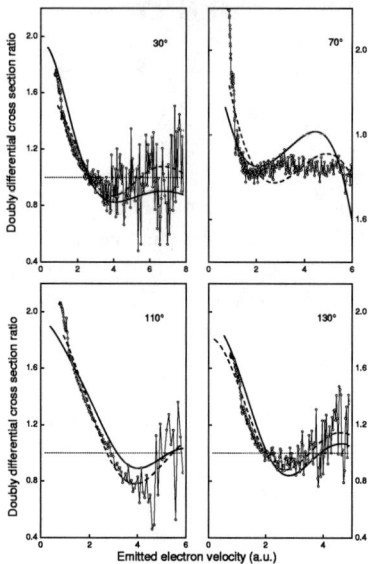

FIGURE 1. Normalized DDCS ratios corresponding to the single ionization of D_2 molecules as a function of the emitted electron velocity, for ejection angles of 30°, 70°, 110°, and 130° at impact energy $E_i = 2.4$ keV. Solid lines, theoretical first-order TEC calculations. Dashed lines, fitting to the experimental ratios using a CF type function. Open circles, experimental data from Ref. [8].

with $\alpha^* = 1.166$, $\rho_0 = 1.406$ and N_{HL} the normalization function. The position vectors of each bound electron with respect to nucleus $j = a, b$ are denoted by $\mathbf{r}_{1j}$ and $\mathbf{r}_{2j}$.

The choice of the approximate final electronic wave function is made within the framework of the molecular three-continuum approximation [4]. In this model, both the two-effective center description (TEC) [5] and the correlated three-body continuum wave function previously used in the atomic case [6] are employed in order to describe the final channel (see reference [4] for details). The molecular three-continuum model (MBBK) was used with success to describe absolute triple differential cross sections (TDCS) for single ionization of H_2 under the same energetic regime of the present interest and also to describe measured TDCS at intermediate incident energies [4].

The exact final wave function may be then approximated as

$$\Psi_f^- \cong \frac{e^{i\mathbf{k}_s \cdot \mathbf{R}}}{(2\pi)^{3/2}} \Phi_f(\rho, \mathbf{r}_2) \xi_c \tag{7}$$

where Φ_f is the wave function corresponding to the bound state of the residual H_2^+ molecular ion represented by a simple linear combination of atomic orbitals,

$$\Phi_f(\rho, \mathbf{r}_2) = N_f(\rho) \{e^{-\alpha r_{2a}} + e^{-\alpha r_{2b}}\} \tag{8}$$

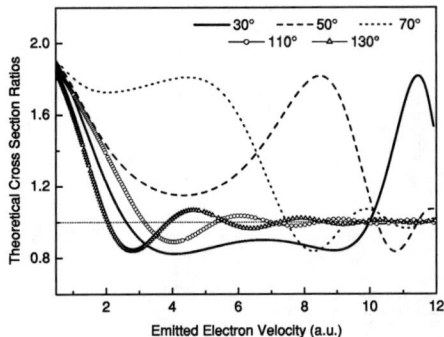

FIGURE 2. Normalized theoretical DDCS ratios for the ionization of H_2 molecules as a function of the ejection energy for several ejection angles. Same kinematical conditions as in Fig. 1

with $\alpha = 1.3918$ and $N_f(\rho)$ the normalization factor evaluated at $\rho_0 = 1.406$. The function ξ_c in equation (7) is chosen as,

$$\xi_c(\mathbf{k}_e,\mathbf{k}_s,\mathbf{R},\mathbf{r}_1,\mathbf{r}_2) = \frac{e^{i\mathbf{k}_e \cdot \mathbf{r}_1}}{(2\pi)^{3/2}} C(\mathbf{k}_e,\mathbf{r}_{1j},\gamma_e) C(\mathbf{k}_s,\mathbf{R}_j,\gamma_s) C(\mathbf{k}_{1p},\mathbf{r}_{1p},\gamma_{ep}) \quad (9)$$

where $j = a$ or b, and the Sommerfeld parameters γ are given by,

$$\gamma_s = -\frac{Z_T}{k_s} \qquad \gamma_e = -\frac{Z_T}{k_e} \qquad \gamma_{ep} = \frac{1}{2k_{1p}}. \quad (10)$$

Here, $\mathbf{k}_{1p} = \frac{1}{2}(\mathbf{k}_s - \mathbf{k}_e)$ is the momentum conjugate to $\mathbf{r}_{1p}$. So, the function ξ_c describes the mutual interactions between the three particles in the final channel of the reaction through the use of the Coulomb wave functions $C(\mathbf{k},\mathbf{r},\gamma)$ given by

$$C(\mathbf{k},\mathbf{r},\gamma) = \Gamma(1-i\gamma) e^{-\pi\gamma/2} {}_1F_1[i\gamma;1;-i(kr+\mathbf{k}\cdot\mathbf{r})]. \quad (11)$$

Following Ref. [7], molecular 5DCS become

$$\sigma^{(5)} = \frac{d^5\sigma}{d\Omega_\rho d\Omega_e d\Omega_s d(k_e^2/2)} \cong 2\left[1+\cos(\chi\cdot\rho_0)\right]\sigma_A^{(3)} \quad (12)$$

where $\chi = \mathbf{k}_e - \mathbf{K}$, and $\mathbf{K} = \mathbf{k}_i - \mathbf{k}_s$ is the momentum transferred to the ionized electron. The function $\sigma_A^{(3)}$ appearing in equation (12) represents a one-center TDCS corresponding to effective H atoms placed at the position of each one of the molecular centers. These effective H atoms TDCS are computed by using the effective charge and the molecular bound energy corresponding to the Heitler-London initial wave function. The interference pattern due to the coherent emission from both molecular centers appears explicitly in equation (12).

Molecular TDCS corresponding to a coplanar geometry are obtained averaging equation (12) over all possible molecular orientations, i.e.,

$$\sigma^{(3)} = \frac{d^3\sigma}{d\Omega_e d\Omega_s d(k_e^2/2)} \cong 2\left[1+\frac{\sin(\chi\rho_0)}{\chi\rho_0}\right]\sigma_A^{(3)}. \quad (13)$$

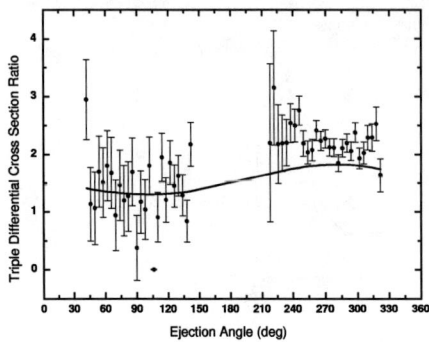
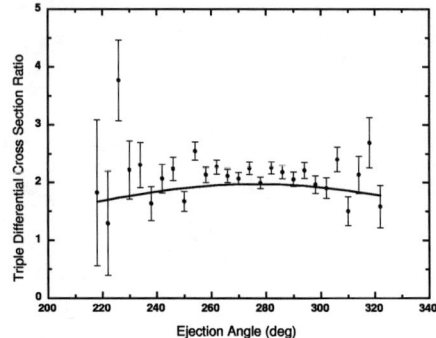

FIGURE 3. TDCS ratios as a function of the ejection angle. $E_i = 4087$ eV. $E_e = 20$ eV. Left panel, $\theta_s = 1.5$ deg. Right panel, $\theta_s = 3$ deg. Full lines, theoretical predictions. Experiments from Ref. [9].

Finally, double differential cross sections (DDCS) may be obtained from equation (13) after performing an additional integration over the solid angle corresponding to the scattered electron,

$$\sigma^{(2)} = \frac{d^2\sigma}{d\Omega_e dE_e} \cong 2\int d\Omega_s \left[1 + \frac{\sin(\chi\rho_0)}{\chi\rho_0}\right] \sigma_A^{(3)} \qquad (14)$$

Eq. (13) is similar to the CF formula [1]. Therefore, interference effects are expected. In order to put them in evidence, the following ratio should be considered,

$$\sigma^{(3)}/2\,\sigma_A^{(3)} \cong \left[1 + \frac{\sin(\chi\rho_0)}{\chi\rho_0}\right]. \qquad (15)$$

It is worthy to note that maximum values of the TDCS ratios (constructive interference) are expected for $\chi = 0$. This condition is satisfied by soft collisions ($\mathbf{k}_e = \mathbf{K} \cong 0$) or in the Bethe region ($\mathbf{k}_e = \mathbf{K}$).

RESULTS

Experimental evidence of the interferences in the case of the DDCS corresponding to the ionization of D_2 molecules by fast electron impact was previously presented [8]. Measured DDCS are divided by twice the effective H atom DDCS. The resulting ratios (open circles) are represented in Fig. 1 as a function of the ejected electron velocity for different ejection angle values. If interference effects were not present, the cross sections ratios at high incident energies would be approximately equal to unity. However, clear oscillations around unity are found except for the case of 70° in which experiments are distributed around the value 1.7. First-order TEC calculations (full lines) are also included in the figures showing a general good agreement with the experimental data. In Fig. 2, we consider theoretical calculations in an extended domain of the emitted electron velocities. Maximum values are found for forward emission angles. These maxima come from the oscillatory term in the integrand of equation (14) which gives a maximum contribution (constructive interference) for $\mathbf{k_e} \cong \mathbf{K}$, i.e., under binary encounter conditions. In particular for 70°, the theoretical ratios oscillate around a value of 1.7, in clear agreement with the measured values. For larger electron velocities, the theoretical ratios present damping oscillations around unity as it occurs for the other ejection angles considered. In Fig. 3, we present theoretical ratios of TDCS obtained within the MBBK model as a function of the ejection angle for impact energy E_i= 4087 eV, ejection energy E_e= 20 eV and scattering angles $\theta_s = 1.5, 3$ deg. Good qualitative agreement is observed between theoretical predictions and experimental ratios obtained from the TDCS of Ref. [9]. For θ_s=1 deg, the theoretical TDCS ratio oscillates around 1.5, exhibiting a maximum value of approximately 1.7 around the binary peak region. The available measurements tend to corroborate this tendency. At θ_s=1.5 deg, theory present a maximum close to 2 again around the binary peak region. Experimental results are distributed around this maximum value.

CONCLUSIONS

Interference effects in multiple differential cross sections have been put in evidence in (e,2e) reactions with D_2 and H_2 molecules. The theoretical predictions present a good qualitative agreement with the available experiments. In particular, constructive interferences have been identified in the binary encounter region.

ACKNOWLEDGMENTS

O.A.F acknowledges fruitful discussions with Prof. Lamham-Bennani and Prof. Birgit Lohmann.

REFERENCES

1. H. D. Cohen and U. Fano, Phys. Rev. **150**, 30 (1966).
2. N. Stolterfoht et al, Phys. Rev. Lett. **87**, 023201 (2001); D. Misra et al, Phys. Rev. Lett. **92**, 153201 (2004); C. Dimopoulou et al, Phys. Rev. Lett. **93**, 123203 (2004).
3. O.A. Fojón, J. Fernández, A. Palacios, R.D. Rivarola, F. Martín, J. Phys. B: At. Mol. Opt. Phys. **37**, 3035 (2004).
4. C.R. Stia, O.A. Fojón, P.F. Weck, J. Hanssen, B. Joulakian, and R.D. Rivarola, Phys. Rev. A **66**, 052709 (2002).
5. P. Weck, O.A. Fojón, J. Hanssen, B. Joulakian and R.D. Rivarola, Phys. Rev. A **63**, 042709 (2001).
6. M. Brauner, J.S. Briggs and H. Klar, J. Phys. B: At. Mol. Opt. Phys. **22** 2265 (1989); G. Garibotti and J.E. Miraglia, Phys. Rev. A **21**, 572 (1980).
7. C. R. Stia, O. A. Fojón, P. Weck, J. Hanssen, and R. D. Rivarola, J. Phys. B: At. Mol. Opt. Phys. **36**, L257 (2003).
8. O. Kamalou, J.-Y. Chesnel, D. Martina, F. Frémont, J. Hanssen, C. R. Stia, O. A. Fojón and R. D. Rivarola, Phys. Rev. A **71**, 010702 (2005).
9. Chérid M, Lahmam-Bennani A, Zurales R W, Lucchese R R, Duguet A, Dal Cappello M C and Dal Cappello C 1989 J. Phys. B: At. Mol. Opt. Phys. **22** 3483.

Double Photoionization of He and H_2

J. Colgan*, M. S. Pindzola† and F. Robicheaux†

*Theoretical Division, Los Alamos National Laboratory, Los Alamos, NM 87545, USA. [1]
†Department of Physics, Auburn University, Auburn, AL 36849, USA.

Abstract. Photoionization cross sections for both atomic helium and molecular hydrogen have recently been calculated using a time-dependent close-coupling method [1]. The total electronic wavefunction for the two electron system is expanded in six dimensions, where four dimensions are represented on a radial and angular lattice and a coupled channels expansion is used to represent the other two dimensions. The double photoionization cross sections obtained for both He and H_2 for a range of photon energies above the complete fragmentation threshold were compared with absolute experimental measurements. Very good agreement is found with experiment. Our method is also capable of being extended to calculations of single and triple differential cross sections of H_2.

Keywords: Double photoionization, molecule
PACS: 32.80.Fb

INTRODUCTION

The double photoionization of two-electron systems has been the subject of intense theoretical and experimental effort in the last 10–15 years. On the theoretical side, the double screened Coulomb [2], convergent close-coupling [3], hyperspherical R-matrix with semi-classical outgoing waves [4, 5], time-dependent close-coupling (TDCC) [6, 7], and exterior complex-scaling (ECS) [8] methods have all calculated total, single, and triple differential cross sections for the double photoionization of helium, the simplest two-electron atom. This theoretical effort has been matched by important experimental advances. Accurate measurements of the total cross section for double photoionization of helium [9] have been obtained and many measurements have now been made on the triple differential cross sections arising from the double photoionization of helium [10, 11, 12, 13].

These experimental measurements of the triple differential cross sections have also been extended to examine the differential cross sections arising from the double photoionization of H_2 (or, equivalently, D_2) [14, 15, 16], i.e. the four-body Coulomb problem. Total cross section measurements for the double photoionization of H_2 were measured in the 1980's first by Dujardin et al [17] and subsequently by Kossman et al [18]. However, it is notable that these experimental efforts have not really been matched by corresponding theoretical work. Early calculations on the double photoionization of H_2 were made by Le Rouzo [19] and the high-energy asympototic limit of the ratio of double-to-single photoionization of H_2 was calculated by Sadeghpour and Dalgarno [20]. However, apart from some qualitative studies of the angular distribution of the elec-

---

[1] E-mail: jcolgan@lanl.gov

trons arising from double photoionization of H_2 by Walter and Briggs [21, 22], there has been little theoretical work to match the experimental measurements. This is no doubt due to the increased complexity of the molecular system, where the non-spherical nature of the potential removes an important symmetry from the problem. Also, there are dynamics in molecular photoionization that do not exist in the atomic case, such as the possible vibrational and rotational motion of the nuclei.

In the last year, some non-perturbative theoretical calculations have started to address this imbalance by calculating the double photoionization cross section for the H_2 molecule using the time-dependent close-coupling method [1] and the exterior complex-scaling method [23]. We also note some promising single-center studies which have begun using the convergent close-coupling method [24]. In this report, we describe the recent progress made using the TDCC method to investigate the double photoionization of H_2. Unless otherwise stated, all quantities are given in atomic units.

THEORY

In the weak-field perturbative limit, the photoionization of a two-electron system may be found by solving the time-dependent Schrödinger equation [6]:

$$i\frac{\partial \psi(\vec{r}_1, \vec{r}_2, t)}{\partial t} = H\psi(\vec{r}_1, \vec{r}_2, t) + V\psi_0(\vec{r}_1, \vec{r}_2)e^{-iE_0 t}, \quad (1)$$

where H is the atomic or molecular Hamiltonian, V is the time-dependent radiation field Hamiltonian, and ψ_0 and E_0 are the exact eigenfunction and eigenenergy of the atomic or molecular ground state. Due to the reduced symmetry of the molecular case, the time-dependent wavefunction for a given MS symmetry is expanded in products of rotation functions:

$$\psi(\vec{r}_1, \vec{r}_2, t) = \sum_{m_1, m_2} \frac{P^M_{m_1 m_2}(r_1, \theta_1, r_2, \theta_2, t)}{r_1 r_2 \sqrt{\sin\theta_1}\sqrt{\sin\theta_2}} \Phi_{m_1}(\phi_1)\Phi_{m_2}(\phi_2), \quad (2)$$

where $\Phi_m(\phi) = \frac{e^{im\phi}}{\sqrt{2\pi}}$ and $M = m_1 + m_2$. For the spherically symmetric atomic case, the time-dependent wavefunction may also be expanded in products of coupled spherical harmonics, as has been used previously with great success [6]. Upon substitution of equation (2) into equation(1) and application of the variational principle, the time-dependent close-coupled partial differential equations are given by:

$$\begin{aligned}
i\frac{\partial P^M_{m_1 m_2}(r_1, \theta_1, r_2, \theta_2, t)}{\partial t} &= T_{m_1 m_2}(r_1, \theta_1, r_2, \theta_2) P^M_{m_1 m_2}(r_1, \theta_1, r_2, \theta_2, t) \\
&+ \sum_{m'_1, m'_2} V^M_{m_1 m_2, m'_1 m'_2}(r_1, \theta_1, r_2, \theta_2) P^M_{m'_1 m'_2}(r_1, \theta_1, r_2, \theta_2, t) \\
&+ \sum_{m''_1, m''_2} W^{M M_0}_{m_1 m_2, m''_1 m''_2}(r_1, \theta_1, r_2, \theta_2, t) \bar{P}^{M_0}_{m''_1 m''_2}(r_1, \theta_1, r_2, \theta_2) e^{-iE_0 t},
\end{aligned} \quad (3)$$

where expressions for the one-electron matrix elements $T_{m_1 m_2}(r_1,\theta_1,r_2,\theta_2)$, the two-electron matrix elements $V^M_{m_1 m_2, m'_1 m'_2}(r_1,\theta_1,r_2,\theta_2)$, and the radiation field coupling operators $W^{MM_0}_{m_1 m_2, m''_1 m''_2}$ may be found in [1]. In this equation, $\bar{P}^{M_0}_{m_1 m_2}(r_1,\theta_1,r_2,\theta_2)$ is the reduced wavefunction for $\psi_0(\vec{r}_1,\vec{r}_2)$. We note that for linear polarization $M = M_0$, while for circular polarization $M = M_0 + 1$. The exact eigenfunction for the He or H_2 ground state is obtained by relaxation of the Schrödinger equation in imaginary time ($\tau = it$):

$$-\frac{\partial \psi_0(\vec{r}_1,\vec{r}_2,\tau)}{\partial \tau} = H\psi_0(\vec{r}_1,\vec{r}_2,\tau) . \tag{4}$$

The total wavefunction is again expanded in products of rotation functions and substituted into equation (4), yielding a set of close-coupled partial differential equations in space and imaginary time.

We solve the TDCC equations using lattice techniques to obtain a discrete representation of the reduced wavefunctions and all operators on a four-dimensional radial and angular grid. Our implementation on massively parallel computers is to partition the radial coordinates (r_1,r_2) over the many processors, so-called domain decomposition. Both explicit and implicit algorithms are used to time propagate the close-coupled equations.

The total cross section for double photoionization is given by:

$$\sigma_{dion} = \frac{\omega}{I}\frac{\partial \mathscr{P}_{dion}}{\partial t}, \tag{5}$$

where $\mathscr{P}_{dion}$ is the double ionization probability [1], ω is the radiation field frequency, and I is the intensity of the radiation field. The double ionization probability can be written in an alternative form from that presented in [1] as

$$\mathscr{P}_{dion} = \sum_{l_1|m_1|,l_2|m_2|} \int_0^\infty dk_1 \int_0^\infty dk_2 |P^M_{l_1|m_1|,l_2|m_2|}(k_1,k_2,t)|^2 , \tag{6}$$

where

$$\begin{aligned} P^M_{l_1|m_1|,l_2|m_2|}(k_1,k_2,t) &= \int_0^\infty dr_1 \int_0^\infty dr_2 \int_0^\pi d\theta_1 \int_0^\pi d\theta_2\, P_{k_1 l_1|m_1|}(r_1,\theta_1) P_{k_2 l_2|m_2|}(r_2,\theta_2) \\ &\times P^M_{m_1 m_2}(r_1,\theta_1,r_2,\theta_2,t) , \end{aligned} \tag{7}$$

and $P_{kl|m|}(r,\theta)$ is a continuum state radial orbital for H_2^+, the construction of which is discussed in [25]. This allows us to define a differential cross section in angle, in analogy with the atomic case [7], as

$$\frac{d\sigma}{d\alpha} = \frac{\omega}{I}\frac{\partial}{\partial t}\int_0^\infty dk_1 \int_0^\infty dk_2\, \delta\!\left(\alpha - \tan^{-1}(\frac{k_2}{k_1})\right) \sum_{l_1|m_1|,l_2|m_2|} |P^M_{l_1|m_1|,l_2|m_2|}(k_1,k_2,t)|^2 , \tag{8}$$

and an ejected-energy single differential cross section as

$$\frac{d\sigma}{dE_1} = \frac{1}{E_1 E_2}\frac{d\sigma}{d\alpha} . \tag{9}$$

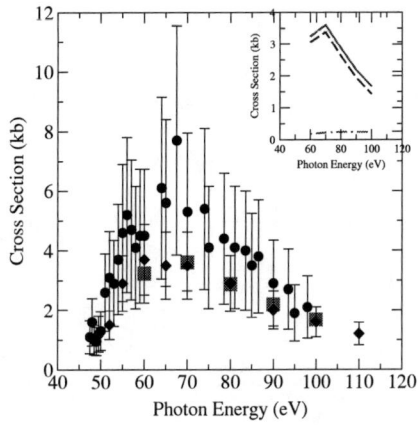

FIGURE 1. Double photoionization cross sections for H_2. The new time-dependent close-coupling calculations (squares) are compared with the experimental measurements of Dujardin et al [17] (circles) and the measurements of Kossman et al [18] (diamonds). (1.0 kb = 1.0×10^{-21} cm^2).

RESULTS

We use the new TDCC method, outlined above, to calculate the double photoionization cross sections for H_2 at its equilibrium internuclear separation of $R = 1.4$ a.u. We employ a $288 \times 288 \times 32 \times 32$ point lattice with a uniform radial mesh spacing of $\Delta r = 0.1$ from 0 to 28.8 in both r_1 and r_2 and a uniform angular mesh spacing of $\Delta \theta = 0.03125\pi$ from 0 to π in both θ_1 and θ_2.

The double photoionization cross section results for H_2, at its equilibrium internuclear separation and averaged over field polarization, are shown in Figure 1. The inset shows the contributions from linear and circular polarized light; as expected for the molecular case, the linear and circular polarized cross sections for H_2 are quite different, with the circular polarized cross sections being very much the larger of the two. Also, the behavior of both contributions as a function of photon energy is different; in this energy range the contribution from circular polarized light peaks and then falls off, whereas the contribution from linear polarized light increases over all this energy range. This large difference indicates that the dominant contribution to double ionization is from the $^1\Pi_u$ state, which implies polarization perpendicular to the molecular axis, in agreement with the conclusions of [18]. We have used two projection methods for calculating the double photoionization cross section; the first method as described in [1], and the second method as outlined by Equations (6,7). The two methods give very similar results. We compare our H_2 results (squares) with the absolute experimental measurements of Dujardin et al [17] (circles) and Kossmann et al [18] (diamonds) and find good agreement between the new TDCC calculations and experiment. Our cross sections are slightly lower than the measurements of Dujardin et al, but well within the large error bars of the experiment. The current calculations are also well within the much smaller error bars of the later experiment of Kossmann et al. We also note that our results are in good agreement with very recent ECS calculations of the total double photoionization cross section for H_2

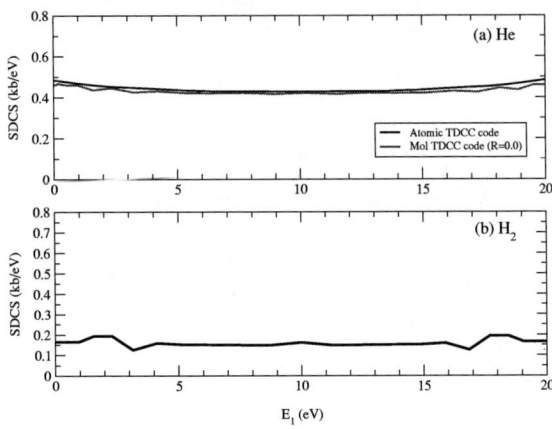

FIGURE 2. Single differential cross sections at an excess energy of 20 eV. In (a) we compare our previous results for He using our atomic time-dependent close-coupling code [9] with the results from our new molecular time-dependent close-coupling code, with $R = 0.0$ a.u. In (b) we present our single differential cross sections for H_2 at $R = 1.4$ a.u. (1.0 kb = 1.0×10^{-21} cm^2).

[23].

Finally, in Figure 2 we present single differential cross sections computed using our molecular TDCC method, for an excess energy of 20 eV. In Figure (2a) we check our new code (with R set to 0.0) against previous results using our atomic TDCC calculations for He [7], which agree well with experiment [26]. Our new calculations with the molecular code are in very good agreement with our previous results using a completely independent code, as are our calculations of the total double photoionization cross section [1]. We note that calculations at $R = 0.0$ using both linear and circular polarized light are very similar to each other, as expected.

In Figure (2b) we present single differential cross sections for H_2 at $R = 1.4$ a.u. In this case, as for the total double photoionization cross section, the contributions from linear and circular polarized light are added appropriately to get the total single differential cross section. In this case, at $R = 1.4$ a.u. the single differential cross section has some small oscillations in the unequal energy-sharing regions. This is likely due to a lack of complete convergence with respect to the number of continuum functions included in Eq. (8).

We also note that our single differential cross sections at 20 eV excess energy (corresponding to a incident photon energy of 71.4 eV) fall between ECS calculations of the single differential cross section at 70.2 eV and 72.9 eV [23]. This indicates that, as for the total double photoionization cross section, the TDCC and ECS methods are in very good agreement for the single differential cross section.

CONCLUSIONS

In this report, we have described our recent progress in calculating double photoionization cross sections for H_2. Our TDCC molecular calculations are an extension of previous atomic calculations which have successfully described double photoionization processes in two-electron atoms. We have shown good agreement with experiment for the total double photoionization cross section of H_2, and have also calculated single differential cross sections which are in very good agreement with recent ECS calculations [23]. We are currently using our techniques to calculate triple differential cross sections for H_2, for which there are many sets of experimental measurements with which to compare. Similar TDCC methods have also been used to calculate the electron-impact ionization cross section of H_2^+ [27], and we look forward to extending our methods to further studies of ionization processes in light molecules.

A portion of this work was performed under the auspices of the US Department of Energy through the Los Alamos National Laboratory. This work was supported in part by grants to Auburn University from the US Department of Energy and the National Science Foundation. Computational work was carried out at the National Energy Research Scientific Computing Center in Oakland, CA, and at the Center for Computational Sciences in Oak Ridge, TN.

REFERENCES

1. J. Colgan, M. S. Pindzola, and F. Robicheaux, *J. Phys. B* **37** L377-384 (2004).
2. D. Proulx and R. Shakeshaft, *Phys. Rev. A*, **48**, R875-R878 (1993).
3. A. S. Kheifets and I. Bray, *J. Phys. B*, **31**, L447-L453 (1998); *Phys. Rev. A*, **62**, 065402 (2000).
4. L. Malegat, P. Selles, and A. K. Kazansky, *Phys. Rev. Lett.*, **85**, 4450-4453 (2000).
5. P. Selles, L. Malegat, and A. K. Kazansky, *Phys. Rev. A*, **65**, 032711 (2002).
6. M. S. Pindzola and F. Robicheaux, *Phys. Rev. A*, **57**, 318-324 (1998).
7. J. Colgan, M. S. Pindzola, and Robicheaux, *J. Phys. B*, **34**, L456-L466 (2001).
8. C. W. McCurdy, D. A. Horner, T. N. Rescigno, and F. Martin, *Phys. Rev. A*, **69**, 032707 (2004).
9. J. A. R. Samson et al, *Phys. Rev. A*, **57**, 1906-1911 (1998).
10. H. Bräuning et al, *J. Phys. B: At. Mol. Opt. Phys.*, **31**, 5149-5160 (1998).
11. J. Roder et al, *Phys. Rev. Lett.*, **79**, 1666-1669 (1997).
12. C. Dawson et al, *J. Phys. B: At. Mol. Opt. Phys.*, **34**, L525-L533 (2001).
13. P. Bolognesi et al, *J. Phys B*, **36**, L241-L247 (2003).
14. J. P. Wightman, S. Cvejanović, and T. J. Reddish, *J. Phys. B*, **31**, 1753-1764 (1998).
15. S. A. Collins et al, *Phys. Rev. A*, **64**, 062706 (2001).
16. D. P. Seccombe et al, *J. Phys. B*, **35**, 3767-3780 (2002).
17. G. Dujardin et al, *Phys. Rev. A*, **35**, 5012-5019 (1987).
18. H. Kossmann et al, *Phys. Rev. Lett.*, **63**, 2040-2043 (1989).
19. H. Le Rouzo, *J. Phys. B*, **19**, L677-L682 (1986); *Phys. Rev. A*, **37**, 1512-1523 (1988).
20. H. R. Sadeghpour and A. Dalgarno *Phys. Rev. A*, **47**, R2458-R2459 (1993).
21. M. Walter and J. Briggs, *J. Phys. B*, **32**, 2487-2501 (1999).
22. M. Walter, A. Meremianin, and J. S. Briggs, *J. Phys. B*, **36**, 4561-4579 (2003).
23. W. I. Vanroose, F. Martin, T. N. Rescigno, and C. W. McCurdy, *Phys. Rev. A*, **70**, 050703 (2004).
24. A. S. Kheifets, *Phys. Rev. A*, **71**, 022704 (2005).
25. M. S. Pindzola et al, *Phys. Rev. A*, submitted (2005).
26. R. Wehlitz et al, *Phys. Rev. Letts.*, **67**, 3764-3767 (1991).
27. M. S. Pindzola, F. Robicheaux, and J. Colgan, *Phys. Rev. A*, submitted (2005).

Photoionization of Xenon Clusters

John D. Bozek[*], Bruce S. Rude[*], and A.L. David Kilcoyne[*]

[*]Advanced Light Source, Lawrence Berkeley National Laboratory, Berkeley, CA 94720 USA

Abstract. High resolution photoelectron spectra of xenon clusters of about 100 atoms, measured with photon energies of 20 to 60 eV are reported. Peaks due to the clusters are observed at lower binding energies than the corresponding atomic lines and exhibit different behaviors. The peak associated with the Xe $5p_{1/2}$ state at higher binding energy is relatively unchanged with photon energy whereas the Xe $5p_{3/2}$ peak changes its shape with increasing photon energy. The change is attributed to differences in the escape depth of the electrons at different kinetic energies and the bulk and surface components of the cluster.

Keywords: photoelectron spectroscopy, clusters, photoionization, xenon, partial cross sections
PACS: 33.60.-q, 36.40.-c, 32.70.Jz

INTRODUCTION

Photoionization studies of clusters of atoms and/or molecules offer unique insight into the evolution of the electronic structure bulk solids from their constituent atomic/molecular building blocks. Amongst the various types of clusters, such as metallic, semiconductor, ionic, etc., rare gas clusters are the most weakly bound, held together by van der Waals forces. Being only weakly bound, clusters of rare gas atoms are difficult to study experimentally since they are easily dissociated. Mass spectroscopic measurements of rare gas clusters, for example, usually results in neutral and ionic evaporation and dissociation of the cluster upon ionization [1].

Photoelectron spectroscopy provides a suitable means to study rare gas clusters owing to the short interaction time of the photoelectron with the residual cluster ion, usually fsecs. The first rare gas clusters to be studied by photoelectron spectroscopy were dimers of Xe [2] followed by dimers of Ar and Kr [3]. Several of the six photoelectron bands expected from molecular orbital theory were observed for each of the rare gas dimers and used to optimize the potential energy surfaces of the ions. More recently, very high resolution PFI-ZEKE photoelectron spectroscopy measurements have been used to construct potential energy surfaces for Ar_2^+, Kr_2^+, and Xe_2^+ with unprecedented accuracy [4,5]. Valence level photoelectron spectra of larger clusters of rare gas atoms have also been reported previously [6,7], although the spectra are dominated by broad unresolved "bands" rather than vibrationally resolved peaks. Inner-valence and inner-shell levels of rare gas clusters have become accessible for photoelectron spectroscopy with the advent of 3rd generation synchrotron light sources, and several examples have been reported recently [8].

Inner-shell and inner-valence photoelectron spectra of rare gas clusters are relatively simple, exhibiting peaks shifted to lower binding energies relative to the corresponding atomic lines. The cluster inner-shell photoelectron peaks are of about the same width as the corresponding atomic lines and can usually be resolved into two peaks due to contributions from atoms on the surface of the cluster and from atoms in the interior or bulk of the cluster [8]. The outer valence photoelectron spectra of rare gas clusters also exhibit peaks shifted to lower binding energies relative to the corresponding atomic levels. In contrast to the inner-shell, however, the outer valence photoelectron peaks are much broader than either the atomic lines or cluster inner-shell photoelectron lines and exhibit different widths for the two spin-orbit split components [6,7].

To further explore the unique behavior of the valence levels of rare gas clusters, we have examined the photon energy dependent behavior of the photoelectron spectra of large clusters of Xe. In addition to probing the photoionization dynamics of the cluster at different photon energies, changes in the excitation energy will result in different kinetic energies of the photoelectrons, which will be sensitive to changes in escape depth for electrons leaving the bulk of the cluster.

EXPERIMENTAL

Clusters of Xe with an average size <N> ~ 100 atoms were produced from pure xenon using a continuous skimmed supersonic expansion gas jet. The skimmer region is fitted with two 1000 l s^{-1} turbo molecular pumps and maintained at pressures of $\leq 10^{-3}$ Torr when the jet is operating. The aperture or nozzle of the source can be cooled with liquid nitrogen and a resistive electric heater is used to maintain the desired temperature. For the results reported here, a 200 μm aperture was used and maintained at a temperature of 170±1 K. A 1mm diameter conical skimmer separates the nozzle and surrounding vacuum from the interaction region and photoelectron spectrometer. Pressures of $\leq 10^{-5}$ Torr were maintained in this region throughout the experiment.

Condensation of clusters from rare gases in expansions is well understood and the resulting cluster size is related to the condensation parameter, Γ^*, which can in turn be parameterized by the stagnation pressure, p_0, nozzle diameter, d, nozzle temperature, t_0, and gas specific constant, k, according to [9]:

$$G^* = k \times p_0 \times d^{0.85} \times T_0^{2.2875} \qquad (1)$$

Using the values given above along with a gas constant, k, for xenon of 5554 [9], an average cluster size <N> of 100 atoms results.

Photoelectron spectra were measured with photons from Beamline 10.0.1 at the Advanced Light Source using a Scienta SES-200 electron spectrometer set at the magic angle [10]. The beamline was operated at 7 meV resolution for photon energies below 35 eV, at 15 meV for 35 eV $\leq$ hυ $\leq$ 50 eV and at 20 meV at higher photon energies. A pass energy of 5 eV was used on the spectrometer with differing slit sizes to approximately match the electron spectrometer resolution with the photon bandwidth.

RESULTS AND DISCUSSION

A photoelectron spectrum of the Xe 5p valence region, including both the atomic lines and the bands from the clusters measured with a total experimental resolution of 8 meV is shown in Fig. 1. The spectrum has been calibrated to the binding energy of the Xe $5p_{3/2}$ line at 12.1298 eV [11]. The spectrum in Fig. 1 is similar to others published previously [7,8] exhibiting sharp lines for the spin-orbit split atom 5p lines, and much broader lines shifted to lower binding energies from each of the atomic peaks. The relative intensity of the cluster peaks is smaller than in recently published spectra due to the smaller cluster size in the present study, 100 atoms versus 1000 atoms, and the greatly enhanced resolution over the previous results [8].

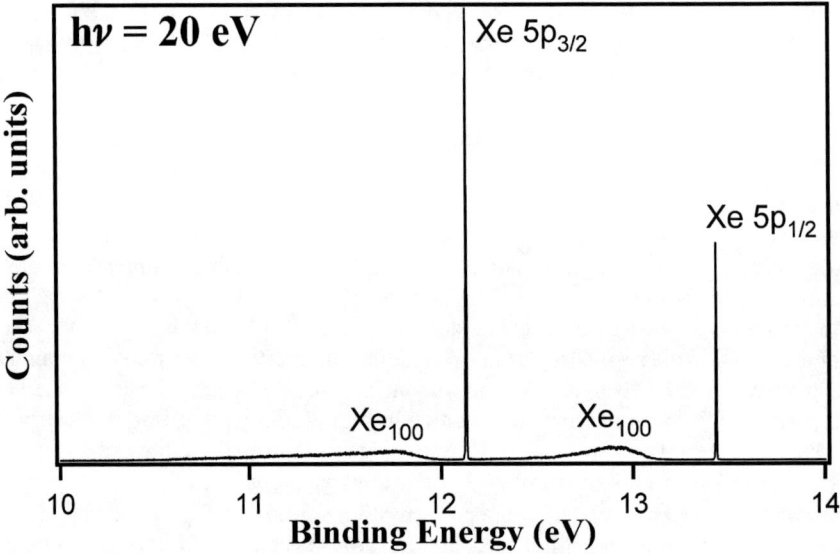

FIGURE 1. Photoelectron spectrum of Xe and clusters with average size of 100 atoms produced in supersonic jet expansion measured with a photon energy of 20 eV. The experimental resolution, as determined from the width of the $5p_{3/2}$ line, is 8 meV

An enhanced view of the cluster peaks measured at the same photon energy but with a somewhat lower experimental resolution of 18 meV is shown in Fig. 2. The two cluster peaks below the Xe $5p_{3/2}$ and Xe $5p_{1/2}$ lines are very different in shape, intensity, and energy separation from the corresponding atomic lines. The spectrum in Fig. 2 also exhibits considerably more structure than has been reported previously for outer valence photoelectron spectra of Xe clusters.

The difference in the shapes of the two cluster peaks is dramatic, with the peak closer to the Xe $5p_{1/2}$ line being much narrower and almost Gaussian in shape with a full width at half maximum of about 300 meV while the Xe $5p_{3/2}$ cluster lines has a trigonal shape with a width at half height of about 750 meV. This disparity in the line shapes has been noted previously and attributed to either a Xe_3^+ chromophore with strongly bound states associated with the $^2P_{3/2}$ ionization limit but only weakly bound or even unbound states for the $^2P_{1/2}$ ionization limit [7], or to the band structure of the condensed Xe cluster particle [8]. These results do not conclusively support either interpretation. Differences between the intensities of the two cluster peaks are somewhat deceptive, since the peak areas are much closer to expected statistical ratio of 2:1 due to the additional breadth of the *5p₃/₂* cluster line.

Structure is apparent in the cluster peaks at binding energies of 12.96, 12.54 and 12.36 eV for the *5p₁/₂* line and 11.76, 11.56, 11.27 and 10.9 eV for the *5p₃/₂* line. It is possible that given the relatively small size of the clusters studied here that a significant percentage of xenon dimer is present in the beam and may give rise to the narrower peaks. Peaks due to dimer were seen in low pressure spectra measured previously, that were quickly overwhelmed by signal from larger clusters as the size of the clusters was increased [7]. The positions of these bands are not in good agreement with previously reported Xe_2^+ ionization energies, however [2,5]. In particular, the lack of intensity at 13.31 eV, where the $C^2\Pi_{(1/2)u}$ state was observed with the highest binding energy in previous photoelectron measurements [2] performed with He I radiation, contradicts the assignment of this structure to dimer contributions.

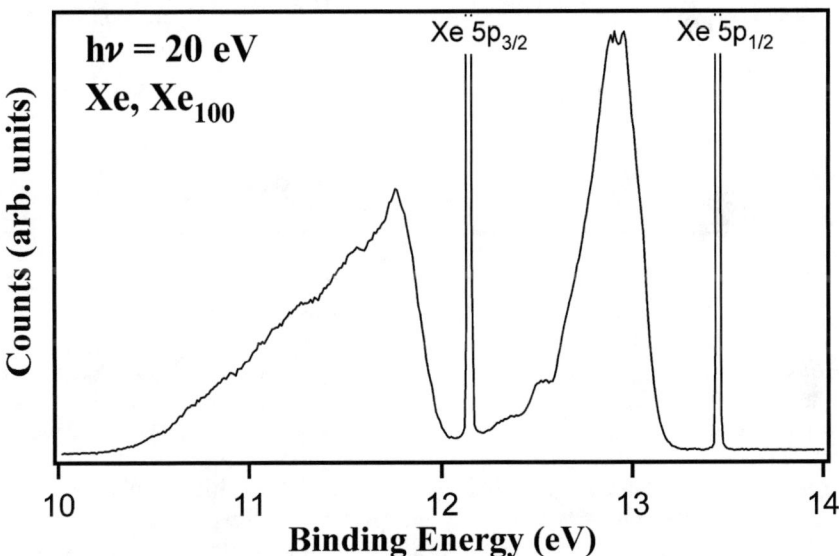

FIGURE 2. Photoelectron spectrum of Xe and clusters measured with a photon energy of 20 eV vertically zoomed to enhance the intensity of the cluster peaks. The experimental resolution, as determined from the width of the *5p₃/₂* line, is 18 meV

Xe 5p photoelectron spectra of the clusters were measured at additional photon energies, and the results are presented in Fig. 3. The higher photon energy spectra have somewhat poorer statistical quality, but several differences from the spectrum measured at 20 eV photon energy emerge. Firstly, the Xe *5p₁/₂* cluster peak is almost identical in shape and position across the sequence of energies, except for the 60 eV spectrum, where the cluster peak is shifted to higher kinetic energy. This shift is due to a change in the stagnation pressure for this spectrum (due to the sample bottle running out) resulting in smaller clusters being produced in the beam. The average cluster size fell quickly from 100 to 80 atoms during the acquisition of this one spectrum. While the main cluster peak shifts to higher kinetic energy, the small peak apparent in several of the spectra at 12.64 eV binding energy does not move. The other small peaks at 12.54 and 12.36 eV appear to persist in the spectra at higher photon energies, although it is difficult to unambiguously identify them due to the noisier data.

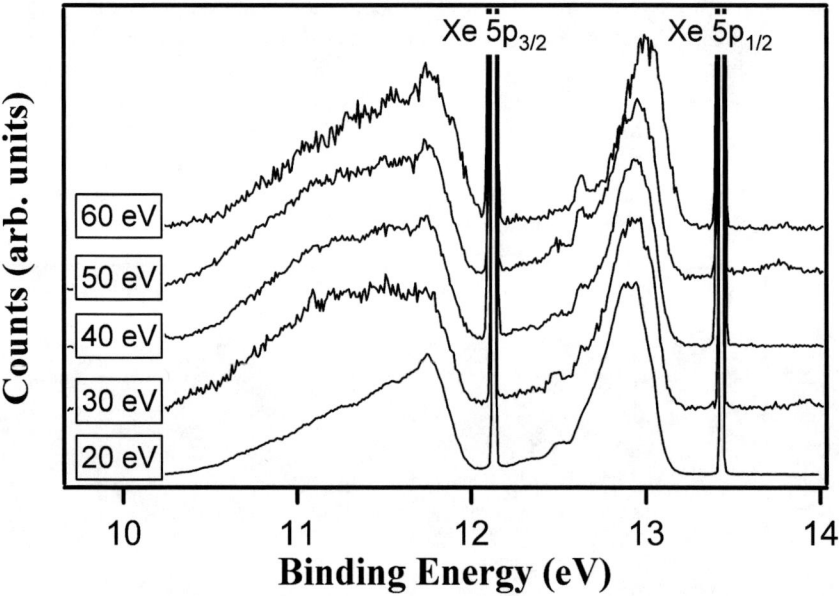

FIGURE 3. Photoelectron spectra of Xe 5p region measured at several different photon energies as indicated on the figure. The intensities were arbitrarily scaled to give the Xe $5p_{1/2}$ cluster lines the same intensities in all of the spectra.

The Xe $5p_{3/2}$ cluster peak changes dramatically at higher photon energies, exhibiting enhanced intensity at binding energies below the peak maximum of 11.76 eV. The peak shape is similar for the spectra measured at 30, 40 and 50 eV although at 60 eV the increased intensity begins to moderate and the shape is returning to that exhibited by the 20 eV spectrum. The maximum of the peak at 11.76 eV is remarkably constant, even for the 60 eV spectrum where the peak broadens to higher binding energy rather than shifting. The origin of the additional intensity on the low binding energy side of the peak could simply be due to reduced escape depth for electrons emerging from the bulk of the cluster at these kinetic energies. The bulk/surface sensitivity ratio of Xe 4d electrons for 300 atom and 1000 atoms clusters was found to exhibit a distinct minimum for kinetic energies between 20 and 40 eV with a sharp maximum at lower kinetic energies and a slow increase at kinetic energies above 40 eV [12]. The present results are consistent with this model if the increased signal between 10.5 and 11.5 eV binding energy is due to signal from the surface of the cluster while the underlying signal represented by the shape of the peak in the spectrum measured at 20 eV photon energy is due primarily to atoms from the bulk of the cluster. Surprisingly, the Xe $5p_{1/2}$ does not exhibit similar behavior, however.

CONCLUSIONS

High resolution outer valence photoelectron spectra of small xenon clusters of about 100 atoms have been measured using photon energies between 20 and 60 eV. Consistent with previous results, the cluster spectra exhibit two broad bands shifted to lower binding energies than the corresponding atomic Xe 5p lines. The cluster peak associated with the atomic Xe $5p_{1/2}$ line is considerably narrower than that associate with the Xe $5p_{3/2}$ line. The spectra were found to exhibit considerably more structure than any that have been reported previously, with distinct peaks apparent within the broader bands. Enhanced intensity in the 10.5 – 11.5 eV binding energy region of the Xe $5p_{3/2}$ cluster peak in the spectra measured at 30, 40 and 50 eV photon energies was assigned to suppressed contributions from the bulk of the cluster due to the reduced escape depth at these kinetic energies in xenon.

ACKNOWLEDGMENTS

The Advanced Light Source is supported by the Director, Office of Science, Office of Basic Energy Sciences, of the U.S. Department of Energy under Contract No. DE-AC02-05CH11231.

REFERENCES

1. Hellmut Haberland, *Clusters of atoms and molecules : theory, experiment, and clusters of atoms.* (Springer-Verlag, Berlin ; New York, 1994).
2. P. M. Dehmer and J. L. Dehmer, Journal of Chemical Physics **68**, 3462 (1978).
3. P. M. Dehmer and J. L. Dehmer, Journal of Chemical Physics **69**, 125 (1978).
4. A. Wuest and F. Merkt, Journal of Chemical Physics **120**, 638 (2004); A. Wuest and F. Merkt, Molecular Physics **103**, 1285 (2005).
5. P. Rupper, O. Zehnder, and F. Merkt, Journal of Chemical Physics **121**, 8279 (2004).
6. Frank Carnovale, J. Barrie Peel, Richard G. Rothwell et al., Journal of Chemical Physics **90**, 1452 (1989).
7. Frank Carnovale, J. Barrie Peel, and Richard G. Rothwell, Journal of Chemical Physics **95**, 1473 (1991).
8. R. Feifel, M. Tchaplyguine, G. Oehrwall et al., European Physical Journal D: Atomic, Molecular and Optical Physics **30**, 343 (2004).
9. R. Karnbach, M. Joppien, J. Stapelfeldt et al., Review of Scientific Instruments **64**, 2838 (1993).
10. N. Berrah, B. Langer, A. A. Wills et al., Journal of Electron Spectroscopy & Related Phenomena, **101-103**, 1, (1999).
11. F. Brandi, I. Velchev, W. Hogervorst et al., Physical Review A **64**, 032505 (2001).
12. M. Tchaplyguine, R. R. Marinho, M. Gisselbrecht et al., Journal of Chemical Physics **120**, 345 (2004).

Ionization of Atoms with Spin Polarized Electrons

J. Lower, S. Bellm, R. Panajotovic[*], E. Weigold, A. Prideaux[**],
D.H. Madison[**], Z. Stegen[**], Colm T. Whelan[+] and B. Lohmann[***]

*Atomic and Molecular Physics laboratories, RSPhysSE,
The Australian National University, Canberra 0200, Australia*
[*]University of Sherbrooke, Québec, Canada J1H 5N4
[**]University of Missouri, Rolla, Missouri, USA
[+]Old Dominion University, Norfolk, Virginia, USA
[***]Griffith University, Queensland 4111, Australia

Abstract. The most detailed insight into the process of electron impact-induced ionization of atomic species is provided by measurements in which both kinematical and quantum mechanical variables are determined. Here we describe recent (e,2e) experimental and theoretical studies involving the ionization of xenon and argon by spin-polarized electrons in which the fine-structure levels of the ion are energetically resolved. Such investigations shed light on the mechanisms driving the ionization reaction and the role of exchange and relativistic processes.

Keywords: Spin Polarized Electrons, Electron Impact-Induced Ionization
PACS: 34.80 Dp, 34.80.Nz

INTRODUCTION

Continuing progress in areas of technological interest, such as plasma formation, gas discharge and laser physics, as well as our understanding of the physics and chemistry of the upper atmosphere, depend crucially on our understanding of the process of electron impact-induced ionization. The most detailed insight into electron impact-induced ionization of atomic species is provided by measurements performed within a kinematically complete framework. In the case of single ionization, this can be achieved through (e,2e) measurements in which pairs of electrons derived from a common ionization event are identified by their correlated arrival times at two detectors [1] or through detection of one scattered electron and the residual ion in an electron-ion coincidence measurement [2]. In recent years a new class of (e,2e) measurements have emerged in which both kinematical and quantum mechanical variables are determined. This has been achieved through preparation of the quantum projection state of the electron-atom system prior to collision. These measurements employ beams of spin-polarized electrons [3-4] and/or spin polarized targets achieved through laser preparation or magnetic selection techniques [5-6]. The addition of quantum state specificity to ionization experiments means that the constituents of the ionization process can be isolated and highlighted.

Numerous refinements in theoretical techniques have accompanied these experimental developments. Electron exchange, relativistic effects and the long-range Coulomb-interaction of the charged particles can now be treated accurately with modern theoretical tools. This paper will focus on calculations performed within the framework of Distorted Wave Born Approximation (DWBA) to interpret the experimental data. Here we describe recent progress in investigations into the electron-atom system and explain how spin-resolved studies can provide insight into the role of electron exchange in electron-impact-induced ionizing collisions.

FINE STRUCTURE EFFECT

Spin dependent effects in the scattering of electrons from atomic targets can arise through three mechanisms, namely electron exchange, spin-orbit interaction of scattered electrons in the field of the atom/ion and through angular-momentum coupling within the target atom and/or residual ion in the case of ionizing collisions [7,8]. For ionization experiments performed with spin polarized electrons and in which the fine-structure levels of the residual ion are resolved, Jones *et al* [9] showed that even in the non-relativistic limit where spin-orbit interaction of scattered electrons in the field of the atom/ion is negligible, a strong dependence of the ionization cross section on the spin projection of the projectile electron could be expected. In their model, the spin dependence results from an interplay of exchange between the projectile and ejected target electron, angular-momentum coupling and the impact-induced orientation of the residual ion. The model predicts zero spin asymmetries in the limit of negligible exchange, and in analogy to the "Fine-structure Effect" in excitation [10,11], zero spin asymmetry for the case where the individual fine-structure transitions are energetically unresolved. Subsequent measurements on xenon [12,13] and later, more sophisticated calculations [14,15], while confirming a strong spin-dependence in the ionization cross sections as originally predicted, have revealed additional and significant contributions from many-body exchange effects which tend to mask the signatures for a "pure" fine-structure effect. Notwithstanding this fact, that [9] predicts zero spin asymmetry in the absence of exchange means that the spin asymmetry parameter is a very sensitive test of the role and nature of electron exchange in electron-impact-induced ionizing collisions, and a sensitive test to the magnitude of contributions from relativistic effects which can also contribute to a non-zero result, even in the absence of exchange.

EXPERIMENT

The experimental apparatus is shown schematically in Figure 1 with details to appear separately [16]. Longitudinally polarized electrons are created by the photo-excitation of valence electrons from a strained gallium arsenide photo-cathode under illumination by circularly polarized laser light [17]. These electrons are extracted, deflected through 90° and focused to produce a beam of transversely polarized electrons. Inversion of the electron beam polarization from into (spin down) to out of (spin up) the scattering plane is achieved by reversing the helicity of the laser light by means of a liquid crystal retarder. Here the scattering plane is defined by the momentum vectors for the incident and measured scattered electrons. After deflection through 90°, the electron beam is accelerated to around 1 keV and transported at high energy through a differential pumping stage before entering the main collision chamber in which the electron spectrometer is housed. Inside this scattering chamber, the electron beam is decelerated to the experimental collision energy E_i by means of a five element electrostatic lens and focused at the grounded interaction volume defined by the overlap of the electron and target beams. The atomic target beam is formed by effusion through a 1mm internal diameter needle orientated orthogonally to the scattering plane. The required beam energy $E_i=eV_c$ is set by adjusting the photocathode potential V_c.

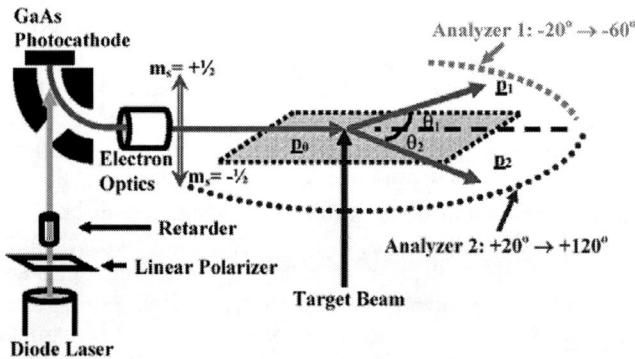

FIGURE 1. Schematic representation of the apparatus used to measure spin-resolved cross sections.

Scattered electrons emitted within the scattering plane are momentum analyzed in one of two toroidal-sector electrostatic-energy-analyzers located on opposite sides of the primary electron beam. Each analyzer is terminated by a micro-channel-plate electron-multiplier pair followed by a crossed delay line detector from which the spatial and temporal arrival coordinates (x_i, y_i, t_i) for each measured electron are determined [18]. From these coordinates, pairs of electrons derived from common (e,2e) ionization events are identified by their correlated arrival times at the two separate detectors and their initial momenta (p_1, p_2) are deduced. One analyzer accepts electrons over the azimuthal angular range $\theta_1 = -20° \rightarrow -60°$, the second over the range $\theta_2 = +20° \rightarrow +120°$ ($-240° \rightarrow -340°$), where θ_1 and θ_2 are measured within the scattering plane and with respect to the direction of the primary electron beam. Due to the limited size of the electron-multipliers (80 mm diameter), parallel measurements can only be performed over a 40° angular capture range in the larger analyzer. However, its multiplier/detector assembly is rotatable which enables access to its full 100° acceptance range. Background resulting from electrons derived from separate ionization events are subtracted using standard statistical techniques [1]. Recently, the spectrometer performance was greatly improved due to suppression of stray-electron background. The small statistical fluctuations of recent valence-shell ionization-data reflect this fact.

COMPARISON OF THEORY WITH EXPERIMENT

The present series of measurements concern the investigation of exchange effects in the electron-atom ionization of the two atomic species xenon and argon. In the case of xenon, we consider the reaction leading to the removal of a target electron in the $5p^6$ valence shell leading to the spin-orbit split Xe^+ $5p^5$ $^2P_{1/2}$ or $5p^5$ $^2P_{3/2}$ final ion states. In the case of argon, a hole is created in the inner-shell $2p^6$ orbital and cross sections have been measured for transitions leading to the Ar^+ $2p^5$ $3p^6$ $^2P_{1/2}$ and $^2P_{3/2}$ ion states. By varying the reaction kinematics and target structure, we aim to obtain a deeper insight into the exchange process.

Xenon was the first closed-shell target atom for which the measurement of non-zero spin asymmetries was established [12,13]. Since that time disparities between experiment and theory have narrowed, leading to an enhanced understanding of scattering dynamics. Calculations performed in the early 90's and subsequent works have shown that while exchange between the incident and ejected target electron, electron impact-induced orientation of the ion, and angular momentum coupling within the residual ion all contribute to the observed asymmetries, as predicted by [9], the form of the asymmetry function is strongly influenced by many-body effects involving exchange of the final-state continuum electrons with the residual electrons in the ion. The fact that significant discrepancies between measurement and theory still remain raises questions about our ability to accurately describe many-body exchange processes in electron-atom scattering. However, the origin for the disparity may lie elsewhere in theory or experiment.

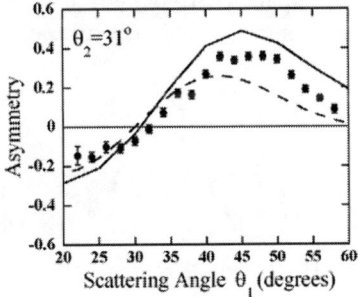

FIGURE 2. Spin asymmetry for the ionization of xenon leading to the $^2P_{1/2}$ final ion state. The incident beam energy is 112 eV and the two outgoing electrons of 49.7 eV leave on opposite sides of the incident beam. One electron is detected at a scattering angle of $\theta_2=31^0$ and the spin asymmetry is measured as a function of the scattering angle θ_1 of the second detected electron. Dashed curve: DWBA calculation, solid curve 3DW calculation.

Figure 2 shows the results of recent measurement and calculation. The reaction kinematics are presented in the figure caption. The experimental results are compared to non-relativistic Distorted Wave Born Approximation (DWBA) calculations in two different forms. In both cases, exchange between the incident and ejected target electron are taken into account as well as exchange between the two final-state continuum electrons and the remaining electrons in the residual ion. This latter many-body exchange effect is accounted for in the calculation of the distorted waves by making a local approximation for the non-local exchange potential. In the calculation labeled DWBA, the Coulomb interaction in the final state is calculated through angle dependent effective charges [19]. In the more sophisticated 3DW calculation [20] the final-state Coulomb interaction is included exactly and constitutes a distorted wave version of the BBK theory [21]. For both cases the agreement with measurement is reasonable, however the remaining discrepancies suggest that a better theoretical treatment of the exchange effect between the continuum electron and passive electrons in the ion may be required.

Given the difficulties in theoretically describing the complex many-body nature of the exchange process for the xenon measurements, kinematics and a target were proposed where the exchange process, to a very good approximation, should reduce to a two-body exchange between projectile and ejected target electron [8]. Thus by performing investigations under those conditions, a clearer understanding might be achieved of the relationship between spin asymmetries and mechanisms of exchange. The proposed experiment involved measuring fine-structure-resolved spin-asymmetries for ionization of the $2p^6$ inner shell of argon with respective energies of 1949 eV and 500 eV for the ejected and scattered electrons. These energies were deemed large enough to ensure a two-body exchange mechanism, but small enough to ensure relativistic spin flip processes, which play a significant role in the K-shell ionization of heavy targets [22], remain small. As the cross section for the proposed kinematics was extremely small, we decided to adopt modified energies of 600 eV and 60 eV for the final-state electron-pair where the cross section is higher [23], hoping that a simplified exchange mechanism would still remain. Preliminary experimental results are shown in Figure 3.

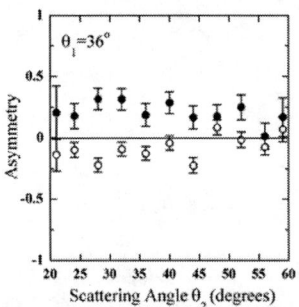

FIGURE 3. Spin asymmetry for the ionization of argon leading to the $^2P_{1/2}$ (solid circles) and $^2P_{3/2}$ (open circles) final ion states. The incident beam energy is 910 eV, the two outgoing electrons of energies 600 eV and 60 eV leave on opposite sides of the incident beam. For this figure, the fast 600 eV electron scatter at an angle $\theta_1=36^0$ and the spin asymmetry is measured as a function of the scattering angle θ_2 of the slow 60 eV electron.

Of particular interest is the relationship between asymmetries for the $^2P_{1/2}$ and $^2P_{3/2}$ ion states, which for a "pure" fine-structure effect should follow the relationship [9]

$$A_{1/2} = -2A_{3/2}. \tag{1}$$

Whilst the error bars in Figure 2 are too large to substantiate this relationship, the spin asymmetry derived from averaging over our full data set (not presented here), corresponding to a range of scattering angles θ_1 and θ_2, suggests that relationship (1) is closely approximated. This suggests that the experiment may indeed describe a simpler scattering process. Comparison of the data with Relativistic Distorted Wave Born Approximation (rDWBA) calculations will appear in a future publication [24].

CONCLUSION

The study of ionization of closed-shell atoms with spin polarized electrons provides insight into target coupling, exchange and relativistic effects in electron-atom collisions. Through a judicious choice of the reaction kinematics and target, contributions from relativistic or exchange processes can be isolated in measurement. Continued improvements in experimental and theoretical tools promise greater insight into the dynamics of electron-atom collisions in the coming years.

ACKNOWLEDGMENTS

J.L and S.B thank the Australian Research Council for their support.

REFERENCES

1. I. E. McCarthy and E. Weigold, *Electron-Atom Collisions*, edited by A. Dalgarno, P. L. Knight, F. H. Read and R. N. Zare, Cambridge: Cambridge University Press, 1995.
2. J. Ullrich *et al.*, Rep. Prog. Phys. **66**, 1463-1545 (2003).
3. C. Mette, T. Simon, C. Herting, G. F. Hanne, and D. H. Madison, J. Phys. B **31**, 4689, (1998).
4. K-H. Besch *et al*, Phys. Rev. A. **58**, R2638-40 (1999).
5. M. Streun *et al.*, Phys. Rev. A. **59**, R4109 (1999).
6. J. Lower, E. Weigold, J. Berakdar and S. Mazevet, Phys. Rev. Lett. **86** (2001) 624-627.
7. J. Kessler, *Polarized Electrons*, edited by G. Ecker, P. Lambropoulos and H. Walther, 2nd edn, Berlin: Springer, 1985.
8. M. Kampp *et al*, Eur. Phy. J. D. **29**, 17 (2004).
9. S. Jones *et al.*, Phys. Rev. Lett. **72**, 2554 (1994).
10. G. F. Hanne, Phys. Rep. **95**, 95 (1983).
11. K. Bartschat and D. H. Madison, J. Phys. B **20**, 5839 (1987).
12. G. F. Hanne, Can. J. Phys. **74**, 811 (1996).
13. X. Guo *et al.*, Phys. Rev. Lett. **76**, 1228 (1996).
14. D. H. Madison *et al*, J. Phys. B: At. Mol. Opt. Phys. 31 (1998) L17-L25.
15. U. Lechner *et al*, in *Electron Scattering from Atoms, Molecules, Nuclei and Bulk Matter*, edited by C. T. Whelan and N. J. Mason, New York: Kluwer/Penum, 2003, pp. 131-142.
16. J. Lower *et al.*, To be submitted to Rev. Sci. Instrum.
17. Nakanishi, T. *et al.*, Phys. Lett. A **158**, 345, (1991).
18. RoentDek GmbH, Kelkheim, Germany (www.roentdek.com).
19. A. Prideaux and D. H. Madison, J. Phys. B: At. Mol. Opt. Phys. 37 (2004) 4423-4433.
20. A. Prideaux and D.H. Madison, Phys. Rev. A **67**, 052710 (2003).
21. M. Brauner *et al*, J. Phys. B **22**, 2265 (1989).
22. W. Nakel and C. T. Whelan, Phys. Rep. **315**, 409 (1999).
23. S. Cavanagh and B. Lohmann, J. Phys. B **30**, L231 (1997).
24. S. Bellm, J. Lower, B. Lohmann and C. T. Whelan (to be published).

Study of coherence in electron-impact excitation of mercury

F. Jüttemann*, G. Außendorf* and G. F. Hanne*

Physikalisches Institut, Universität Münster, 48149 Münster, Germany

Abstract. An adequate description of electron-impact excitation of mercury (Hg) 6s6p levels usually demands the application of the intermediate coupling scheme, leading to a singlet-triplet mixing between the LS-coupled $^{1,3}P_1$ configurations. However, for small scattering angles and impact energies well above the excitation threshold (i.e. 15 eV), the singlet part of the mixture is predicted to be dominant. Hence, for Hg 6s6p $^1P_1$ an almost 'coherent excitation' by electron-impact can be expected for the energetic and angular range stated above. Contrary to this expectation previous measurements by Murray et al [1] and Masters et al [2] seemed to show a significant loss of coherence. We present results from our recent electron-photon coincidence measurements and compare them with the results of theoretical calculations using R-matrix and RDWBA methods. Overall a good agreement is shown, thus confirming our expectations.

Keywords: <electron impact excitation mercury Hg coherence>
PACS: 34.80.Dp,34.80.Nz,34.80.Pa

INTRODUCTION

Electron scattering from heavy atoms such as mercury represents a typical example for the study of relativistic effects both on the continuum electron and the target electrons. Nevertheless simplified theoretical descriptions, which typically neglect relativistic effects for the continuum electron, can provide quantitative agreement with experimental data. To show the motivation for simplifications leading to the expectation of an almost 'coherent excitation' of the 6s6p $^1P_1$ state we start with intermediate coupled target states. The description of excited 6s6p $^{1,3}P_1$ mercury states in the intermediate coupling scheme by McConnell and Moiseiwitsch [3] provides

$$\Psi(^3P_1) = \alpha\Psi^0(^3P_1) + \beta\Psi^0(^1P_1)$$
$$\Psi(^1P_1) = \alpha\Psi^0(^1P_1) - \beta\Psi^0(^3P_1)$$

where Ψ^0 are LS-coupled states and the mixing coefficients are $\alpha = -0.987$ and $\beta = 0.171$ given by Lurio [4]. The mixing coefficients naturally show a dominant singlet part for the $^1P_1$ state, further strengthened by the characteristics of the differential cross sections (DCS) for electron impact excitation of the $^1P_1$ and $^3P_1$ states of mercury; i.e. at 15 eV impact energy and 10°-30° the $^1P_1$ DCS exceeds the $^3P_1$ DCS by one order of magnitude (see Zubek [5]).
In order to study the relevance of the triplet admixture to the $^1P_1$ state, electron-photon coincidence measurements were carried out.

CP811 *Ionization, Correlation, and Polarization in Atomic Collisions*, edited by A. Lahmam-Bennani and B. Lohmann
© 2006 American Institute of Physics 0-7354-0303-1/06/$23.00

SCATTERING GEOMETRY AND OBSERVABLES

The excitation process of interest, Hg $6s^2\ {}^1S_0 \to 6s6p\ {}^1P_1$, is described in the 'natural frame', where the wave vector $\mathbf{k}_0$ of the incident electrons defines the x-axis. The wave vector $\mathbf{k}_1$ of the outgoing electron together with $\mathbf{k}_0$ defines the scattering plane. The quantization axis (z-axis) and the polarization vector of the incident electrons $\mathbf{P}_e$ are perpendicular to this plane. With m_0 and m denoting the spin projections of the incident and outgoing electrons and M_J denoting the magnetic substate of the excited atom, the process can be described by six independent scattering amplitudes $f(M_J, m, m_0)$ (see Andersen et al [6]).

$$f(1, \tfrac{1}{2}, \tfrac{1}{2}) = a_1, \quad f(-1, \tfrac{1}{2}, \tfrac{1}{2}) = a_3, \quad f(0, \tfrac{1}{2}, -\tfrac{1}{2}) = b_1$$
$$f(1, -\tfrac{1}{2}, -\tfrac{1}{2}) = a_2, \quad f(-1, -\tfrac{1}{2}, -\tfrac{1}{2}) = a_4, \quad f(0, -\tfrac{1}{2}, \tfrac{1}{2}) = b_2 \quad (1)$$

Note that all fixed quantum numbers ($J_0 = M_0 = 0$ and $J = 1$) have been omitted. As a consequence of the invariance of the interaction with respect to reflection in the scattering plane, the substates $|M_J = \pm 1\rangle$ can only be excited by non-spin-flip processes ($a_1, ..., a_4$), whereas $|M_J = 0\rangle$ can only be excited by spin-flip processes (b_1, b_2).

The excited atomic state is then analyzed by determination of the polarization components of the decay photons. For the given experimental setup with the polarization vector of the incident electrons $\mathbf{P}_e$ perpendicular to the scattering plane, a set of four Stokes parameters fully describes the excited p-state (see Andersen et al [6]).

$$P_1^z = P_1 = \frac{I(0°) - I(90°)}{I(0°) + (90°)} \qquad P_3^z = P_3 = \frac{I(\sigma^-) - I(\sigma^+)}{I(\sigma^-) + (\sigma^+)}$$

$$P_2^z = P_2 = \frac{I(45°) - I(135°)}{I(45°) + (135°)} \qquad P_1^y = P_4 = \frac{I(0°) - I(90°)}{I(0°) + (90°)}$$

$I(\gamma)$ denotes the light intensity transmitted by a linear polarizer aligned at an angle γ with respect to the wave vector $\mathbf{k}_0$ of the incident electrons. In addition $I(\sigma^+)$ and $I(\sigma^-)$ are the intensities of left- and right-hand circularly polarized light, respectively. Furthermore, the superscripts z and y denote the direction for the light detector.

POLARIZATION AND COHERENCE

Let us first consider the asymptotic entire state for a given initial spin projection m_0 of the incident electron (i.e. $m_0 = \tfrac{1}{2}$), with $|JM_J\rangle$ denoting atomic states and $|m\rangle$ denoting electronic states.

$$\Psi_{r \to \infty} \longrightarrow |00\rangle|\tfrac{1}{2}\rangle e^{ikx}$$
$$+ \{f(1, \tfrac{1}{2}, \tfrac{1}{2})|11\rangle|\tfrac{1}{2}\rangle + f(0, -\tfrac{1}{2}, \tfrac{1}{2})|10\rangle|-\tfrac{1}{2}\rangle + f(-1, \tfrac{1}{2}, \tfrac{1}{2})|1-1\rangle|\tfrac{1}{2}\rangle\} \frac{e^{ik'r}}{r}$$

Studying the radiation from atoms excited in this way, one should usually observe a total polarization $P_{\text{total}} = (P_1^2 + P_2^2 + P_3^2)^{1/2} < 1$, since the scattered electrons are detected without simultaneous spin analysis, thus averaging over different spin projections of the scattered electrons. Yet this is not valid for all light detection directions. To understand why, let us consider the special case of light detection in z-direction, as it is realized for measurements of the Stokes parameters $P_{1,2,3}$.

Light emission from an excited state requires that the wave functions of the excited state meet the selection rules for dipole radiation. As a consequence the $|M_J = 0\rangle$ part cannot contribute to light detected in z-direction. Note that this means a 'natural' spin selection because there is no contribution of spin-flip scattering processes to light emitted in z-direction. This in turn cancels the averaging over different spin projections of the scattered electrons when observing the decay photons. Thus one should see $P_{\text{total}} = 1$.

Summarizing we can say that for a given initial spin projection m_0 of the incident electrons (totally polarized electron beam) the detection of the decay photons in z-direction always leads to $P_{\text{total}} = 1$.

However, for unpolarized-electron-impact excitation $P_{\text{total}} < 1$ can result from differences in the three Stokes parameters $P_{1,2,3}$ for spin-up and spin-down electrons. Since the spin-orbit effects contribute to theses differences we would like to define the term 'coherent excitation' as excitation, which does not depend at all on spin effects and therefore leads to $P_{\text{total}} = 1$ even for unpolarized-electron-impact.

A straighter observation of a relevance of spin-orbit effects, particularly of the triplet admixture, for electron-impact excitation can be provided by means of P_4-measurements. Here the radiation from $|M_J = 0\rangle$ substate can be observed directly to reveal a relevance of the triplet admixture in the excitation of the $|M_J = 0\rangle$ substate.

For the sake of clarity let us additionally consider the radiation from a pure LS-coupled $^1P_1$ state. With a focus on small angle scattering, the spin-orbit interaction of the continuum electron (i.e. Mott scattering) can be neglected and spin conservation yields $\Delta M_S \equiv 0$. This leads to vanishing scattering amplitudes for excitation of the $|M_J = M_L = 0\rangle$ substate, because this excitation would require $\Delta M_S = \pm 1$ thus violating the selection rule $\Delta M_S \equiv 0$ arising from spin conservation. So the $|M_L = 0\rangle$ part (see Figure 1) does not appear at all in the expression for the asymptotic entire state leading to $P_{\text{total}} = 1$ for light detection in arbitrary directions.

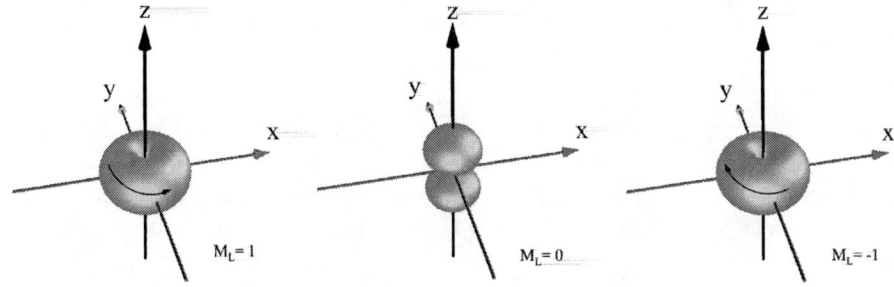

FIGURE 1. Atomic basis set for a p-state in the 'natural frame'

RESULTS AND DISCUSSION

The natural isotope mix of mercury demands corrections in form of perturbation coefficients in the theoretical calculation of the Stokes parameters due to the occurrence of non-vanishing nuclear spin projections. This leads to a reduced P_{total}, because the measurement process averages over a mixture of different nuclear spin projections. In cases, where $P_{total} = 1$ is anticipated, this yields $0.84 < P_{total} < 0.95$ (see for example Wolcke et al [7]). For better orientation this data range is marked by dashed lines in the following plots for P_{total}.

For scattering energies of 50eV and 100eV we expect neither exchange nor relativistic effects for the continuum electron to play an important role in small angle scattering. According to FIGURE 2, the comparison between different theoretical calculation ap-

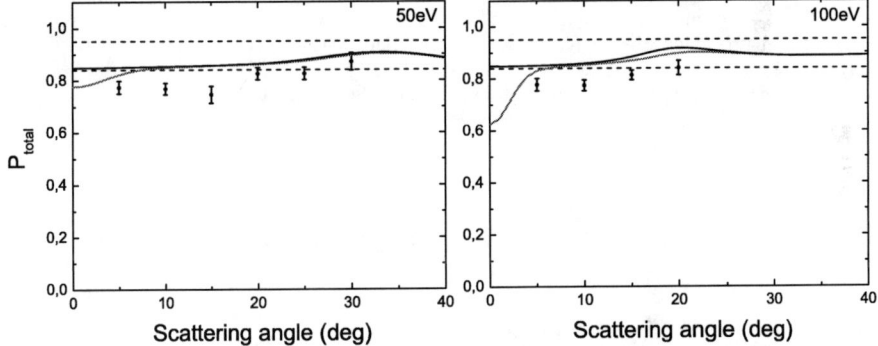

FIGURE 2. P_{total} for unpolarized electron-impact. • experimental data for unpolarized electron impact at 50eV and 100eV impact energy. Left: R-matrix calculation by K. Bartschat for unpolarized electrons at 50eV impact energy. Right: RDWBA calculation by R. Srivastava for unpolarized electrons at 100eV impact energy. The grey lines each indicate the theoretical calculations adapted to the solid angle of the electron spectrometer. The dashed lines show the anticipated data range of $0.84 < P_{total} < 0.95$.

proaches shows a good agreement between experiment and theoretical calculation. The effect of the solid angle of the electron spectrometer was found to be relevant for small scattering angles (grey plots). Please note as well that the simple anticipation of the data range $0.84 < P_{total} < 0.95$ meets both theoretical calculations.

P_{total} values for 100eV impact energy of ≈ 0.27 at $10°$ and ≈ 0.44 at $20°$ as presented by Masters et al [2] could not be reconfirmed.

More experimental data was acquired at 15eV impact energy to determine whether exchange effects, which are more likely to occur at scattering energies closer to the excitation threshold, could lead to a more important role of the triplet admixture. Again, both theoretical calculations meet the anticipated data range of $0.84 < P_{total} < 0.95$ for small scattering angles (see FIGURE 3). However, as predicted by both calculations for larger scattering angles, the differences in P_{total} for spin-up and spin-down electrons could not be seen under the given statistical uncertainty. The effect of the solid angle of the electron spectrometer was found not to be relevant at this scattering energy.

Therefore, it seems very unlikely that P_{total} values of ≈ 0.3 and below at $10°$ and $20°$ can be reconfirmed, as presented by Murray et al [1] for a scattering energy of 16eV.

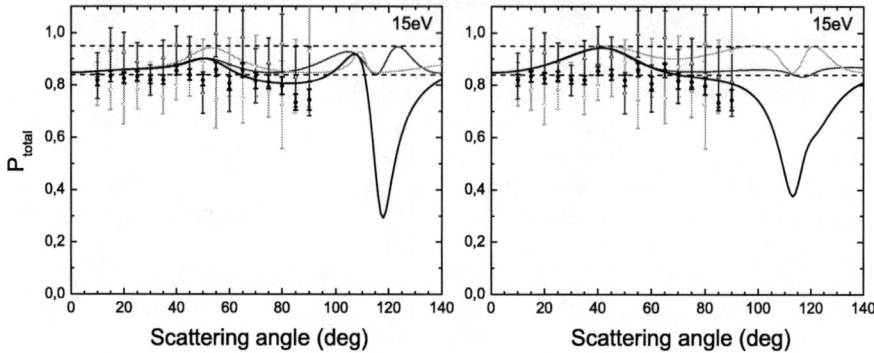

FIGURE 3. P_{total} for unpolarized (●), spin-up (△) and spin-down (▽) electron-impact at 15eV impact energy. Left: R-matrix calculation by K. Bartschat for unpolarized (black), spin-up (dark grey) and spin-down (light grey) electrons. Right: RDWBA calculation by R. Srivastava for unpolarized (black), spin-up (dark grey) and spin-down (light grey) electrons. An influence of the solid angle of the electron spectrometer on the calculated data was not visible at 15eV impact energy. The dashed lines show the anticipated data range of $0.84 < P_{total} < 0.95$.

FIGURE 4 shows our results for P_4 at 15eV impact energy, again compared to both theoretical calculation approaches. The agreement between calculation and experiment is good, although slightly better for the R-matrix method. Only for small scattering angles a deviation of up to 10% is visible for both calculations.

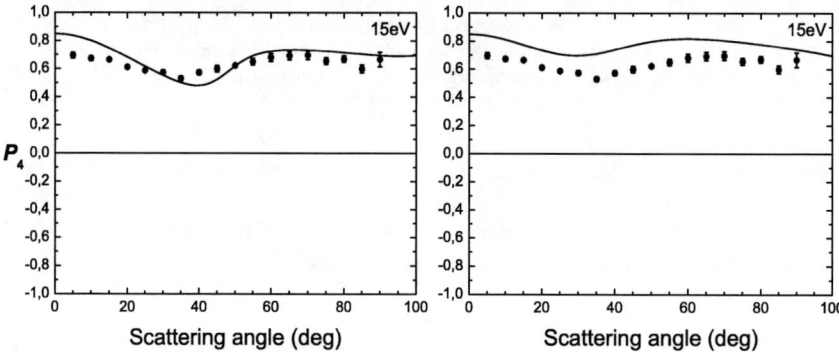

FIGURE 4. P_4 for unpolarized electron-impact at 15eV impact energy. Left: R-matrix calculation by K. Bartschat for unpolarized electrons. Right: RDWBA calculation by R. Srivastava for unpolarized electrons.

CONCLUSIONS

An overall good agreement between theoretical calculations and experiment was achieved. This holds for P_{total} as well as for P_4. Hence the assumption of an almost coherent excitation of the Hg $^1P_1$ state can be considered valid.

ACKNOWLEDGMENTS

The authors thank K. Bartschat and R. Srivastava for stimulating discussions and for providing the calculated data used in this article.

REFERENCES

1. A. J. Murray, R. Pascual, W. R. MacGillivray and M. C. Standage, *J. Phys. B* **25** 1915–1930 (1992).
2. A. T. Masters, W. R. MacGillivray, M. C. Standage and I. Humphrey, *Abstracts of contributed papers Vol. I*, ICPEAC 1997, Vienna Austria, 1997, p. TH032
3. J. C. McConnell and B. L. Moiseiwitsch, *J. Phys. B* **1**, 406-413 (1968).
4. A. Lurio, *Phys. Rev. A* **140**, 1505–1508 (1965).
5. M. Zubek, N. Gulley, A. Danjo and G. C. King, *J. Phys. B* **29**, 5927–5936 (1996).
6. N. Andersen, K. Bartschat, J. T. Broad, and I. V. Hertel, *Phys. Rep.* **279**, 251-396 (1997).
7. A. Wolcke, K. Bartschat, K. Blum, H. Borgmann, G. F. Hanne and J. Kessler, *J. Phys. B* **16**, 639-655 (1983).

Theoretical Treatment of Electron-Impact Ionization of Molecules

Junfang Gao, J.L. Peacher, and D.H. Madison

Department of Physics, University of Missouri-Rolla, Rolla, MO, USA 65409

Abstract. There is currently no reliable theory for calculating the fully differential cross section (FDCS) for low energy electron-impact ionization of molecules. All of the existing experimental FDCS data represent averages over all molecular orientations and this can be an important theoretical complication for calculations that are computer intensive. We have found that using an averaged molecular orbital is an accurate approximation for ionization of ground states. In this paper, we will describe the approximation, discuss its expected range of validity and show some FDCS results using the approximation for ionization of H_2 and N_2.

Keywords: Ionization, Molecules, Electron-impact, Charged particle ionization.
PACS: 34.10.+x, 34.80.Gs

INTRODUCTION

Over the last two decades, there has been considerable interest in the fully differential cross section (FDCS) for electron-impact ionization of molecules. However most of this work has concentrated on either high incident electron energies or small molecules. For high enough energies, all the continuum electrons can be represented as plane waves and, in the plane wave impulse approximation (PWIA), the cross section becomes proportional to the momentum space wave function of the active electron. The PWIA was developed by McCarthy and his co-workers [1-4] and it has been extremely useful for studying molecular structure.

As the energy of the incident electron is decreased or the ejected electron has a low energy, it is clear that approximating the continuum electrons as plane waves will fail. Although there has been some improved theoretical models proposed for ionization of H_2 [5-6], very little work has been reported for ionization of larger molecules. One of the practical problems associated with any theoretical calculation of molecular ionization lies in the fact that existing experimental data represent an average over all molecular orientations and this can be a significant obstacle for approaches that are computer intensive. For example, we have introduced the 3DW (3 body distorted wave) model for charged particle ionization of atoms. This model contains the interactions between all two-particle subsystems to all orders of perturbation theory and practical calculations for a single point on a graph can take a day or more on the fastest computers. As a result, calculating enough points to take an accurate numerical 3-dimensional average becomes impractical (at least with the computing resources that we have available for use). Consequently, finding an alternative to brute force numerical averages is highly desirable.

In this paper, we report a simple approximation which eliminates the necessity for averaging cross sections over all molecular orientations. This approximation is to form an orientation averaged molecular orbital (OAMO) first and then use the averaged orbital for the calculation of the FDCS. We will demonstrate that this elementary approximation is valid for ionization of ground gerade states which are dominated by s-basis functions. Plane wave impulse approximation (PWIA) and distorted wave impulse approximation (DWIA) results will be presented for ionization of H_2 and N_2.

THEORY

We start with the plane wave impulse approximation (PWIA) which has been very successful for high energy collisions. In the PWIA, the FDCS is given by

$$\sigma^{PWIA}(\hat{\mathbf{R}}) = \frac{4}{(2\pi)^5} \frac{k_a k_b}{k_i} F(\mathbf{k}_i, \mathbf{k}_a, \mathbf{k}_b) |T(\hat{\mathbf{R}})|^2 \quad (1)$$

where $(\mathbf{k}_i, \mathbf{k}_a, \mathbf{k}_b)$ are the momenta of the (incident, fast-final, slow-final) electrons, respectively. F is an elementary function of these momenta [4] and

$$T(\hat{\mathbf{R}}) = \int e^{i\mathbf{q}\cdot\mathbf{r}} \Phi(\mathbf{r},\mathbf{R}) d\mathbf{r} \quad (2)$$

Here $\mathbf{q} = \mathbf{k}_i - \mathbf{k}_a - \mathbf{k}_b$ is the momentum transferred to the residual ion and $\Phi(\mathbf{r},\mathbf{R})$ is the molecular orbital (MO) which depends on the orientation of the molecule determined by the internuclear vector $\mathbf{R}$. To perform a proper average over orientations, one must average eq. (1) over all possible orientations.

$$\sigma^{PWIA} = \frac{1}{4\pi} \int \sigma^{PWIA}(\hat{\mathbf{R}}) d\hat{\mathbf{R}} \quad (3)$$

For the orientation averaged molecular orbital (OAMO) approximation which we have proposed, the orientated MO is replaced by

$$\phi^{OAMO}(\mathbf{r}) = \frac{1}{4\pi} \int \Phi(\mathbf{r},\mathbf{R}) d\hat{\mathbf{R}} \quad (4)$$

Now eq. (2) does not depend upon the orientation of the molecule

$$T^{OA} = \int e^{i\mathbf{q}\cdot\mathbf{r}} \phi^{OAMO}(\mathbf{r}) d\mathbf{r} \quad (5)$$

and the OA cross section is given by

$$\sigma^{PWIAOA} = \frac{4}{(2\pi)^5} \frac{k_a k_b}{k_i} F(\mathbf{k}_i, \mathbf{k}_a, \mathbf{k}_b) |T^{OA}|^2 \quad (6)$$

Using the OAMO for the calculation of the FDCS represents a significant simplification since: (1) the OAMO is relatively easy to calculate and (2) it only has to be calculated once since it is independent of the incident electron energy.

As a first attempt to improve the PWIA, we have proposed the distorted wave impulse approximation (DWIA). The idea of the DWIA is to take advantage of the factorization simplification of the PWIA and then to replace the plane waves in eq. (2) by distorted waves. As a result, the T-factor of eq. (2) becomes

$$T^{DWIA} = \int \chi_a^{-*}(\mathbf{k}_a,\mathbf{r}) \chi_b^{-*}(\mathbf{k}_b,\mathbf{r}) \chi_i^{+}(\mathbf{k}_i,\mathbf{r}) \phi^{OAMO}(\mathbf{r}) d\mathbf{r} \quad (7)$$

where (χ_i, χ_a, χ_b) are (incident, fast-final, slow-final) distorted waves, respectively. The distorted waves are calculated in the normal way as eigenfunctions of the Schrödinger equation using a spherically symmetric potential obtained from the charge density for the molecule averaged over all orientations. For the incident channel distorted wave, the charge density for a neutral molecule is used and for the final channel distorted waves, the charge density for a molecular ion is used. The DWIA cross sections are given by

$$\sigma^{DWIA} = \frac{4}{(2\pi)^5} \frac{k_a k_b}{k_i} F(\mathbf{k}_i, \mathbf{k}_a, \mathbf{k}_b) \left| T^{DWIA} \right|^2 \qquad (8)$$

RESULTS

The relative simplicity of the PWIA allows us to test the validity of using the OAMO approximation. If the approximation were valid, eqs. (3) and (6) should give identical results. We have investigated the validity of the approximation for ionization of H_2 and N_2. We have formed Hartree-Fock MO's for both molecules using the GAMES software [7]. For H_2, the ground state wave function was formed from three s-basis functions and for N_2, the ground state wave function was formed from two s-basis functions and one p-basis function.

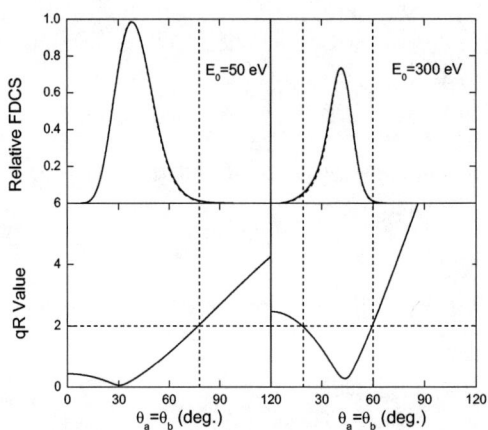

FIGURE 1. Upper part of the figure is the relative FDCS for electron impact ionization of H_2 in the coplanar symmetric scattering geometry. For coplanar symmetric scattering, the electrons are detected with equal energies and at equal angles on opposite sides of the beam direction. The energy of the incident electron is E_0 and each outgoing electron has an energy of $(E_0 - E_{ion})/2$ where E_{ion} is the ionization energy of the ground state orbital. The horizontal axis is the angular location for the two electron-detectors and the corresponding qR value is shown in the lower part of the figure where R=1.4 a_0.

In fig. 1, σ^{PWIA} and σ^{PWIAOA} results are compared for ionization of H_2 by 50 and 300 eV incident-energy electrons for coplanar symmetric scattering. Although it is hardly visible on the figure, there are actually two cross sections in the top part of the figure. It can be shown by analytical integration that, if a gerade state s-basis function is used to calculate σ^{PWIA} and σ^{PWIAOA}, the results are the same providing qR (momentum transfer times internuclear distance) is less than about two. In the bottom half of fig. (1), the qR values corresponding to the scattering angles are shown and the vertical dashed line indicates where qR crosses two. It is seen that the cross sections are very small for $qR > 2$. However, the OAMO approximation is still valid even for $qR > 2$.

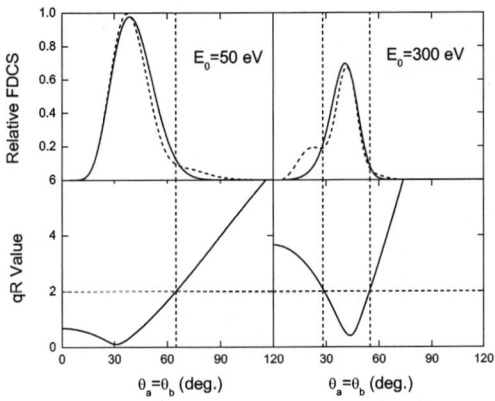

FIGURE 2. Same as Fig. 1 except for ionization of N_2 where $R=2.1\, a_0$.

Figure 2 shows the same comparison for electron-impact ionization of N_2. Here we see that using the OAMO to calculate the OA cross sections is reliable for $qR < 2$ as expected. For $qR > 2$, the approximation is still reasonably good except for high energy and small scattering angles.

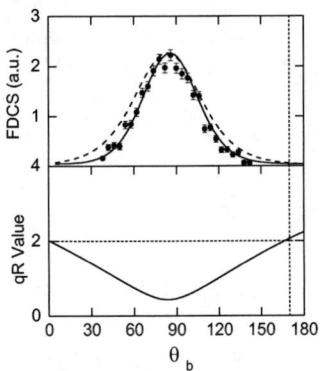

FIGURE 3. Upper part of the figure are the FDCS for electron-impact ionization of H_2 for coplanar asymmetric scattering. The energy of the incident electron is 4087 eV, the faster final state electron is observed at 3^0 and the energy of the ejected electron is 20 eV. The experimental data are those of Cherid et al. [8], the solid curve are the DWIA results; and dashed curve are PWIA results. The horizontal axis is the observation angle for the ejected electron and the corresponding qR value is shown in the lower part of the figure.

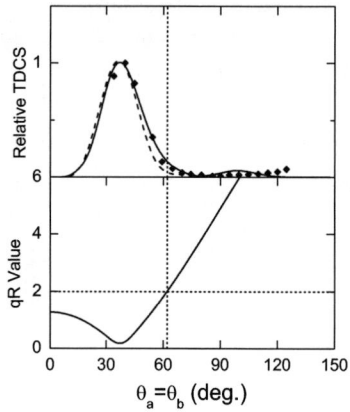

FIGURE 4. Upper part of the figure is the relative FDCS for electron-impact ionization of N_2 for coplanar symmetric scattering. The energy of the incident electron is 75.6 eV and both final state electrons have energy of 30 eV. The experimental data are those of Hussey and Murray [9], the solid curve are the DWIA results; and the dashed curve are PWIA results. The horizontal axis is the angular location for the two electron-detectors and the corresponding qR value is shown in the lower part of the figure.

Figures 3 and 4 compare experimental and theoretical results for ionization of H_2 and N_2. For H_2, experimental data are absolute and we have normalized our relative cross sections to experiment at the maximum value. It is seen that there is excellent agreement between experiment and the DWIA at this high energy (of course there is also little difference between the PWIA and DWIA at this energy). Figure 4 compares experiment and theory for low energy electron-impact ionization of N_2 for coplanar symmetric scattering. For this case, the PWIA does not predict a measurable cross section for larger angles. Although the large angle behavior does not agree with experiment, the DWIA, on the other hand, does predict a measurable large angle peak in the cross section. Large angle peaks in the FDCS are normally attributed to the atomic electron being back scattered from the ion. Since the physics contained in a distorted wave is elastic scattering from the molecule (initial state) and ion (final state), the fact that the DWIA contains a large angle peak while the PWIA does not clearly indicates that this peak results from elastic back scattering from the ion. The lack of good quantitative agreement for the largest angles may also be related the fact that qR is quite a bit larger than 2 in this angular region which means that using the OAMO to calculate orientation averaged FDCS becomes questionable.

CONCLUSIONS

In conclusion, we have shown that using orientation averaged molecular orbital (OAMO) to calculate the FDCS averaged over all molecular orientations is reliable for ionization of H_2 at all energies and all qR values where the cross section is large enough to be measured. For ionization of N_2, the approximation was valid for $qR < 2$

at all energies and also for $qR > 2$ except for small angles and high energies. This elementary approximation for evaluating orientation averaged cross sections will greatly simplify the evaluation of an orientation averaged FDCS for electron-impact ionization of molecules using better and more sophisticated theoretical approaches.

We also used the OAMO to calculate the DWIA FDCS for ionization of H_2 and N_2. We found that the DWIA represents an improvement over the PWIA particularly for low energies where the PWIA fails to predict the secondary large angle peaks. The DWIA was in qualitative, but not quantitative, agreement with the secondary large angle peak seen in the experimental data.

ACKNOWLEDGMENTS

The authors would like to acknowledge the support of the NSF under grant number PHY-0070872.

REFERENCES

1. E. Weigold and I.E. McCarthy, Electron Momentum Spectroscopy, Kluwer Academic/Plenum Publishers, New York,1999.
2. I.E. McCarthy and E. Weigold, Phys. Rep. Vol. 27, N.6, 275-371(1976).
3. I.E. McCarthy and A.M. Rossi, Phys. Rev. A, Vol.49, N.6, 4645-4652(1994).
4. I.E. McCarthy and E Weigold, Reports on the Progress in Physics, Vol.51, 299-392(1988).
5. C.R. Stia, O.A. Fojón, P.F. Weck, J. Hanssen, B. Joulakian and R.D. Rivarola, Phys. Rev. A. 66, 052709(2002).
6. P. Weck, O.A. Fojon, J. Hannsen, B. Joulakian, and R.D. Rivarola, Phys. Rev. A 63, 042709 (2001).
7. M.W. Schmidt, K.K. Baldridge, J.A. Boats, S.T. Elbert, M.S. Gordon, J.H. Jensen, S. Koseki, N. Matsunaga, K.A. Nguyen,S.J. Su, T.L. Windus, M. Dupuis and J.A. Montgomery,Jr., J. Comp. Chem. Vol.14, N.11, 1347-1363(1993)
8. M. Cherid, A. Lahmam-Bennani, R.W. Zurales, R.R. Lucchese, A. Duguet, M.C. Dal Capello, and C. Dal Cappello, J. Phys. B At.Mol.Opt . 22, 3483 (1989).
9. M.J.Hussey and A.Murray, J.Phys. B: At.Mol.Opt. 35, 3399-3409(2002).

New Results On Excitation, Ionization And Ionization-Excitation By Electron Impact

Albert Crowe

School of Natural Sciences (Physics), University of Newcastle upon Tyne, NE1 7RU, United Kingdom

Abstract. A brief outline is given of the information which can be obtained from correlation studies of each of these three processes. A sample of recent results, mostly from this laboratory, are analysed critically with the aim of identifying the combination of experimental conditions necessary to provide the ultimate test of theoretical models in each case.

Keywords: Excitation, Ionization Excitation-Ionization, Correlations.
PACS: 34.80.Dp, 34.80.Pa

INTRODUCTION

The topics in this very wide ranging title are tackled by presenting some recent experimental data, mostly from this laboratory, and analyzing them in terms of their achievements and their weaknesses, particularly with regard to providing the most stringent test of theory. In this way it is hoped that clear directions can be given regarding the type of future experiments which will give the most sensitive tests of theory. The discussion will be restricted to correlation studies which give the maximum information on the dynamics of each of the processes.

IONIZATION OF A SINGLE ATOMIC ELECTRON

Experiments conducted using the (e,2e) technique in asymmetric geometry are discussed here. These yield what are commonly referred to as triple differential cross sections (TDCS). Even though these experiments totally define the kinematics of the incident, scattered and ejected electrons, the information available is restricted relative to that available from excitation and potentially ionization-excitation experiments, as discussed later.

Figure 1 shows very recent [1] results on ionization of the outer p-electron in argon at an incident electron energy of 200 eV. A similar study at an incident energy of 113.5 eV has been reported by Haynes and Lohmann [2]. The experimental data are compared with two calculations, one a standard distorted wave Born (DWBA) in which exchange distortion (ED) is deliberately left out, and a second (DWBA-RM2) in which ED is added using the *R*-matrix method [1]. Both theories reproduce the shapes of the TDCS when the slow electrons are emitted near the momentum transfer (**K**) directions, although the relative sizes of the two lobes are best predicted by DWBA-RM2. Similarly, for electrons ejected close to the opposite direction the recoil

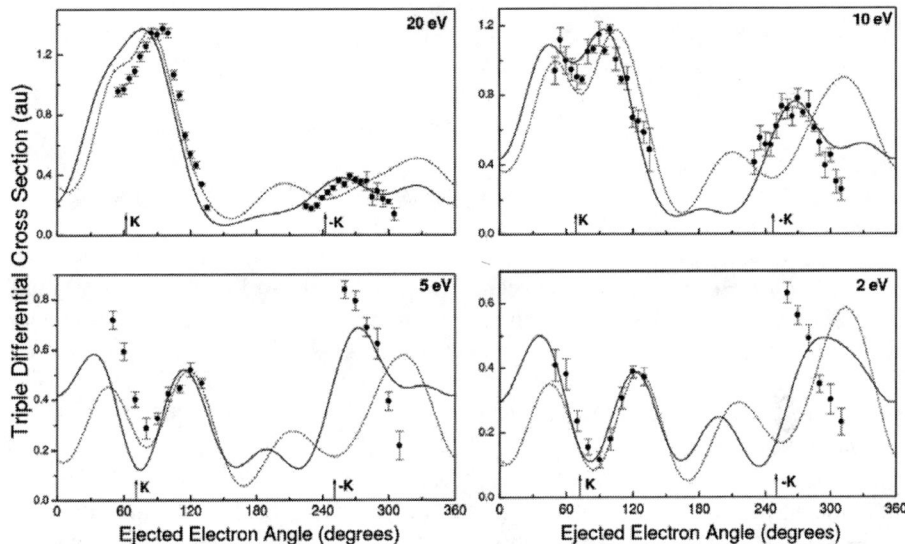

FIGURE 1. Experimental and theoretical TDCS for the ejection of a 3p-electron from argon at an incident electron energy of 200 eV and a fast (scattered) electron angle of -15° as a function of the slow (ejected) electron angle for the ejected electron energies shown. ●, experimental data; full line, DWBA-RM2; dashed line, DWBA.

peak is best predicted by DWBA-RM2 with the recoil/binary peaks ratio being very well reproduced for the two highest ejected electron energies.

Although this comparison between experiment and theory emphasizes the importance of a proper treatment of exchange in theoretical models of ionization, a number of improvements should be made to experimental data of this type to provide a greater challenge for theory.

(i) Put cross sections on an absolute scale.

In Fig. 1 both the experimental data and the DWBA curves have been arbitrarily normalized to the DWBA-RM2 at the higher angle binary lobe for each ejected electron energy. This masks large discrepancies between the cross section magnitudes predicted by the two calculations and gives no information on the reliability of either. At an ejected electron energy of 20 eV, for example, the TDCS predicted by the DWBA model lies 27% below the DWBA-RM2 at the point of normalization, whereas it lies 121% above at 5 eV.

(ii) Extend angular range of data.

The data range in Fig. 1 is fairly typical of conventional (e,2e) spectrometers where the angular range is restricted by the finite size of the electron gun and analysers. It is also obvious from Fig. 1 that the two theories predict quite different behaviour of the TDCS in regions where it is untested by experiment, including very significant ejected electron signal in the same direction as the fast scattered electron, -15° (345°in Fig. 1). This can now be overcome using a magnetic angle changing device [3] and the first (e,2e) experiment of this type has just been reported [4].

(iii) Improve the energy resolution of the experiment.

The overall energy resolution used in this experiment, ~0.6 eV, is not ideal for the lowest ejected electron energy studied, 2 eV.

The implementation of all of these improvements in a single experiment would provide a much more rigorous evaluation of theoretical methods.

EXCITATION OF A SINGLE ATOMIC ELECTRON

In this case the scattered electron-polarized photon correlation method and the equivalent laser excited-superelastic scattering technique provide information beyond the cross section level. They are often referred to as quantum mechanically complete experiments, yielding both the *magnitudes and phases* of the complex excitation amplitudes. Alternatively, the data can be analysed in terms of the shape of the excited state charge cloud and the expectation value of the angular momentum transfer ($L_\perp$) in the collision [5].

Close interplay between experimental and theoretical correlation studies of excitation have led to major theoretical advances over the last decade. Reliable experimental results are reproduced in detail by theory for simple atoms over a wide range of both scattering energies and angles. The convergent close coupling (CCC) and *R*-matrix with pseudo states (RMPS) methods have been at the forefront of these developments. Processes showing excellent agreement between experiment and theory include 1S-nP excitation in hydrogen and the pseudo-one-electron atoms, helium and the pseudo-two-electron atoms, including heavier elements such as barium. Experimental data for ^{1}S to ^{1}D and ^{3}D (resolved) transitions in helium are also described in detail by these theories.

Attempts have also been made to understand many aspects of the data from these experiments in terms of simple physical models ([6], and references therein). These have included the variation in sign of the angular momentum transfer, $L_\perp$, and the shape and alignment of the excited state charge cloud. Here, attention is focused on a general observation concerning the behaviour of $L_\perp$ and the differential cross section (DCS) for the excited state, first reported by Teubner and Scholten [7] for the $3^2P_{3/2}$ state of sodium.

Figure 2 shows recent comparisons from this laboratory for the 3^1P state of magnesium [8, 9]. It can be clearly seen, at both energies, that the first minimum in the DCS corresponds to a zero value of $L_\perp$. It should also be noted that the variation of the experimental $L_\perp$ and the shape of the measured relative DCS at both incident electron energies are well reproduced by the CCC (shown) and RMPS (not shown) theories, demonstrating that theory and experiment both confirm the existence of this relationship. $L_\perp$ and the magnetic sub-level cross sections are related [6] (in the natural co-ordinate frame) by,

$$L_\perp \sim \frac{\sigma_{+1} - \sigma_{-1}}{\sigma_{+1} + \sigma_{-1}}$$

Hence $L_\perp = 0$ requires $\sigma_{+1} = \sigma_{-1}$. Unless there is a propensity for this condition at DCS

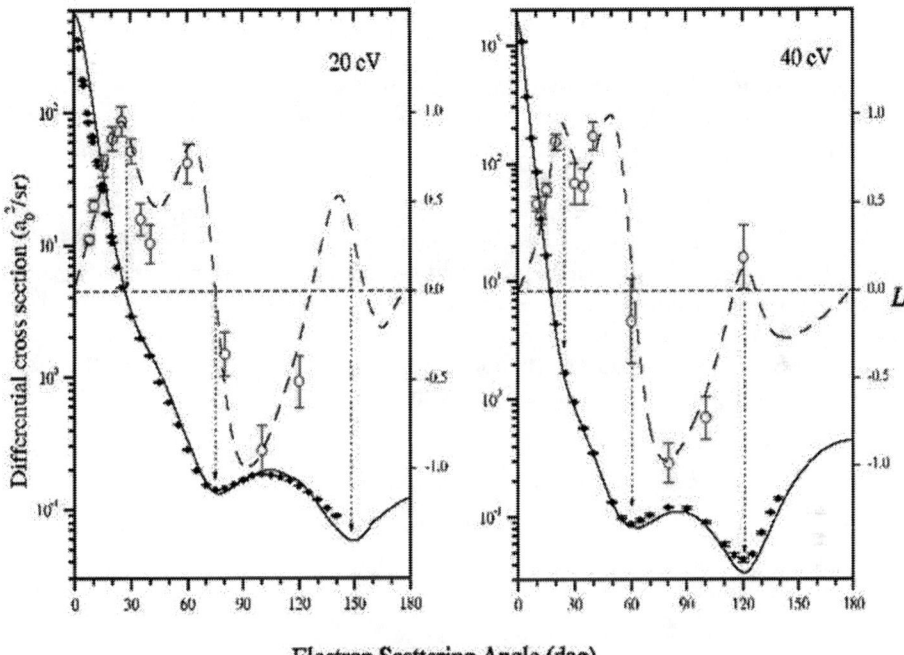

FIGURE 2. A comparison between the differential cross sections and the angular momentum transfer, $L_\perp$, for excitation of the 3^1P state of magnesium. ○, experimental $L_\perp$ at 20 eV [8], 40 eV [9]; •, corresponding experimental DCS; dashed line, CCC theory for $L_\perp$ at 20 eV [8], 40 eV [10]; full line, corresponding theoretical DCS.

minima where both cross sections are small, this simple approach is not helpful in understanding the relationship. It may be that the best physical explanation comes from an extension of an analysis [11] which concluded that the angular behaviour of $L_\perp$ is due to quantum mechanical interference effects. The presence of maxima and minima in DCS is often accounted for qualitatively in terms of diffraction effects.

Other features concerning the relative behaviour of the DCS and $L_\perp$ can be seen in Fig. 2. At 40 eV, the near-zero $L_\perp$ around 120° corresponds with the second DCS minimum, while at 20 eV, theory shows close, but not exact, agreement between the position of the third $L_\perp$ zero and the second DCS minimum. At both energies there are also similarities in the positions of the first maximum in $L_\perp$ and a change of slope in the DCS.

Future excitation studies are likely to concentrate on more complex systems where relativistic effects are more important. Here spin-polarized electron studies may play an important role. At present only the Munster group are performing spin-polarized electron-photon correlation studies, see [12] for example. It may also be timely to extend some early correlation studies of simple molecules.

IONIZATION-EXCITATION

Various aspects of this two-active-atomic-electron process were discussed in detail at the 2003 meeting in Königstein [13, 14] and so a limited discussion is given here. In principle the (e,2e), (e,eγ) and (e,2eγ) methods can be used to study these processes, but only (e,2eγ) with a measurement of the polarization state of the emitted radiation gives the complete information described for single electron excitation.

Figure 3 shows an example of an (e,2e) study of He^+ ($n = 2$) which has received considerable theoretical attention and is one of the few cases where more than a single experiment has been carried out. Discrepancies between the experiments may well be accounted for by the influence of doubly excited states for these kinematics [15]. Nevertheless, the second Born (two-step) [16] and second-order RMPS [17] calculations show good agreement in shape with the present data.

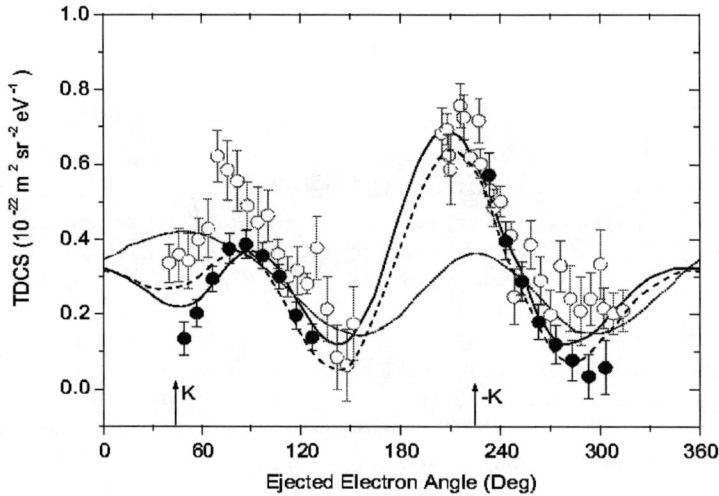

FIGURE 3. He^+ ($n = 2$) TDCS (arbitrarily normalized to the absolute scale of [18]) at an incident electron energy of 645.4 eV, ejected electron energy of 10 eV and a scattering angle of -4°. Experiment: ●, present data; ○, Avaldi et al [18]. Theory: dash-dot line, first Born [16]; full line, second Born [16]; dashed line, second-order RMPS [17]. (Note that the TDCS scale shown should be decreased by a factor of 100).

The most challenging aspect of ionization-excitation studies is the low coincidence signal rates arising from their low cross sections. The He^+ ($n = 2$) TDCS are typically more than a factor of 100 down on the corresponding $n = 1$ TDCS. Clearly the way ahead for these studies is through the use of single-particle multi-detectors capable of providing timing information on individual events. An outstanding example of the use of these is shown in Fig. 4, with this coincidence spectrum [19] having statistics more reminiscent of a single particle energy-loss spectrum. This spectrometer also has the facility to carry out spin-polarized electron studies.

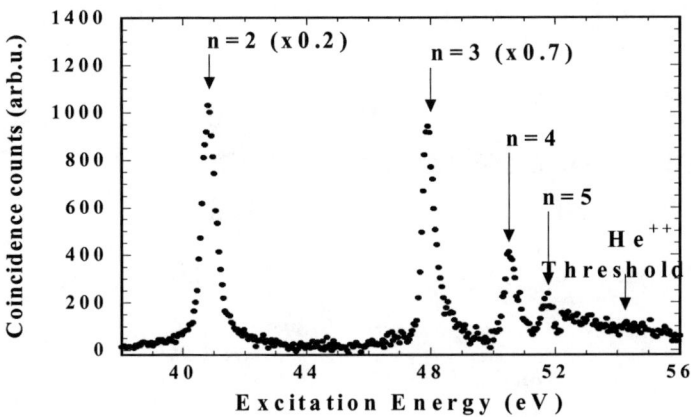

FIGURE 4. He$^+$ ($n \geq 2$) binding energy spectrum. The energy scale is relative to the He$^+$ ($n = 1$) energy [19].

ACKNOWLEDGMENTS

This work is supported by the United Kingdom Engineering and Physical Sciences Research Council (EPSRC). I am grateful to Julian Lower for providing Fig. 4 in electronic form.

REFERENCES

1. M. Stevenson, G. J. Leighton, A. Crowe, K. Bartschat, O. K. Vorov and D. H. Madison, *J. Phys. B: At. Mol. Opt. Phys.* **38**, 433-440 (2005).
2. M. A. Haynes and B. Lohmann, *Phys. Rev. A* **64**, 044701 (2001).
3. F. H. Read and J. M. Channing, *Rev. Sci. Instrum.* **67**, 2372-7 (1996).
4. M. A. Stevenson and B. Lohmann in *XXIV ICPEAC abstracts*, edited F. D. Colavecchia et al., 2005, p. Fr 035.
5. N. Andersen, J. W. Gallagher and I. V. Hertel, *Phys. Rep.* **165**, 1-188 (1988).
6. N. Andersen and K. Bartschat, *Polarization, Alignment, and Orientation in Atomic Collisions*, Springer-Verlag, New York, 2001.
7. P. J. O. Teubner and R. E. Scholten, *J. Phys. B: At. Mol. Opt. Phys.* **25**, L301-6 (1992).
8. D. O. Brown, A. Crowe, D. V. Fursa, I. Bray and K. Bartschat, *J. Phys. B: At. Mol. Opt. Phys* (submitted).
9. D. O. Brown, D. Cvejanovic and A. Crowe, *J. Phys. B: At. Mol. Opt. Phys.* **36**, 3411-23 (2003).
10. D. V. Fursa and I. Bray, *Phys. Rev. A* **63**, 032708 (2001).
11. D. H. Madison, G. Csanak and D. C. Cartwright, *J. Phys. B: At. Mol. Phys.***19**, 3361-6 (1986).
12. C. Herting, G. F. Hanne, K. Bartschat, K. Muktavat, R. Srivastava and A. D. Stauffer, *J. Phys. B: At. Mol. Opt. Phys.* **36**, 3877-87 (2003).
13. K. Bartschat in *Correlation and Polarization in Photonic, Electronic and Atomic Collisions*, edited by G. F. Hanne et al, AIP Conference Proceedings 697, Melville, NY, 2003, pp. 213-8.
14. M. Stevenson, M. Dogan and A. Crowe in *Correlation and Polarization in Photonic, Electronic and Atomic Collisions*, edited by G. F. Hanne et al, AIP Conference Proceedings 697, Melville, NY, 2003, pp. 219-24.
15. Y. Fang and K. Bartschat, *Phys. Rev. A* **64**, 020701 (2001).
16. P. J. Marchalant, J. Rasch, C. T. Whelan, D. H. Madison and H. R. J. Walters, *J. Phys. B: At. Mol. Opt. Phys.* **32**, L705-10 (1999).
17. Y. Fang and K. Bartschat, *J. Phys. B: At. Mol. Opt. Phys.* **34**, L19-25 (2001).
18. L. Avaldi, R. Camilloni, R. Multari, G. Stefani, J. Langlois, O. Robaux, R. J. Tweed and G. Nguyen Vien, *J. Phys. B: At. Mol. Opt. Phys.* **31**, 2981-97 (1998).
19. J. Lower, R. Panajotovic and E. Weigold, in *XXIII International Conference on the Physics of Electronic and Atomic Collisions*, edited by R. Schuch, et al, *Phys. Scr.* **T110**, 216-221 (2004).

Calculation of excitation-autoionization of helium with three active electrons: Sharp Resonance features in the SDCS

C. W. McCurdy*,†, D. A. Horner** and T. N. Rescigno†

Departments of Applied Science and Chemistry, University of California, Davis, CA 95616
†*Lawrence Berkeley National Laboratory, Chemical Sciences, Berkeley, CA 94720*
**Department of Chemistry, University of California, Berkeley, CA 94720*

Abstract.
A clear signature in the form of sharp features in the singly differential cross section for the (e,2e) process due to excitation of autoionizing states of helium is demonstrated in calculations in the S-wave limit. All three active electrons are treated on the same footing in calculations using the exterior complex scaling (ECS) method that applies the correct outgoing boundary conditions for all breakup channels in the problem. At the energies treated here, within a few eV of the threshold for the excitation-autoionization process, the effects of post-collision interactions are clearly demonstrated. It is suggested that such behavior should be seen generally in (e,2e) experiments on atoms that measure the single differential cross section.

Keywords: electron-impact ionization
PACS: 34.80.Dp

INTRODUCTION

When an electronic collision is capable of exciting a doubly excited autoionizing state of an atom or molecule, there are two mechanisms for electron-impact ionization at play. The first of those is direct ionization, and the second is the process of excitation-autoionization. Very early experiments saw this effect [1], and it has long been recognized that the two processes involved can interfere [2, 3]. Theoretical treatments of this interesting phenomenon have generally assumed that, for situations where one of the exiting electrons is near an autoionizing level, the overall amplitude for ionization can be written as a sum of direct and resonant terms, frequently using a plane-wave or distorted-wave representation for the direct component [4, 5] and an independent calculation of the resonant autoionizing contribution. The marked sensitivity of these approaches to the model used for the direct ionization is well known [6]. Recent progress has also been made by methods that treat only the fast electron perturbatively [7].

Our purpose here is to present a completely non-perturbative treatment of excitation-autoionization of helium in the S-wave model. Since the S-wave model simplifies the full problem by treating only states with zero orbital angular momentum, it of course cannot give a quantitatively accurate description of the full e^--He ionization problem. It does, however, represent a true four-body Coulomb problem and therefore shows much of the complexity associated with the full problem, of which it is an important part. In particular these calculations show that doubly excited states of the target leave a clear signature in the single differential cross section (SDCS) for ionization. Pairs of resonance peaks appear, symmetrically related by the energies of either the scattered or ejected electron, signaling the decay of an autoionizing state. Moreover those peaks also provide a sensitive measure of post-collision interaction effects. We expect that these aspects of the results of these calculations will carry over to the full problem. To date, the interference of the direct and autoionizing components of the process have only been observed in measurements of differential cross sections for particular angles of ejection. The possibility that such a clear signature of the physics might be seen in a more integrated quantity, namely the SDCS, suggests that observations that effectively integrate over ejected electron directions could provide a direct measure of these effects.

By using the method of exterior complex scaling (ECS) to compute the required wave functions, we are able to compute accurate ionization cross sections without having to make any *a prioi* assumptions about the form of the ionization amplitudes. Thus, these calculations provide a way of investigating the essential correlation effects involving three electrons in this process [8, 9].

THEORY

The S-wave limit for two electrons is known as the Temkin-Poet model, and for three electrons the main feature of the Hamiltonian is the same. The interaction between electrons i and j is just the spherical average of the Coulomb potential, namely $1/r_>$, where $r_>$ denotes the greater of the radial coordinates r_i and r_j. Given that simplification, we have three electrons moving in three dimensions, and any two (or three) of them can be in the continuum in the final state.

Exterior Complex Scaling. The method of exterior complex scaling uses a transformation on the underlying coordinates of the electrons to apply purely outgoing boundary conditions, regardless of the number of particles that exit in the final channel. Therefore we begin with an equation that determines the purely outgoing part of the full wave function. To that end, we partition the full wave function Ψ^+ into two parts:

$$\Psi^+ = \Phi_0 + \Psi_{SC}, \qquad (1)$$

where the unperturbed function Φ_0 specifies the initial conditions and the scattered wave Ψ_{SC} contains only outgoing waves. Substituting Eq. (1) into the time-independent Schrödinger equation gives a driven equation for the scattered wave:

$$(E-H)\Psi_{SC} = (H-E)\Phi_0. \qquad (2)$$

Eq. (2) must be solved with purely outgoing boundary conditions, and the scattered wave Ψ_{SC} carries information about all the dynamical processes of interest.

The ECS method allows one to determine the scattered wave on a finite volume without having to specify its detailed asymptotic form. The essential idea of the method is that the solution of the Schrödinger equation is an analytic function of the coordinates in the asymptotic region. Like every analytic function, it can be viewed along any ray in the complex coordinate plane. In some directions it is simpler than in others, and we can choose to apply the boundary conditions in a direction in which they are simpler, *but rigorously equivalent*, to the correct boundary conditions along the line of real-valued coordinates.

The method uses an analytic transformation where the electron coordinates are rotated into the complex plane beyond some point R_0. This is accomplished by replacing each radial electron coordinate r with a scaled coordinate $R(r)$, defined by

$$R(r) = \begin{cases} r & ,r < R_0 \\ R_0 + (r - R_0)e^{i\theta} & ,r \geq R_0. \end{cases} \tag{3}$$

Purely outgoing functions decay on the complex portion of the coordinate $R(r)$, and on that portion of the contour the simple boundary condition $\Psi_{SC} \to 0$ is exactly equivalent to the physical outgoing boundary conditions. Critically, the function at distances less than R_0 are unaffected by the scaling, and it is at those distances that we extract the scattering information from Ψ_{SC}.

We can represent Eq. (2) using the finite element method with the discrete variable representation (FEM-DVR) that has proven effective in many other appliccations [10] when applied to the radial variables. In this case it is applied in three dimensions, and converts Eq. (2) to a large set of linear equations. We can avoid solving them directly by reformulating those equations so that the scattered wave function in Eq. (2) is computed as the Fourier transform of a time-dependent wavepacket,

$$\Psi_{SC} = -i \int_0^\infty e^{iEt} \chi(t)\, dt, \tag{4}$$

with $\chi(t) = e^{-iHt}\chi(0)$, and where the initial "wavepacket" is simply given by $\chi(0) = (H - E)\Phi_0$, the right hand side of Eq. (2).

Evaluating the ionization amplitude. We now turn to the evaluation of scattering cross sections in the S-wave electron-He problem. The most practical approach to calculating both excitation and breakup cross sections is to formulate the problem in terms of integral expressions for the underlying scattering amplitudes [11]. For a two-electron problem the breakup amplitude, say for ionization in e^--H collisions, can be written formally as $f(k_1, k_2) = 2\langle \sin(k_1 r_1) \sin(k_2 r_2) | E - T | \Psi_{SC} \rangle$. This expression, while formally correct, since all the physics is in Ψ_{SC}, is not computationally useful in this context, because on a finite volume it fails to distinguish the ionization from the discrete excitation channels at the same energy — an infinite volume is necessary to produce the momentum delta functions that exclude those channels. We have shown that instead $f(k_1, k_2) = 2\langle \varphi_{k_1} \varphi_{k_2} | E - T - V_1 | \Psi_{SC} \rangle$, where V_1 is the distorted wave potential corresponding to the final state "testing functions", can be converted to a surface integral that does in fact give only the breakup amplitude. That is true provided that the functions φ_{k_i} are orthogonal to the target states (e.g. the bound states of the H atom).

The three-electron case is substantially more complicated, because at present no correspondingly simple amplitude expression exists. The analogous expression for this case, for the (e,2e) process, is

$$f(k_1, k_2) = 2\langle \varphi_n \varphi_{k_1} \varphi_{k_2} | E - T - V_1 | \Psi_{SC} \rangle, \tag{5}$$

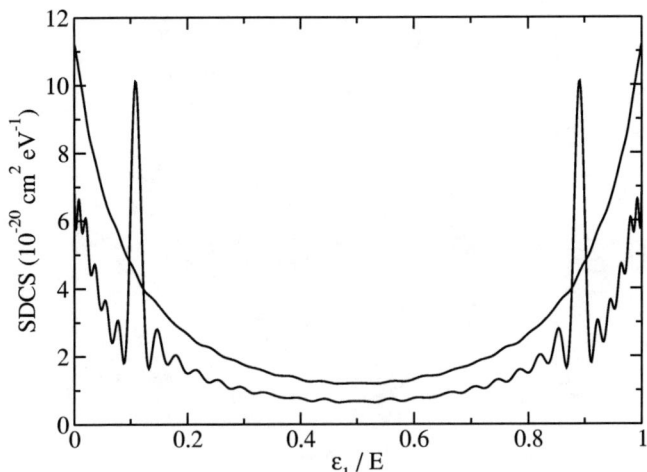

FIGURE 1. SDCS for ionization from the 2^3S state at 37.0 eV (upper curve) and 44.0 eV (lower curve) incident electron energy, showing the characteristic symmetric appearance of autoionizing features in the SDCS, in this case due to the $2s^2(^1S)$ state.

where φ_n is the state of He$^+$ left behind after ionization, and V_1 is a static-exchange potential. Using Green's theorem we can convert this expression to a surface integral

$$f(k_1,k_2) = \int_S \left[\varphi_n(r_1)\varphi_{k_1}(r_2)\varphi_{k_2}(r_3)\nabla \Psi_{SC}(r_1,r_2,r_3) \right. \\ \left. - \Psi_{SC}(r_1,r_2,r_3)\nabla \varphi_n(r_1)\varphi_{k_1}(r_2)\varphi_{k_2}(r_3) \right] \cdot \hat{n}\, dS, \quad (6)$$

that encloses a finite volume, bounded by R_0, the exterior scaling radius. But now the simple excitation channels, leaving behind excited states of the He atom must be subtracted from Ψ_{SC} in order to make this expression numerically accurate [10, 8]. This is not a perfect solution, but it is numerically practical in this case. The general problem of extracting the ionization amplitudes from an exact three-electron wave function known on an arbitrarily large, but finite, volume still has no simple solution.

THE SDCS IN THE CASE OF EXCITATION-AUTOIONIZATION

The description of the helium atom in the S-wave limit is well known, and displays a number of autoionizing states, a few of which we give in Table 1. Figure 1 shows the SDCS for scattering from the 2^3S state at incident electron energies of 37.0 and 44.0 eV. The lower energy is $\sim$2.5 eV below the energy required to excite the $2s2s(^1S)$ autoionizing state (39.49 eV), and we find the usual smile-shaped SDCS, but above that threshold, at 44.0 eV, the SDCS shows two sharp peaks ("teeth in the smile") that are symmetrically positioned with respect to $E/2$ as they must be, at energies for either scattered or ejected electron corresponding to the decay of the $2s2s(^1S)$ autoionizing state. Because of post-collision interactions, there is a shift of $\sim$0.25 eV between the

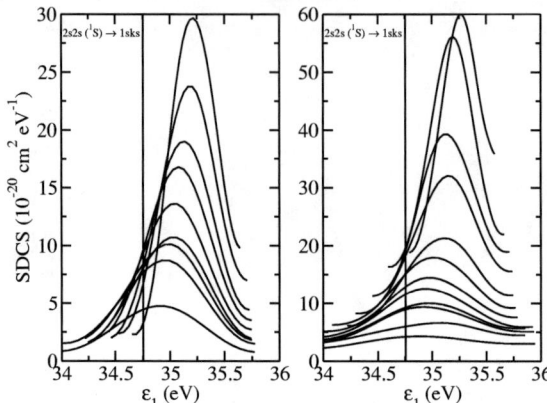

FIGURE 2. SDCS for ionization from the $2\,^3S$ (left) and $2\,^1S$ (right) states, for various incident electron energies, at ejected electron energies near the decay of the $2s^2(^1S)$ resonance state. For the $2\,^3S$ case, incident energies are (top to bottom) from 41.0 to 45.0 eV in increments of 0.5 eV. For the $2\,^1S$ case, the energies are from 39.5 to 45.0 eV in increments of 0.5 eV.

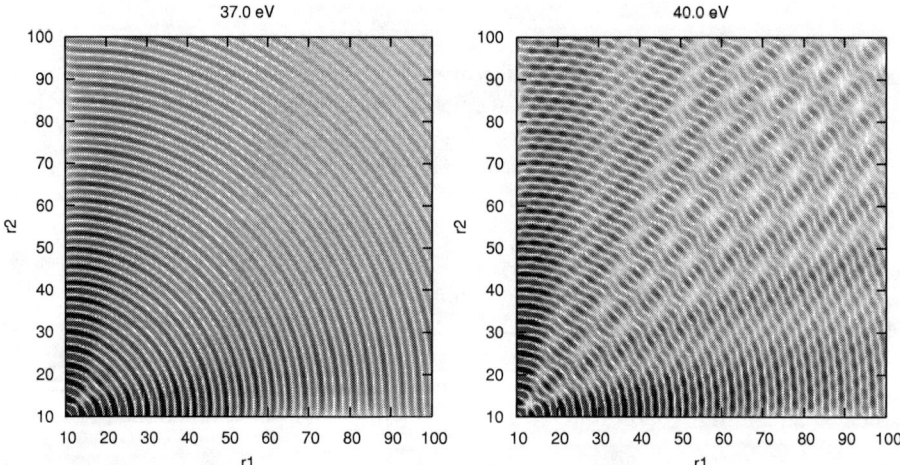

FIGURE 3. Ψ_{SC} (with $r_3 = 1.0a_0$) just below (left) and above (right) the autoionization resonance.

unperturbed energy of the autoionizing state and the energy at which the peak appears in the SDCS. The post-collision effect is further illustrated in Fig. 2 where we show the SDCS, from both the $2\,^3S$ and $2\,^1S$ initial states, for ejected electron energies near those corresponding to the decay of the $2s2s(^1S)$ autoionizing state, as a function of incident electron energy. We see clearly in that figure that as the incident energy increases, the magnitude of the peaks decrease as they shift closer to the unperturbed energy of the doubly excited target state. The effects of post-collision interaction are expected to decrease with increasing incident energy above the excitation energy for the autoionizing

TABLE 1. Energy levels for S-wave helium from two-electron ECS calculations.

State	Energy(a.u.) = $E_r - i\Gamma/2$	
$ksk's$	0	
$2sks$	-0.5	
$2s3s(^1S)$	-0.571923	$-0.28473(-3)i$
$2s3s(^3S)$	-0.584855	$-0.90332(-6)i$
$2s2s(^1S)$	-0.722837	$-0.11992(-2)i$
$1sks$	-2	
$1s3s(^1S)$	-2.06079	
$1s3s(^3S)$	-2.06849	
$1s2s(^1S)$	-2.14420	
$1s2s(^3S)$	-2.17426	
$1s1s(^1S)$	-2.87903	

state, and Fig. 2 shows exactly that effect.

It is interesting to see the wave functions at energies above and below the threshold for excitation autoionization. $\Psi_{SC}(r_1, r_2, r_3)$ can be visualized by fixing one of the coordinates ($r_3 = 1.0$ a$_0$) and plotting the wave function as a function of the remaining coordinates. Suppressed in this discussion, but of course correctly present in the calculation, are the spin couplings. In Fig.3 we show the wave function for scattering from the $2s^2(^1S)$ state just below and above the resonance energy. Below the resonance we see the outgoing spherical waves characteristic of the direct (e,2e) process, and just above the resonance we see a pattern that is the signature of interference between the autoionizing contribution and direction ionization.

ACKNOWLEDGMENTS

This work was performed at the University of California Lawrence Berkeley National Laboratory under the auspices of the U. S. DOE under Contract DE-AC02-05CH11231and was supported by the Office of Basic Energy Sciences, Division of Chemical Sciences.

REFERENCES

1. J. A. Simpson, G. E. Chamberlain, and S. R. Mielczarek, *Phys. Rev.*, **139**, A1039 (1965).
2. J. P. van den Brink, J. van Eck, and H. G. M. Heideman, *Phys. Rev. Lett.*, **61**, 2106 (1988).
3. D. G. McDonald, and A. Crowe, *J. Phys. B*, **25**, 4313 (1992).
4. V. V. Balashov, S. S. Lipovetsky, and V. S. Senashenko, *Zh. Eksp. Theo. Fiz.*, **63**, 1622 (1972), [Sov. Phys. JETP **36**, 858 (1973)].
5. A. Pochat, R. J. Tweed, M. Doritch, and J. Peresse, *J. Phys. B*, **15**, 2269 (1982).
6. I. E. McCarthy, and B. Shang, *Phys. Rev. A*, **47**, 4807 (1993).
7. Y. Fang, and K. Bartschat, *Phys. Rev. A*, **64**, 020701 (2001).
8. D. A. Horner, C. W. McCurdy, and T. N. Rescigno, *Phys. Rev. A*, **71**, 012701 (2005).
9. D. A. Horner, C. W. McCurdy, and T. N. Rescigno, *Phys. Rev. A*, **71**, 010701 (2005).
10. C. W. McCurdy, M. Baertschy, and T. N. Rescigno, *J. Phys. B*, **37**, R137 (2004).
11. C. W. McCurdy, and T. N. Rescigno, *Phys. Rev. A*, **62**, 032712 (2000).

Interference Of Exchange And Processes In The Case Of Double Ionisation Of Argon

F. Catoire[*], C. Dal Cappello[+], A. Lahmam-Bennani[*] and A. Duguet[*]

Laboratoire des Collisions Atomiques et Moléculaires, LCAM, (UMR 8625) and Fédération Lumière et Matière (FR 2764), Bât. 351, Université Paris XI, 91405 Orsay Cedex, France
+ Institut de Physique, LPMC, 1 Bd Arago, 57078 Metz Cedex 3, France

Abstract. The (e,3e) angular distributions for the double ionisation of Ar where at least two of the three outgoing electrons have the same kinetic energy (205 eV) as the Auger electron issued from a 2p-ionisation and the subsequent electronic relaxation have been theoretically investigated at three different impact energies within a first Born model and including electron exchange and interference of processes. The results are compared with recent (e,3-1e) measurements under the same kinematics. Possible dominant ionisation mechanisms are tentatively identified in each case.

Keywords: Coincident angular distributions, Auger transition, Interference, Four-body problem.
PACS: 32.80.Dp, 34.80.Pa

INTRODUCTION

We investigate in this work the double ionisation of Ar from the 3p shell ($3p^{-2}$) that occurs *via* two main processes:
- the direct double ionisation (DDI) written as

$$e_0 + Ar \rightarrow e_a + e_b + e_c + Ar^{++}(3p^{-2}).$$

Theoretically, this DDI will be assumed to be the result of a shake off (SO) process [1].
- the Auger process (AP) which will be considered as a two step process [2]. The reaction can be written as:

$$e_0 + Ar \rightarrow e_a + e_b + Ar^{+*}$$
$$Ar^{+*} \rightarrow Ar^{++}(3p^{-2}) + e_c (205 \text{ eV}),$$

where e_0 is the incident electron with the energy E_0, while e_a, e_b, e_c are the three outgoing electrons with respective energies E_a, E_b and E_c.

The experimental data were obtained in a series of (e,3-1e) experiments [3] where only two of the three outgoing electrons are observed. The toroidal energy analysers were both tuned to detect electrons of the same energy, 205 eV, that is corresponding to the energy of the Auger electron. In this situation, exchange has to be taken into account in the theoretical approach in both DDI and AP cases. The experimental procedure is described in more details in [3]. After a brief review of our experimental results we will describe the model used for comparison with the data.

EXPERIMENTAL APPROACH AND RESULTS

Double ionisation of the 3p shell of Argon is investigated at three different incident energies E_0 = 956 eV, 658 eV and 466 eV. Due to energy conservation and to the fact that we detect two electrons of 205 eV, these incident energies correspond to an unobserved electron energy of 503 eV, 205 eV and 13 eV, respectively.

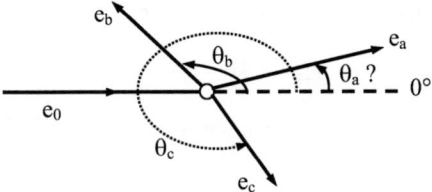

FIGURE 1. Diagram of the accessible angles for each detected electron.

The two observed electrons are detected in a coplanar geometry, in an angular range covering one half plane, as shown in figure 1, that is $0° \leq \theta_b \leq 180°$ and $180° \leq \theta_c \leq 360°$ while the θ_a-angle is unknown since electron-a is undetected. Figure 2 presents the experimental results for the different cases.

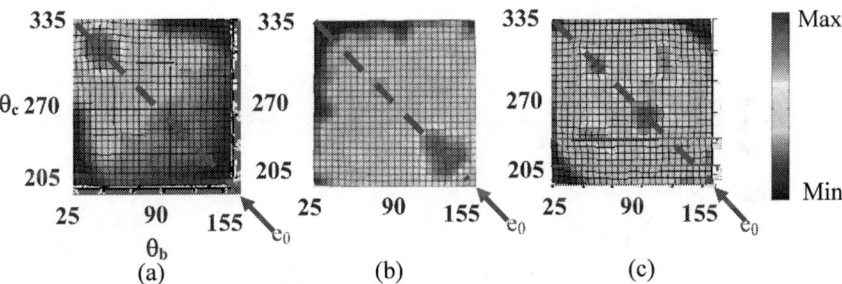

FIGURE 2. Experimental angular distributions for the coincidence detection of two 205 eV electrons where the incident energy is : (a) 466 eV, (b) 658 eV and (c) 956 eV. The energy of the unobserved electron is 13 eV, 205 eV and 503 eV, respectively. The intensity scale is given on the right. All plots are relative measurements.

The description of these experimental results is detailed in [3], hence we only remind here the main point, i.e. the observed strong dependence of the angular distributions as a function of the incident energy. E.g., the case (b) exhibits a peak at the position (120°,240°) and the case (a) at the position (50°,310°), while the case (c) has a much more complicated structure. This dependence will tentatively be explained with a theoretical investigation.

THEORETICAL MODEL

In this section we present the theoretical model used for the description of the DDI and the AP. The DDI is described within the first Born approximation (FBA). We use the frozen-core approximation to reduce the matrix element to a two active electrons transition. In the model used in [1], each outgoing electron was described by a coulomb wave (3CW model). Here, due to the fact that the Auger electron is described by a plane wave and due to the use of exchange, all electrons have to be described identically (3PW model). We thus neglect the interaction of all electrons with the residual ion. In comparison with the model of [3], we have taken Z→0 in each coulomb wave of the SO process. In fact, the calculated angular distributions using the 3CW or the 3PW models are not significantly different, except for the magnitudes. This justifies the use of this approximation. However, we include the Gamow factor to take into account the pair-wise electron interaction.

The Direct Double Ionisation

The DDI is assumed to result from a SO process (figure 3), whose amplitude is given by (see details in [3]):

$$t^{SO} = \langle \Psi_f | \frac{-2}{r_a} + \frac{1}{r_{ab}} + \frac{1}{r_{ac}} | \Psi_i \rangle$$

while the cross section is : $\sigma^{SO} = (2\pi)^4 \frac{k_s k_b k_c}{k_a} |t^{SO}|^2$

A Clementi [4] wave function is used for the initial state of the two electrons in Argon. The final state wave function is expressed as:

$$\langle \vec{r}_c, \vec{r}_b, \vec{r}_a | \Psi_f \rangle = g(\eta) \prod_{j=a}^{c} \frac{1}{(2\pi)^{3/2}} e^{\frac{\pi}{2}\alpha_j} \Gamma(1+i\alpha_j) e^{i\vec{k}_j \cdot \vec{r}_j} {}_1F_1(-i\alpha_j, 1, -i(k_j r_j + \vec{k}_j \cdot \vec{r}_j))$$

where $\alpha_j = \frac{2}{k_j}$, $\eta = \frac{1}{k_{ab}} = \frac{1}{\|\vec{k}_b - \vec{k}_a\|}$ and $g(\eta) = e^{\frac{\pi}{2}\eta} \Gamma(1-i\eta)$ is the Gawow factor.

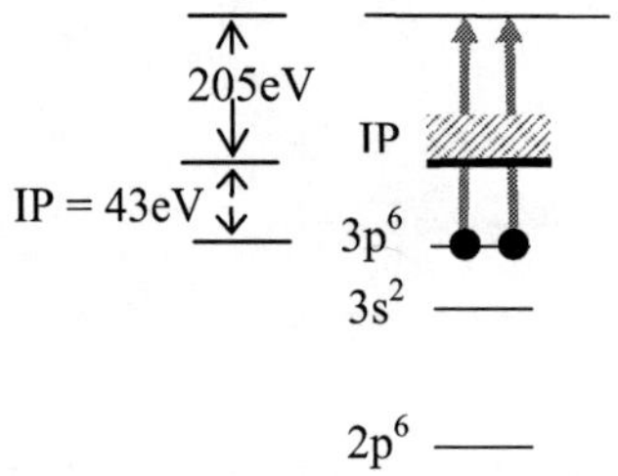

FIGURE 3. Schematic diagram of the direct valence double ionisation via a SO process.

The Auger Process

The Auger process is assumed to be a two step process. First, a simple (e,2e) ionisation of the 2p shell which leaves the residual ion in an excited state (one of the three 2p^{-1} states). The hole created in the 2p shell is filled by recombination of a 3p-electron followed by the ejection of an electron from the same shell. The resulting transition is described by an amplitude corresponding to the product of the two processes, i.e. the Auger transition is written as:

$$t^{Au} = t^A t^{e2e} = \langle \Psi_f | \frac{1}{r_{cd}} | \Psi^+ \rangle \langle \Psi^+ | \frac{-1}{r_a} + \frac{1}{r_{ab}} | \Psi_i \rangle$$

where a and b are the electrons of the (e,2e) reaction and c and d are those of the Auger process. We are dealing here with a problem where four particles are involved. We use Clementi wavefunctions for the description of the electrons in the 2p and 3p shells. The ejected and Auger electrons are described by plane waves. In this specific reaction, $|\Psi^+\rangle$ describes the two electrons involved in the Auger process.

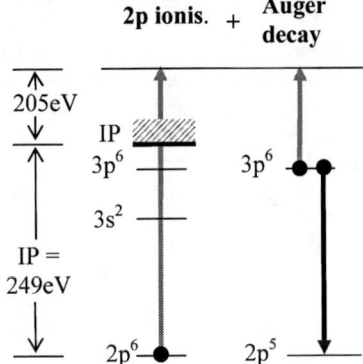

FIGURE 4. Schematic diagram of the Auger transition described as a two step process.

In order to calculate the Auger amplitude, we expand the term $e^{i\vec{k}\cdot\vec{r}}$ appearing in $|\Psi_f\rangle$ on the basis of spherical harmonics, yielding

$$t^A = 4\pi\sqrt{27} \sum_{l,l'} (-1)^{l'-x_f+y} (i)^{l'-x_f+x_i+y} \sqrt{2l'+1} \begin{pmatrix} 1 & 1 & l' \\ 0 & 0 & 0 \end{pmatrix} \begin{pmatrix} l' & l & 1 \\ 0 & 0 & 0 \end{pmatrix}$$

$$\times \begin{pmatrix} 1 & 1 & l \\ x_i & -x_f & x_f-x_i \end{pmatrix} \begin{pmatrix} l' \\ -x_i+x_f-y \end{pmatrix} \begin{pmatrix} l & l \\ x_i-x_f & y \end{pmatrix} Y_{l'}^{-x_f+x_i+y}(\hat{k}_{AU}) I(l,l',k_{AU})$$

where x_f, x_i and y are respectively the projection of the orbital momentum of the electron involved in the transition, the Auger electron and the hole. The term $I(l,l',k_{AU})$ only depends of l, l' and k_{AU} the modulus of $\hat{k}_{AU}$. The vector $\hat{k}_{AU}$ is oriented such as the z-direction is the direction of the incident electron. Firstly, we can observe that we verify the conservation of the LS coupling. Secondly, for specific angles, some transitions are forbidden.

For a given final state $3p^{-2}$, the global amplitude of the Auger transition is a sum over all the intermediate ion states. So, the Auger amplitude is written as:

$$t^{Au} = \sum_{m_l} \frac{\langle \Psi_f | V^{Au} | \Psi_1^+(m_l) \rangle \langle \Psi_2^+(m_l) | V^{SO} | \Psi_i \rangle}{E_0 - E_a - E_b - E_c - E^{++} + i\frac{\Gamma_{m_l}}{2}}$$

where Γ_{m_l} is the life-time of the m_l intermediate state. $\Gamma_{m_l} \to 0$ in our approximation, so that:

$$\frac{1}{E_0 - E_a - E_b - E_c - E^{++} + i\frac{\Gamma_{m_l}}{2}} \underset{\Gamma_{m_l} \to 0}{=} -i\pi\delta(E_0 - E_a - E_b - E_c - E^{++})$$

As a result, the amplitude of the Auger transition reduces to a simple sum of products of amplitudes corresponding to the (e,2e) reaction and the Auger one.

Interference Of Processes

Two of the three outgoing electrons have the same energy. In the particular case of $E_0 = 658$ eV, three outgoing electrons have the same energy, hence they are fully indistinguishible. To describe correctly this possibility of exchange of electrons, we have to construct eigenvectors of L^2 and S^2. In the case of SO process, this procedure is given in [3]. We have now included the Auger process so that the formulation of exchange in the case of SO has to be adapted to the specific case of Auger.

The first point concerns the non-symmetry with respect to exchange of ejected electrons when these electrons are issued from the same outer shell. The potential is symmetric with respect to this exchange in the case of SO, but it is not symmetric in the Auger case. So, we have to change the global expression of the cross section to adapt it to the Auger process.

The second point is related to the fact that the Auger process deals with the problem where four electrons are involved. But the expression of the cross section is obtained for the case of three active electrons. The procedure we used is the following: we have to calculate different amplitudes corresponding to specific initial and final states, but only for three active electrons. Nonetheless, the fourth active electron introduced for the Auger transition is only a spectator for what concerns the exchange in the continuum. Hence, we added coherently the Auger amplitude (i.e. summing over the intermediate states where the fourth involved electron is spectator) and the SO amplitude.

The Auger transition amplitude corresponds to a transition where the initial state is described by a product of two wave functions of the 3p shell and a final state described by a product of a plane wave and a wave function of the 2p shell. If we denote by e_1 and e_2 electrons of the 3p shell, either e_1 or e_2 could fill the hole in the 2p shell. To account for this, the spin states of these two electrons are introduced by using determinants in the initial and final state of the Auger transition. Hence, the amplitude of this transition is a sum (or a difference, depending on whether the spin configurations are up-up or up-down) of the direct amplitude and the exchange one.

RESULTS AND DISCUSSION

The Auger and SO cross sections are calculated in the three different cases. We observe some similarities between the theoretical and experimental distributions.

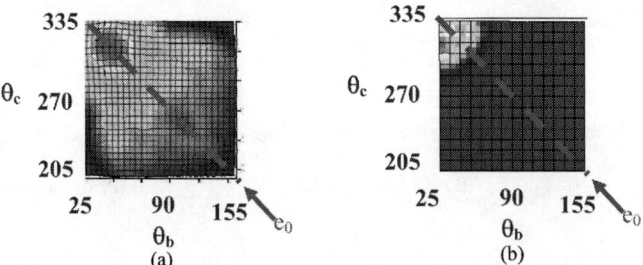

FIGURE 5. The case with $E_0 = 466$ eV. (a) Experimental results. (b) (e,3e) calculations where the scattering angle is -3°, and only the SO contribution is included.

In particular, in the case $E_0 = 466$ eV, the SO process seems to be dominant and shows a peak at the position (50°, 310°) as it is shown in figure 5.

In the case of SO process the main contribution is due to ^3P states. This could be explained by the fact that these ^3P final states represent 9 of the 15 possible final states. The contribution of all these states is an incoherent sum of each amplitude. On the contrary, the situation is much more complex in the Auger case due to the introduction of intermediate ion states and interference of exchange. As a result, the ^3P final states do not contribute as strongly as they do in the SO process.

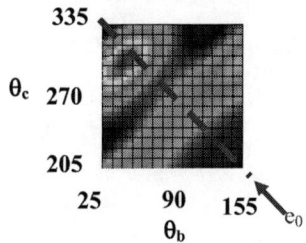

FIGURE 6. (e,3e) calculated angular distribution due to the Auger process for $E_0 = 956$ eV and a scattereing angle of -3°.

In conclusion, the SO process seems to be dominant in the case $E_0 = 466$ eV. As shown in [3], the position of the main forward peak could be explained with a simple analysis based on a billiard ball model.

The situation is more complex for $E_0 = 956$ eV. No agreement with experiments is obtained when introducing only the SO contribution. Figure 6 where only the Auger process is included exhibits peaks more or less resembling the experimental ones. Possibly, the introduction of interference of processes would bring the theoretical distribution in closer agreement with the experimental one.

For $E_0 = 658$ eV, it was shown in [3] that the main contribution is an (e,2e) process on the 2p-shell which is energetically accessible only in this case.

Finally, in this work, the analysis was done by comparing the (e,3-1e) experiments to an (e,3e) model where the outgoing electrons are described by plane waves. Hence, we still have on the one hand to introduce a coulomb wave in the Auger process description, and on the other hand to perform an (e,3e) experiment for a more detailed comparison. This work is in progress. To this purpose, we have modified our (e,3e) spectrometer to include a third toroidal analyser for multi-angle detection of all three outgoing electrons. The system is presently under final tests and will be used in the immediate future to complete this study.

REFERENCES

1. E. G. Berezhko, N. M. Kabachnik and V. V. Sizov *J. Phys B.* **11**, 1819 (1978).
2. S.A. Sheinerman and V. Schmidt *J. Phys B*, 30,1677 (1997).
3. F. Catoire, C. Dal Cappelo, A. Lahmam-Bennani and A. Duguet, in Electron and Photon Impact Ionization and related Topics 2004, edited by B. Piraux, IOP Conferences Series 183, Bristol, 2004, pp. 53-62.
4. E. Clementi and C. Roetti, *At. Data Nucl. Data Tables* **14**, 177 (1974).

(e,2e) and (e,3-1e) studies on double processes of He near the Bethe ridge

N. Watanabe[1,2,*], Y. Khajuria[1], M. Takahashi[1,2], Y. Udagawa[2],
P.S. Vinitsky[3], Yu. V. Popov[3], O. Chuluunbaatar[4] and K. A. Kouzakov[5]

1) Institute for Molecular Science, Okazaki 444-8585, Japan
*2) Institute for Multidisciplinary Research for Advanced Materials, Tohoku University.,
Sendai 980-8577, Japan*
3) Nuclear Physics Institute, Moscow State University, Moscow 119992, Russia
4) Join Institute for Nuclear Research, Dubna 141980, Moscow Region, Russia
5) Physics Department, Moscow State University, Moscow 119992, Russia

Abstract. We report (e,2e) and (e,3-1e) experiments on the double processes of He, i.e. single ionization with simultaneous excitation and double-ionization. The symmetric noncoplanar geometry combined with high incident electron energies has made it possible to study at large momentum transfers. The results are compared with plane wave impulse approximation (PWIA) calculations using He wavefunctions of various levels of sophistication. It is shown that shapes of the momentum dependent (e,2e) and (e,3-1e) cross sections are well reproduced by the PWIA calculations when highly correlated wavefunctions are employed, but noticeable discrepancies between experiment and theory remain in magnitude. The discrepancies are, however, reduced with increasing impact energies, suggesting higher excitation energies may be required to analyze these double processes in terms of the PWIA.

Keywords: (e,2e), (e,3-1e), ionization-excitation, double ionization, electron correlation
PACS: 34.80.Dp

INTRODUCTION

The electron impact single-ionization experiment under the high-energy Bethe ridge conditions is referred to as binary (e,2e) or electron momentum spectroscopy and provides direct information on the one-electron momentum density of the ionized orbital [1]. The key concept in connecting the (e,2e) cross section with the one-electron momentum density is the plane wave impulse approximation (PWIA).

Double processes of the two-electron system He, i.e. single-ionization with simultaneous excitation to n=2 orbital (hereafter called ionization-excitation) and double-ionization, are particularly attractive in view of (e,2e) spectroscopy. Because of the absence of electron correlation in the final ion state, they provide ideal opportunities to examine PWIA as well as electron correlation in the initial target state. As to the ionization-excitation processes of He, the PWIA allows us to probe the one-electron momentum density of excited orbital component involved in the target initial state wavefunction. In the case of double-ionization, the two-electron momentum density can be derived by so-called (e,3e) experiments, where all the three outgoing electrons should be detected in coincidence [2,3]. Although direct determination of the two-electron momentum density is not possible, experimentally less demanding (e,3-1e) method, proposed by Popov et al.[4], is very sensitive to electron correlation in the target initial state. In spite of the fundamental importance, however, very few studies have been made on the double processes of He under the kinematics that seek to satisfy the high-energy Bethe ridge conditions [5-10]. Extremely small cross sections of the double processes at large momentum transfer are responsible for the scarcity of such studies.

Under these circumstances, we report (e,2e) and (e,3-1e) experiments on He using a recently developed multichannel (e,2e) spectrometer [9] that features high sensitivity. An impact energy of 2080 eV was used in the symmetric noncoplanar geometry and we have thus achieved a large momentum transfer of 9 a.u., a value that has never been realized so far for study on double-ionization. Thanks to multichannel detectors, primary ionization, where the residual electron remains in the 1s orbital, can also be measured simultaneously with double processes. Comparisons with the primary ionization have made it possible to accurately determine relative intensities of the double processes. Furthermore, measurements at impact energies of 1240 and 4260 eV were also performed in order to examine the incident electron energy dependence of the momentum profiles of the ionization-excitation process. The results are compared with PWIA calculations using various wavefunctions in terms of both shape and absolute magnitude of the cross sections. The present work is the examination of the PWIA for the double processes under the kinematics that are the closest to the high-energy Bethe ridge conditions compared with those employed in the previous (e,2e) and (e,3-1e) studies [5-8].

EXPERIMENTAL METHOD

For electron-impact single-ionization and double-ionization processes of He conservation of linear momentum and energy requires:

$$p_{He+} = p_0 - p_1 - p_2, \quad (1)$$

$$E_{bind} = E_0 - E_1 - E_2 \quad (2)$$

and

$$p_{He2+} + p_3 = p_0 - p_1 - p_2, \quad (3)$$

$$E_3 = E_0 - E_1 - E_2 - IP^{2+}. \quad (4)$$

Here p_j's and E_j's (j = 0,1,2,3) are momenta and kinetic energies of the incident and outgoing electrons, respectively. p_{He+} and p_{He2+} represent the recoil momentum of the residual ion He^+ and that of He^{2+}, respectively. E_{bind} and IP^{2+} are ionization energy and the double-ionization threshold of He. Since the present experiment involves coincidence detection of two outgoing electrons, p_{He+} and E_{bind} are fully determined for the (e,2e) processes. On the other hand, for the (e,3-1e) process the obtainable quantities are $(p_{He2+} + p_3)$ and E_3. For the sake of simplicity both p_{He+} and $(p_{He2+} + p_3)$ are denoted as momentum q here, and in the same sense (e,2e) and (e,3-1e) momentum profiles refer to $|p_{He+}|$-dependent (e,2e) cross section and $|p_{He2+} + p_3|$-dependent (e,3-1e) cross section respectively.

In the symmetric noncoplanar geometry, two outgoing electrons having equal energies ($E_1 = E_2$) and polar angles ($\theta_1 = \theta_2 = 45°$) with respect to the incident electron beam axis are detected in coincidence. Then, the magnitude of the momentum q is expressed by

$$q = \sqrt{(p_0 - \sqrt{2}p_1)^2 + (\sqrt{2}p_1 \sin(\Delta\phi/2))^2}, \quad (5)$$

where $\Delta\phi$ ($= \phi_1 - \phi_2 - \pi$) is the out-of-plane azimuthal angle difference between the two outgoing electron detected. If the incident electron energy and momentum are fixed, a given ionization transition can be selected simply by the choice of detection energy ($E_1 = E_2$) and then q can be determined only by $\Delta\phi$. The same is true for (e,3-1e) experiments, if we detect two fast outgoing electrons with equal energies in the symmetric noncoplanar geometry while leaving one slow ejected electron undetected.

In this work, a recently developed multichannel (e,2e) spectrometer has been employed to carry out (e,2e) and (e,3-1e) measurements simultaneously. Details of the spectrometer have been described elsewhere [11].

THEORY

Within the PWIA, the triple differential cross section for (e,2e) process on He is described by

$$\frac{d^3\sigma}{d\Omega_1 d\Omega_2 dE_1} = (2\pi)^4 f_{ee} \frac{p_1 p_2}{p_0} G(q), \quad (6)$$

$$G(q) = \left| \left(\frac{1}{2\pi}\right)^{3/2} \sqrt{2} \int \varphi_f^*(r_1) e^{iq \cdot r_2} \Phi(r_1, r_2) dr_1 dr_2 \right|^2, \quad (7)$$

where $d\Omega_j$ denotes the element of solid angle for the jth outgoing electron and f_{ee} is the half-off-shell Mott scattering cross section [1]. The structure factor $G(q)$ is proportional to the square modulus of the Fourier transform of the overlap between the initial ground state $\Phi(r_1, r_2)$ and the final ion state $\varphi_f(r_1)$. For (e,3-1e) reaction of He, the four-fold differential cross section based on the PWIA model is given by [4]

$$\frac{d^4\sigma}{d\Omega_1 d\Omega_2 dE_1 dE_2} = (2\pi)^4 f_{ee} \frac{p_1 p_2 p_3}{p_0} F(p_3, q), \quad (8)$$

$$F(p_3, q) = \int \left| \left(\frac{1}{2\pi}\right)^{3/2} \sqrt{2} \int \varphi_c^*(p_3, r_1) e^{iq \cdot r_2} \Phi(r_1, r_2) dr_1 dr_2 \right|^2 d\Omega_3. \quad (9)$$

Here $\varphi_c(p_3, r_1)$ is a Coulomb wave with momentum p_3. Eqs.(7) and (8) tell us that within the PWIA the structure factors involved in the cross sections are independent of E_0. Underlying assumptions leading to these equations are that electron energies have to be high enough to describe the incoming and fast two outgoing electrons by plane waves. To make extensive comparisons with experiment, five kinds of models for the He ground state wavefunction have been employed in the present study. The simplest model is the Hartree-Fock (HF) wavefunction [12], in which electron correlation is neglected. Radial correlation is taken into account by the Hylleraas-Eckart-Chandrasehkar (HEC) wavefunction [13]. To examine both radial and angular correlations on momentum profiles we have used the following three wavefunctions; a 12-component variation of the Chuluunbaatar, Puzynin and Vinitsky (CPV) wavefunction [14], a configuration interaction (CI) wavefunction of Mitroy et al. [15], and the Bonham and Kohl (BK) wavefunction [16].

RESULTS AND DISCUSSION

Figures 1(a) and (b) show experimental momentum profiles for the primary ionization and ionization-excitation processes. Likewise, the experimental (e,3-1e) momentum profiles at E_3=10 and 20eV are plotted in Figs. 1(c) and (d). Also included in Fig. 1 are the associated PWIA calculations using the HF, HEC, CPV, CI, and BK wavefunctions. Although absolute cross sections can not be determined with the present experiment, relative magnitudes of the individual transitions are maintained. Thus, by normalizing a certain experimental momentum profile to an associated theoretical one, one can place all other experiments on an absolute scale. Here we have chosen the PWIA cross section using the CI wavefunction (PWIA/CI) for the primary ionization as a standard for the normalization procedure. Furthermore, in order to compare the theoretical momentum profiles using different wavefunctions in a common scale, all the calculations are multiplied by scaling factors, which have been obtained so that the area of the theoretical momentum profile for the primary transition is the same as that of the PWIA/CI n=1 momentum profile.

It is immediately clear from Fig. 1(a) that all the PWIA calculations, including HF, are in general agreement with each other and with the experiment. Clearly, there are no noticeable effects of electron correlation in the primary ionization. In contrast, it is evident from Fig. 1(b) that the PWIA calculation using the uncorrelated HF wavefunction shows marked deviation from the experimental momentum profiles. The difference in shape is greatly reduced by taking radial correlation (HEC) or radial and angular correlation (CPV, CI, BK) into account. Although the shape of the experimental momentum profile is reproduced, discrepancies remain in magnitude; calculations always show about 30 % smaller cross sections.

As to the (e,3-1e) momentum profile, HF completely fails to reproduce the observed one, and HEC predicts an order of magnitude smaller cross section. The other three highly correlated wavefunctions result in similar momentum profiles with each other, which are in reasonably good agreement with the experimental ones in terms of shape. However, discrepancies again remain in magnitude; calculated ones underestimate the experimental cross sections by a factor of about 2.4 and 2.9 at E_3=10 and 20eV, respectively.

To see effects of distortion of the electron waves on the observed intensity differences, we have calculated (e,2e) momentum profiles within the distorted-wave Born approximation (DWBA) using the CI wavefunction. It is clear from Fig. 1(b) that the DWBA/CI momentum profile is very similar to the PWIA calculations except for at large momentum and can not explain the observed disagreements between experiment and theory. Thus, the intensity difference can not be attributed to distorted-wave effects.

Two possible sources of the difference in magnitude can be conceived; one is an incomplete description of the target ground-state wavefunction, which has been

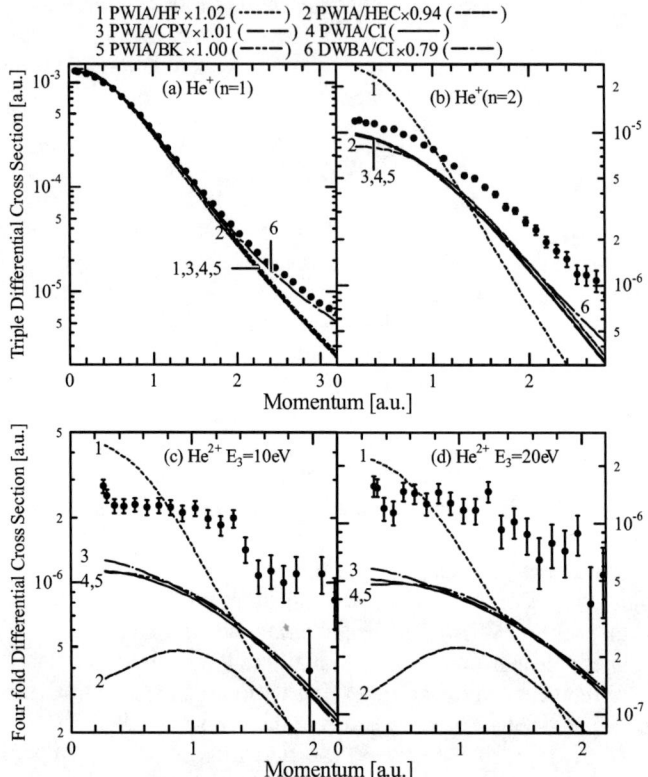

FIGURE 1. Upper panel: comparison of experimental (e,2e) momentum profiles of He for (a) the primary ionization and (b) the ionization-excitation processes and associated theoretical calculations. Lower panel: comparison of experimental (e,3-1e) momentum profiles of He for the doubly ionized He^{2+} with E_3= 10 (c) and 20 eV (d). PWIA calculations are also shown. All the experimental and theoretical momentum profiles are provided as normalized intensities to the PWIA/CI cross sections for the primary transition. For details, see ref. [17].

extensively examined in a separate paper [17]. Another is a failure of the PWIA description of the double processes due to contributions of higher order Born terms, which are not involved in the PWIA model. In what follows we focus our attention on contributions of higher order Born terms for the ionization-excitation process [18].

As mentioned earlier, the structure factor involved in the PWIA cross section should be independent of impact energy (cf. Eqs. (7) and (9)). Hence the validity of the PWIA can be experimentally assessed by a comparison of momentum profiles measured at different impact energies. Figure 2 shows experimental momentum profiles of the ionization-excitation transition at impact energies of 1240, 2080, and 4260eV. In order to compare the momentum profiles at different impact energies, the cross sections have been divided by the kinematical factors. Also illustrated in the figure are the PWIA/CI calculations folded with momentum resolutions of the spectrometer at the energy employed.

The experimental momentum profiles vary considerably with incident electron energy, decrease monotonically with an increase in E_0, and are approaching the PWIA

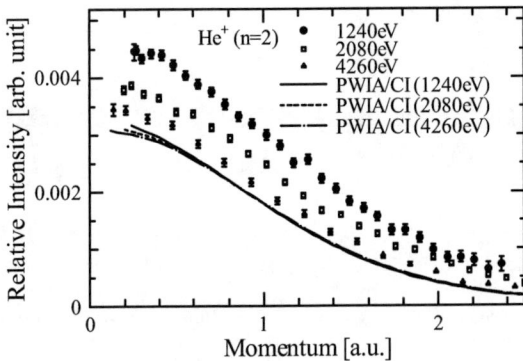

FIGURE 2. Experimental and PWIA/CI momentum profiles of He for the ionization-excitation transition at impact energies of 1240, 2080, and 4260eV. All the momentum profiles are divided by the kinematical factors.

prediction. This suggests that the ionization-excitation experiment should be performed at impact energy of about 5 keV or higher in order to make correct interpretation in terms of PWIA.

In summary, the following conclusions can be drawn from the present findings. First, the primary (e,2e) process of He is very accurately described by PWIA/HF calculation. Second, electron correlation plays a crucial role in ionization-excitation and double ionization processes. Finally, it is suggested that a higher electron impact energy is required than those employed in the present study in order to come to a correct interpretation of the double processes in terms of PWIA.

REFERENCES

* corresponding author. e-mail address noboru@tagen.tohoku.ac.jp
1. I. E. McCarthy and E. Weigold, *Rep. Prog. Phys.* **54**, 789-879 (1991).
2. V. G. Neudatchin, Y. F. Smirnov, A. V. Pavlitchenkov, and V. G. Levin, *Phys. Lett.* A **64**, 31 (1977).
3. Y. F. Smirnov, A. V. Pavlitchenkov, and V. G. Levin, and V. G. Neudatchin, *J. Phys.* B **11**, 3587 (1978).
4. Yu. V. Popov, C. Dal Cappello, and K. Kouzakov, *J. Phys.* B **29**, 5901 (1996).
5. J. P. D. Cook, I. E. McCarthy, A. T. Stelbovics, and E. Weigold, *J. Phys.* B **17**, 2339 (1984).
6. A. D. Smith, M. A. Coplan, D. J. Chornay, J. H. Moore, J. A. Tossell, J. Mrozek, V. H. Smith Jr., and N. S. Chant, *J. Phys.* B **19**, 969 (1986).
7. N. Lermer, B. R. Todd, N. M. Cann, C. E. Brion, Y. Zheng, S. Chakravorty, and E. R. Davidson, *Can. J. Phys.* **74**, 748 (1996).
8. P. Bolognesi, C. C. Jia, L. Avaldi, A. Lahmam-Bennani, K. A. Kouzakov, and Yu. V. Popov, *Phys. Rev.* A **67**, 034701 (2003).
9. A. Lahmam-Bennani, C. C. Jia, A. Duguet, and L. Avaldi, *J. Phys.* B **35**, L215 (2002).
10. A. Dorn, A. Kheifets, C. D. Schröter, B. Najjari, C. Höhr, R. Moshammer, and J. Ullrich, *Phys. Rev.* A **65**, 032709 (2002).
11. M. Takahashi, N. Watanabe, Y. Khajuria, K. Nakayama, Y. Udagawa, and J. H. D. Eland, *J. Electron Spectrosc.* **141**, 83 (2004).
12. E. Clementi and C. Roetti, *At. Data Nucl. Data Tables* **14**, 177 (1974).
13. E. A. Hylleraas, *Z. Phys.* **54**, 347 (1929).
14. O. Chuluunbaatar, I. V. Puzynin and S. I. Vinitsky, *J. Phys.* B **34**, L425 (2001).
15. J. Mitroy, I. E McCarthy and E. Weigold, *J. Phys.* B **18**, 4149 (1985).
16. R. A. Bonham and D. A. Kohl, *J. Chem. Phys.* **45**, 2471 (1966).
17. N. Watanabe, Y. Khajuria, M. Takahashi, Y. Udagawa, P. S. Vinitsky, Yu. V. Popov, O. Chuluunbaatar, and K. A. Kouzakov, *Phys. Rev.* A, in press.
18. N. Watanabe et al. (to be published).

(e,3-1e) Reactions at Large Momentum Transfer: The Plane-Wave Second Born Approximation

Pavel S. Vinitsky*, Yuri V. Popov*, Konstantin A. Kouzakov[†], Noboru Watanabe** and Masahiko Takahashi**

Institute of Nuclear Physics, Moscow State University, Moscow 119992, Russia
[†]*Department of Nuclear Physics and Quantum Theory of Collisions, Moscow State University, Moscow 119992, Russia*
**Institute of Multidisciplinary Research for Advanced Materials, Tohoku University, Sendai 980-8577, Japan*

Abstract. We consider theoretically symmetric (e,3-1e) reactions in He atom at large momentum transfer. For evaluating the corresponding four-fold differential cross sections, a theory based on the renormalized plane-wave second Born approximation (SBA) is developed. The numerical SBA calculations for both coplanar and noncoplanar symmetric geometries are performed and compared to the experimental values.

INTRODUCTION

In recent years the first symmetric (e,3-1e) measurements on He atom at large momentum transfer have been realized using both coplanar [1] and noncoplanar [2] geometries. These studies have been motivated by the idea of obtaining new information on a helium wave function, since, as was shown in [3] using the plane-wave impulse approximation (PWIA) theory, the four-fold differential cross sections for symmetric (e,3-1e) reactions at large momentum transfer strongly depend on the spatial correlations of electrons in a target. However, marked discrepancies between the experimental values and the PWIA calculations with various models of helium wave function have been determined. Moreover, the calculations in the distorted-wave Born or impulse approximation (DWBA or DWIA, respectively) employing different models for distorting potentials [4, 5] only slightly diminish the observed discrepancies. Thus, it is on agenda to examine the role of dynamical mechanisms which are ignored by the PWIA and/or DWIA theories. These (typical) mechanisms are inelastic collisions of the fast incoming/outgoing electrons, which participate in a binary electron-electron collision, with the rest of the system. A natural starting point for theoretical analysis of such processes is the SBA.

This work aims at developing the plane-wave SBA theory for the electron-impact double ionization of helium in symmetric kinematics involving large momentum transfer. To our knowledge, no such theory has been published so far. In this respect it should be noted that specific feature of the plane-wave Born series for the scattering of charged particles consists in divergencies of higher-order terms if the particles' energies and momenta obey the dispersion relations (see, for instance, [6]). Therefore, when developing the SBA theory, one must treat the divergencies due to the second-order Born terms by employing a proper renormalization procedure. Below we present the plane-wave SBA formalism and discuss in detail the issue of renormalization. Then, the pilot numerical results are presented and discussed. The atomic units (a.u.) $e = \hbar = m_e$ are used throughout.

THEORY

General formulation

We specify the energies and momenta of the fast incoming and outgoing electrons with $(E_0, \bm{p}_0)$, $(E_1, \bm{p}_1)$ and $(E_2, \bm{p}_2)$, respectively. It is assumed that the two fast outgoing electrons form a symmetric pair, i.e. $E_1 = E_2$ and $\theta_1 = \theta_2$ (the polar angles are measured relative to the direction of the incident electron momentum $\bm{p}_0$). The energy and momentum of a slow ejected electron, which is not detected in the (e,3-1e) experiment, are $(E_3, \bm{p}_3)$. The four-fold differential cross section (4DCS) is thus given by

$$\frac{d^4\sigma}{dE_1 dE_2 d\Omega_1 d\Omega_2} = \frac{p_1 p_2 p_3}{(2\pi)^8 p_0} \int d\Omega_3 \left(\frac{1}{4}|2T_1 - T_2 - T_3|^2 + \frac{3}{4}|T_2 - T_3|^2 \right). \quad (1)$$

Here T_i ($i = 1, 2, 3$) is the amplitude of the process where the scattered electron has the momentum $\bm{p}_i$. In what follows we focus on the amplitude T_1 noticing that T_2 derives from T_1 upon exchanging the momenta $\bm{p}_1 \leftrightarrow \bm{p}_2$ and the amplitude T_3 can be neglected in the kinematical regimes under consideration.

The amplitude T_1 is evaluated as follows:

$$T_1 = \langle \bm{p}_0 \Phi_0 | V_i [1 + G(E) V_i] | \bm{p}_1 \Phi_f \rangle, \quad (2)$$

where Φ_0 is the helium wave function, Φ_f is the doubly ionized helium state with asymptotic electron momenta $\bm{p}_2$ and $\bm{p}_3$, V_i is the interaction potential between the incident electron and He atom, and

$$G(E) = (E - H + i0)^{-1} \quad (E = E_1 + E_2 + E_3 = E_0 + \mathscr{E}_0 \text{ on the energy shell})$$

is the Green's function with H being a Hamiltonian of the four-body system. Using spectral expansion of the Green's function we arrive at

$$T_1 = \langle \bm{p}_0 \Phi_0 | V_i | \bm{p}_1 \Phi_f \rangle + \sum_\alpha \frac{\langle \bm{p}_0 \Phi_0 | V_i | \Psi_\alpha \rangle \langle \Psi_\alpha | V_i | \bm{p}_1 \Phi_f \rangle}{E - E_\alpha + i0}, \quad (3)$$

where Ψ_α are the eigenstates of the four-body Hamiltonan, $H\Psi_\alpha = E_\alpha \Psi_\alpha$.

The plane-wave SBA amplitude

Consider the first term in (3) with a final-state wave function Φ_f in the form of a symmetrized product

$$\Phi_f^{(0)}(\bm{r}_1, \bm{r}_2) = \frac{1}{\sqrt{2}} [e^{i\bm{p}_2 \bm{r}_1} \varphi_{p_3}(\bm{r}_2) + e^{i\bm{p}_2 \bm{r}_2} \varphi_{p_3}(\bm{r}_1)], \quad (4)$$

where φ_{p_3} is the continuum one-electron state with asymptotic momentum p_3 in a Coulomb field of He^{2+} ion. This yields the amplitude in the first Born approximation (FBA)

$$T_1^{FBA} = \frac{4\sqrt{2}\pi}{(p_0 - p_1)^2} \langle \Phi_0 | q \varphi_{p_3} \rangle, \qquad q = p_1 + p_2 - p_0. \tag{5}$$

The correction to (4) due to the interaction of the fast electron with the slow electron and He^{2+} ion amounts to the two-step 1 (TS1) mechanism. The corresponding second Born term is given by

$$TS1 = \frac{4\sqrt{2}\pi}{(p_0 - p_1)^2} \sum_\lambda \int \frac{dx}{(2\pi)^3} \frac{\langle \Phi_0 | q - x, \varphi_\lambda \rangle \langle \varphi_\lambda | e^{ixr} - 2 | \varphi_{p_3} \rangle}{E - E_1 - \varepsilon_\lambda - \frac{(p_2 - x)^2}{2} + i0} \frac{4\pi}{x^2}, \tag{6}$$

where φ_λ and ε_λ are the one-electron eigenstates and eigenenergies in a Coulomb field of He^{2+} ion.

Let us turn to the second term in (3) that is usually referred to as the two-step 2 (TS2) mechanism. Using (4) and leaving only the lowest-order terms in V_i we obtain

$$TS2 = \sum_n \int \frac{dk}{(2\pi)^3} \frac{\langle p_0 \Phi_0 | V_i | k \Phi_n \rangle \langle k \Phi_n | V_i | p_1 \Phi_f^{(0)} \rangle}{E - \mathscr{E}_n - \frac{k^2}{2} + i0}, \tag{7}$$

where Φ_n and $\mathscr{E}_n$ are the eigenstates and eigenenergies of the helium Hamiltonian. The TS2 mechanism (7) can be divided into the TS2$_1$ and TS2$_2$ contributions which correspond to the intermediate (virtual) transitions to the bound and continuum (with one or two unbound electrons) helium states, respectively. In the present study, when treating the TS2$_1$ contribution, we account only for the elastic intermediate transition ($n = 0$)

$$TS2_1 = \int \frac{dx}{(2\pi)^3} \frac{4\sqrt{2}\pi}{(p_0 - p_1 + x)^2} \frac{\langle \Phi_0 | e^{ixr_1} + e^{ixr_2} - 2 | \Phi_0 \rangle \langle \Phi_0 | q - x, \varphi_{p_3} \rangle}{E - \mathscr{E}_0 - \frac{(p_0 + x)^2}{2} + i0} \frac{4\pi}{x^2}. \tag{8}$$

The TS2$_2$ contribution is evaluated as follows:

$$TS2_2 = TS2_{21} + TS2_{22}, \tag{9}$$

where

$$TS2_{21} = \sum_\lambda \int \frac{dx}{(2\pi)^3} \frac{4\sqrt{2}\pi}{(p_0 - p_1 + x)^2} \frac{\langle \Phi_0 | q - x, \varphi_\lambda \rangle \langle \varphi_\lambda | e^{ixr} - 2 | \varphi_{p_3} \rangle}{E - E_2 - \varepsilon_\lambda - \frac{(p_1 - x)^2}{2} + i0} \frac{4\pi}{x^2} \tag{10}$$

and

$$TS2_{22} = \langle \Phi_0 | q \varphi_{p_3} \rangle \int \frac{dx}{(2\pi)^3} \frac{4\sqrt{2}\pi}{(p_0 - p_1 + x)^2} \frac{1}{E - E_3 - \frac{(p_1 - x)^2}{2} - \frac{(p_2 + x)^2}{2} + i0} \frac{4\pi}{x^2}. \tag{11}$$

Thus, the SBA amplitude is given by

$$T_1^{SBA} = T_1^{FBA} + \delta T_1^{SBA}, \qquad \delta T_1^{SBA} = TS1 + TS2_1 + TS2_{21} + TS2_{22}. \tag{12}$$

The renormalized SBA amplitude

A generalization of the plane-wave Lippmann-Schwinger integral equations to the scattering of a few charged particles has been considered in [7]. Below we outline the main result which is exploited in practical calculations. Examining the structure of integrals in (6), (10) and (11), we deduce that the $TS2_{22}$ contribution is divergent and the TS1 and $TS2_{21}$ contributions diverge if $\varphi_\lambda = \varphi_{p_3}$. To "circumvent" this situation and obtain a convergent (physical) SBA amplitude, let us consider the plane-wave Born series for the amplitude T_1 on the out-of-energy shell assuming that $E - E_1 - E_2 - E_3 = \Delta > 0$:

$$T_1(\Delta) = T_1^{FBA} + \delta T_1^{SBA}(\Delta) + \ldots \qquad (13)$$

Here the SBA and all other higher-order Born terms diverge if $\Delta \to +0$. It is possible to single out the divergencies and collect them in a phase (exponential) factor (see [6] for details), so that

$$T_1(\Delta) = e^{i\eta \ln \Delta} N[T_1^{FBA} + \delta \tilde{T}_1^{SBA} + \ldots], \qquad (14)$$

where all the values, excepting the phase factor, are convergent if $\Delta \to +0$. The renormalized (physical) amplitude is given by [8]

$$\tilde{T}_1 = \frac{e^{-\pi\eta/2 + iA}}{\Gamma(1 - i\eta)} \lim_{\Delta \to +0} e^{-i\eta \ln \Delta} T_1(\Delta), \qquad (15)$$

where

$$\eta = -\frac{1}{p_1} - \frac{1}{p_2} + \frac{1}{|\mathbf{p}_1 - \mathbf{p}_2|}$$

is a total Coulomb number and

$$A = -\frac{1}{p_1} \ln(4p_1^2) - \frac{1}{p_2} \ln(4p_2^2) + \frac{1}{|\mathbf{p}_1 - \mathbf{p}_2|} \ln |\mathbf{p}_1 - \mathbf{p}_2|^2$$

is a Dollard phase [9]. For deriving the renormalized SBA amplitude $\tilde{T}_1^{SBA}$ we must consider a first-order expansion of (15) into series over η. Thus, according to (12) we get

$$\tilde{T}_1^{SBA} = \lim_{\Delta \to +0} \{[1 + R(\Delta)]T_1^{FBA} + TS1(\Delta) + TS2_1 + TS2_{21}(\Delta) + TS2_{22}(\Delta)\} \qquad (16)$$

with

$$R(\Delta) = -i\eta \ln \Delta - (\pi/2 + i\gamma)\eta + iA,$$

where $\gamma = 0.5772$ is the Euler constant. The result (16) is convergent and unique.

It should be remarked that the form of the convergent factor in the right-hand side of (14) is not unique, for it depends on the procedure utilized in factoring out the divergencies. This fact is reflected in the presence of the coefficient N, which depends on the Coulomb numbers and thereby can be redefined, and has the following important consequence: there is no *a priori* choice for the Gamow factor in the PWIA theory.

NUMERICAL RESULTS

In the numerical realization of the present SBA theory we treat the TS1 and TS2$_{21}$ mechanisms in the closure approximation by replacing the one-electron energies ε_λ in (6) and (10) with an average value $\bar{\varepsilon}_\lambda = -(\bar{E}+2)$ a.u. Since in the closure approximation the TS1 and TS2$_{21}$ contributions are convergent, we drop the one-electron terms in η and A when calculating (16). Further, the three-dimensional integral in (11) is presented as a sum of the convergent and divergent integrals following the procedure of [6]. The convergent integral is carried out numerically, while the divergent integral is performed analytically and compensates for divergency of the FBA term in (16). Below we present the numerical results for two symmetric setups: (1) the coplanar with $E_0 = 580$ eV and $E_3 = 1$ eV [1] and (2) the noncoplanar with $E_0 = 2080$ eV, $E_3 = 10$ eV, and $\theta_1 = \theta_2 = 45°$ [2]. In the coplanar case the electron momenta $\boldsymbol{p}_0$, $\boldsymbol{p}_1$ and $\boldsymbol{p}_2$ are in the same plane and the 4DCS is studied as a function of angle $\theta = \theta_1 = \theta_2$. In the non-coplanar case the 4DCS is studied as a function of momentum $q = |\boldsymbol{p}_1 + \boldsymbol{p}_2 - \boldsymbol{p}_0|$.

FIG. 1 shows the results using a Hylleraas model [10] for the helium wave function:

$$\Phi_0(\boldsymbol{r}_1, \boldsymbol{r}_2) = \frac{Z_{eff}^3}{\pi} e^{-Z_{eff}(r_1+r_2)}, \qquad (17)$$

where $Z_{eff} = 27/16$. In the numerical SBA calculations we have used the value $\bar{E} = -100$ eV. The PWIA results have been generated in accordance with the formula

$$d^4\sigma^{PWIA} = G d^4\sigma^{FBA}, \qquad (18)$$

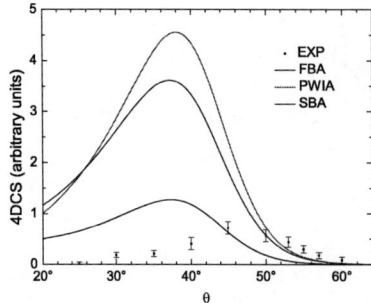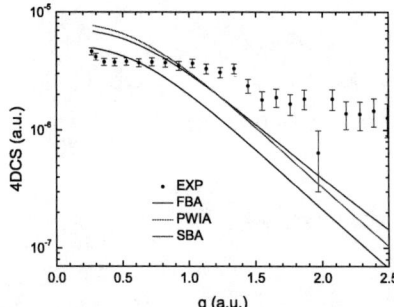

FIGURE 1. The 4DCS in the symmetric coplanar (left panel) and non-coplanar (right panel) setups as a function of $\theta_1 = \theta_2 = \theta$ and q, respectively. The experimental values are due to [1] and [2]. In the coplanar case, the SBA results are normalized to the experimental value at $\theta = 50°$, and the FBA and PWIA results are scaled according to the normalization coefficient. In the noncoplanar case, theoretical results have been convoluted with momentum resolution function [2] and the experiment is normalized to the SBA results for (e,2e) reaction following the procedure of [2].

where

$$G = \frac{2\pi\eta}{e^{2\pi\eta} - 1} \qquad (19)$$

is a Gamow factor which takes into account all final-state Coulomb interactions on equal footing. It should be emphasized that the Gamow factor given by (19) differs from the traditional one that involves only the electron-electron variable $\eta = 1/|\boldsymbol{p}_1 - \boldsymbol{p}_2|$ (see, for instance, [3]). The reason for such a choice is due to the fact that it provides a close description of the SBA results, as can be seen in Fig. 1, while the traditional Gamow factor reduces the FBA results in magnitude.

Comparing the present SBA results with experiments in Fig. 1, we attribute the marked discrepancies to employing in the calculations the Hylleraas function (17), which ignores spatial correlation of electrons in a He atom, and the closure approximation, which treats the virtual SBA transitions in a rather crude way.

CONCLUDING REMARKS

In this work we have developed a novel approach to the calculations of the SBA and PWIA amplitudes which correctly accounts for Coulomb singularities pertinent to the plane-wave Lippmann-Swinger integral equations. We have performed the pilot numerical SBA calculations using the Hylleraas function and closure approximation in order to simplify the six-dimensional integrals (see [11] for more details). A new Gamow factor that relates the FBA and PWIA cross sections has been specified.

This work provides a basis for further studies on the mechanisms of the (e,3e) and (e,3-1e) reactions at large momentum transfer. Note that the 4DCS treated within the context of the first-order theories, such as FBA and/or PWIA, is very sensitive to the electronic correlations in a target. Therefore, a topical issue is to determine the kinematical regimes where the effects of higher-order mechanisms are subdominant. In general the SBA calculations are less sensitive to the initial-state electronic correlations, however the present calculations using (17) indicate that the role of the electronic correlations in the process is not less important than that of the second-order mechanisms.

REFERENCES

1. P. Bolognesi et al., *Phys. Rev. A*, **67**, 034701 (2003).
2. N. Watanabe et al., *Phys. Rev. A*, (in press).
3. Yu. V. Popov, C. Dal Cappello, and K. Kuzakov, *J. Phys. B: At. Mol. Opt. Phys.*, **29**, 5901 (1996).
4. I. E. McCarthy, *Aust. J. Phys.*, **48**, 1 (1995).
5. K. A. Kouzakov, and Yu. V. Popov, *J. Phys. B: At. Mol. Opt. Phys.*, **35**, L537 (2002).
6. Y. Popov, *J. Phys. B: At. Mol. Opt. Phys.*, **14**, 2449 (1981).
7. V. L. Shablov, V. A. Bilyk and Yu. V. Popov, *J. Phys. IV France*, **9**, Pr6-59 (1999); V. L. Shablov, V. A. Bilyk, and Yu. V. Popov, "An Application of the Coulomb Scattering Theory to Ionization Processes", in *Many-Particle Spectroscopy of Atoms, Molecules, Clusters and Surfaces*, edited by J. Berakdar and J. Kirshner, Kluwer Academic/Plenum Publishers, New York, 2001, pp. 71–80.
8. V. L. Shablov, V. A. Bilyk, and Yu. V. Popov, *Phys. Rev. A*, **65**, 042719 (2002).
9. C. Chandler, *Nucl. Phys. A*, **353**, 129c (1981).
10. E. A. Hylleraas, *Z. Phys.*, **54**, 347 (1929).
11. P. S. Vinitsky, Yu. V. Popov, and O. Chuluunbaatar, *Phys. Rev. A*, **71**, 12706 (2005).

Angular Momentum Partitioning in the Dissociation of Diatomic Molecules

T.J.Gay*, J.D.Bozek[†], J.E.Furst[‡], G.A.Gallup*, A.S.Green*[+],
A.L.D.Kilcoyne[†], J.R.Machacek*, J.W.Maseberg*, K.W.McLaughlin[#],
and M.A.Rosenberry*[$]

*Behlen Laboratory of Physics, University of Nebraska, Lincoln, NE 68588-0111 USA
[†]Advanced Light Source, Lawrence Berkeley Laboratory, Berkeley, CA 94720, USA
[‡]University of Newcastle-Ourimbah, Ourimbah, NSW 2258, Australia
[+]University of St. Thomas, 2115 Summit Ave., St. Paul, MN 55105-1080 USA
[#]Loras College, Department of Physics and Engineering, Dubuque, IA 52001 USA
[$]Siena College, Department of Physics, Loudonville, NY 12211 USA

Abstract. We discuss recent experiments that study the transfer of angular momentum from a projectile to the residual target in collisions between the simple diatomic molecules H_2 and N_2 and spin-polarized electrons or circularly-polarized photons. We observe the fluorescence of both the atomic fragments and excited molecular states, and measure the circular polarization fraction of this light, P_3. The incident electron energies range from 10 to 100 eV; the incident photon energies from 33 to 38 eV.

Keywords: angular momentum; spin polarization; circular polarization; molecular dissociation
PACS: 33.15.Vb; 33.80.Gj; 34.30.+h; 34.80.Ht; 34.80.Nz

INTRODUCTION

An important problem in atomic collisions is the distribution, or partitioning, of angular momentum in an excited or ionized atomic target produced by photon or electron bombardment [1]. A complete picture of the angular momentum dynamics must include the ionized or scattered electrons as well [2]. An interesting extension of this problem involves molecular targets, which have the additional complication of rotational angular momentum. One experimental approach to the general problem of angular momentum dynamics in such collisions is to use incident photon or electron beams that are spin polarized, and to detect the polarization of the fluorescence emitted by the target or its fragments following the collision. The advent of third-generation light sources and GaAs polarized electron sources has made such experiments much easier. Earlier experiments had used unpolarized electrons and detected the polarization (both circular and linear) of emitted light in coincidence with the scattered electron [3,4], or used linearly-polarized incident light and detected linear polarization of the subsequent molecular or atomic fluorescence[5-7]. We report here the results of several recent experiments done by our group and others which have begun to elucidate how angular momentum, inserted into the collision complex by the use of either polarized electrons or photons, is distributed in a collision with a simple diatomic molecule or molecular ion. We will concentrate on the measurement of P_3, the circular polarization fraction (or Stokes parameter), as a direct measure of the angular momentum transferred along a specific axis to the target by the incident polarized particle.

EXPERIMENTS WITH ELECTRONS

A particularly simple example of angular momentum transfer in the electron-impact excitation of atoms is one in which a longitudinally-polarized electron excites, via exchange, an upper state of the atom which subsequently emits a photon that is circularly polarized along the axis of the incident electron. Ultimately, for an atom to emit circularly-polarized light, its *orbital* angular momentum must have a magnetic dipole along the direction of photon emission. The incident electron spin provides this through the atom's internal spin-orbit coupling. An example of this occurs in the exchange excitation of the 1s3p 3^3P state of He. For 100% incident electron spin polarization, the emitted photon in the 2^3S-3^3P 389 nm transition has a P_3 of 50% near threshold [8]. Similar angular momentum transfer has been observed in a variety of atomic systems [9,10].

Problems arise with molecules, however. The Münster group bombarded N_2 in its singlet closed-shell ground state with beams of polarized electrons, and measured P_3 of the resultant $C^3\Pi_u - B^3\Pi_g$ 337 nm fluorescence [11]. The excitation of a triplet state with light targets guarantees that exchange excitation, and thus angular momentum transfer, has occurred. Within their statistical uncertainty of 2×10^{-3}, however, they found P_3 to be nil at all the incident electron energies they investigated (see Figure 1).

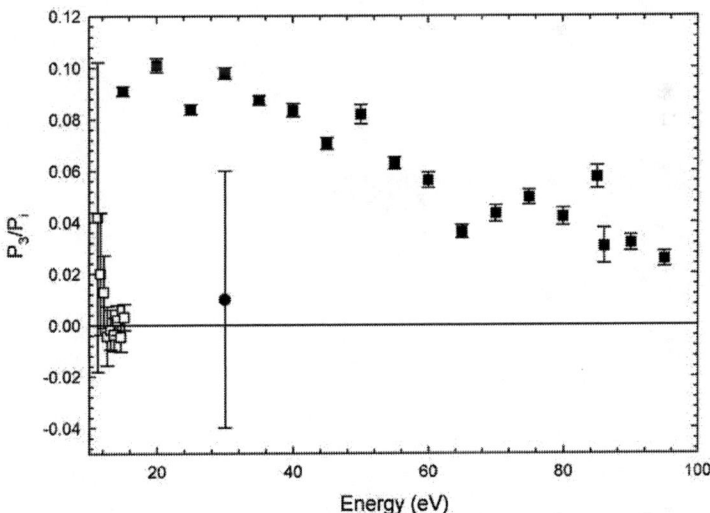

FIGURE 1. Circular polarization fraction P_3 normalized to incident electron polarization *vs.* incident electron energy. Open squares: data of ref. 11; solid circle: measurement with N_2 target and 388 nm filter (see text); solid squares: data of ref. 12 taken with a 600 ± 5 nm filter.

A similar, more crude measurement was recently made in our lab, in which we used an interference filter with a bandpass of 388±5 nm. This filter isolates light from the $A^3\Sigma_u^+ - X^1\Sigma_g^+$, $C^3\Pi_u - B^3\Pi_g$, $A^3\Sigma_u^+ - X^1\Sigma_g^+$, $C^3\Pi_u - B^3\Pi_g$, and $C'^3\Pi_u - B^3\Pi_g$ transitions in N_2, and the $B^2\Sigma_u^+ - X^2\Sigma_g^+$ transitions in N_2^+. Since the N_2 transitions involve triplet states, and the N_2^+ excited state is a doublet, exchange collisions dominate the production of the light we observe. Nonetheless, as with the Münster data, we find P_3 to be consistent with zero (Fig. 1).

We can start to understand these data by remembering the relevant time scales for molecular processes. The impact excitation occurs in times of the order of 10^{-16}s, whereas N_2 rotational motion occurs on a scale of $\sim 10^{-13}$s, with fluorescence lifetimes being more typically 10^{-8}s. The spin-orbit coupling time for excited states of N_2, i.e., the time required for the electron to "spin-up" the orbital angular momentum of the excited state, is $\sim 10^{-13}$s. Thus while the N_2 target develops an orbital orientation over the course of several rotational periods, its internuclear axis subsequently rotates thousands of times before it decays, causing its orbital orientation to be essentially randomized in space. Thus P_3 is nil.

One can, however, expect that an exchange collision followed by a prompt dissociation of the molecule would result in atomic fragments with a non-zero expectation value of spin along the incoming electron axis. The atomic "memory" of this spin direction would not be lost. Thus it should be possible to investigate angular momentum transfer to the molecular fragments by investigating the circular polarization of atomic fluorescence. Using this idea, our group and the Perth group

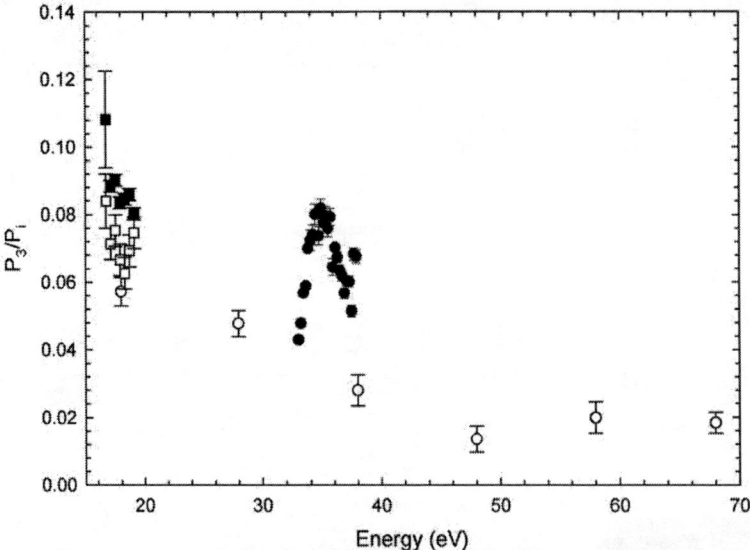

FIGURE 2. P_3 of Hα (656 nm) atomic fluorescence, normalized to the incident photon or electron polarization P_i, vs. incident electron or photon energy. Photon results are also divided by two to take into account the different angular momentum deposited by electrons vs. photons. Solid and open squares: electron data of ref. 13 with different incident electron polarizations; open circles: electron data of ref. 12; solid circles: photon data of this work.

have measured P_3 of H(n=3), Hα (656 nm) fluorescence resulting from electron-impact dissociation of H$_2$ [12,13]. These data are shown in Figure 2. Not surprisingly, significant polarizations near the n=3 production threshold of ~17 eV are apparent in these data, with a slow drop off as the electron energy increases.

It is instructive to compare these polarization values with those resulting from the direct impact excitation of atoms by polarized electrons. To do this, we define a "spin transfer efficiency," T, equal to the initial spin polarization of the excited system divided by the electron spin polarization, P_e. Thus, if we excited a pure molecular triplet state by exchange, $T = 2/3$ [1]. Following dissociation, we assume that the atomic 3s, 3p, and 3d states all have equal spin polarization T', and that this polarization is completely coupled to the orbital angular momentum in the case of the 3p and 3d states, while we ignore hyperfine depolarization [12]. Taking into account the branching ratio between Hα and Lyα radiation (see Figure 3), assuming that light from the 3s state is unpolarized, and using published data for the relative production cross sections for 3s, 3p, and 3d states as the result of electron impact dissociation of H$_2$ [14], we can infer a threshold value of $T' = 0.37$. Assuming that all three n=3 initial populations are equal, $T' = 0.47$. These values are surprisingly comparable to those for direct excitation of, e.g., alkali atoms, from Na ($T = 0.25$) to Cs ($T = 0.45$) [15].

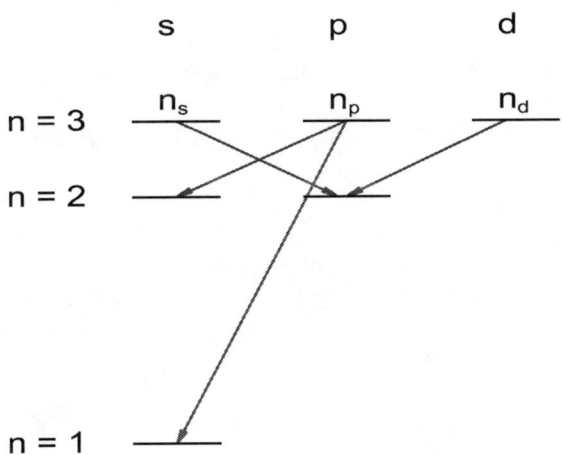

FIGURE 3. Grotrian diagram of the first three energy levels of the H atom. The circular polarization we observe is due to transitions from the 3p and 3d states; photons from the 3s state are unpolarized. In the calculations of T', initial populations of the n=3 states are taken from ref. 14. Only ~1/7th of the 3p states decay to the 2s state; the rest decay via Lyα emission.

EXPERIMENTS WITH PHOTONS

We can learn more about how angular momentum is distributed in a dissociating molecular complex by using circularly-polarized incident photons to provide the dissociation energy. This method has the advantage that a full $\hbar$ of angular momentum is dumped into the target, as opposed to $\hbar/2$ for electrons. Moreover, spin-orbit coupling is not needed to convert spin to orbital orientation; the coupling between the photon and the orbital angular momentum of the molecule is direct. In a recent experiment done at the Advanced Light Source (ALS) at Lawrence Berkeley Laboratory, we have measured Hα P_3 values when light with energy between 33 and 38 eV was used to dissociate H_2. The linearly-polarized synchrotron radiation was turned into circularly-polarized light by passage through a four-reflection quarter-wave retarder [2]. These data are shown in Figure 2. They have been divided by two to account for the larger amount of angular momentum carried by the photons, and adjusted to correspond to photon emission directly along the incident photon axis. Interestingly, the P_3 values produced by the incident photons are comparable to the results for electron bombardment, even though the angular momentum coupling is much more efficient. It is important to keep in mind, however, that the internal molecular spin-orbit coupling, while serving to *produce* orbital orientation in the case of electron bombardment, acts only to reduce it (by spinning up the electrons at the expense of orbital angular momentum) in the case of photon bombardment.

POLARIZED MOLECULAR FLUORESCENCE

Upon closer examination, the simple model of rotational destruction of polarization for excited molecular fluorescence fails. We used a broad (600±5 nm) interference filter to monitor fluorescence in the Fulcher band of H_2 excited by polarized electron impact. These results are shown in Figure 1 together with the N_2 and N_2^+ data. Our

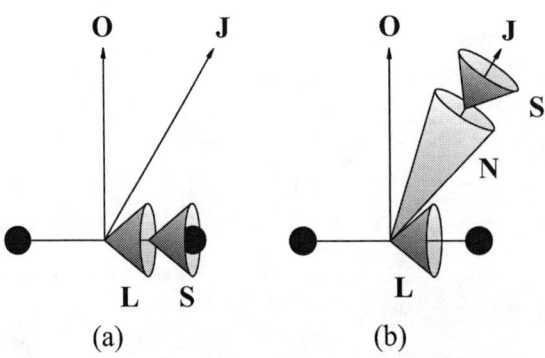

FIGURE 4. Hund's cases a) and b). The nuclear rotational angular momentum is **O**; total molecular angular momentum = **J**. In the case of H_2, **L** + **O** = **N**.

600 nm filter passes light which is due to a variety of H_2 transitions; we estimate that approximately 94% of the transmitted intensity is due to emission by triplet states which can, in principle, produce circularly-polarized light. Naively, one might expect that H_2 fluorescence would be suppressed even more than that of N_2, since, classically speaking, the H_2 molecule at a given temperature rotates somewhat more rapidly than does a nitrogen molecule, while both have comparable fluorescence lifetimes. This neglects the relative strength of the couplings between the various angular momenta in the molecule, however. Nitrogen is a Hund's case a) molecule, in which the spin is essentially coupled directly to the internuclear axis (see Figure 4). This is because the spin-orbit coupling time is comparable to the internuclear rotational period, $\sim 10^{-13}$s. Hydrogen, on the other hand, is a Hund's case b) molecule, meaning that the spin is much more loosely tied to the internuclear axis. While its rotational period is still $\sim 10^{-13}$ s, its spin-orbit coupling time is closer to $\sim 10^{-10}$s. Thus hydrogen can retain a better "memory" of the initial spin direction [12]. We have investigated several other transitions in molecular hydrogen, and find them to be polarized as well, although with generally lower polarization than that shown in Figure 1. Clearly a systematic study of this problem, with wavelength selection for specific vibrational and rotational levels is warranted.

ACKNOWLEDGMENTS

We would like to thank Kevin Dooley, Orhan Yenen, and Duane Jaecks for useful conversations and contributions to the photodissociation data reported here. This work has been funded by the U.S. Department of Energy through use of the ALS, and the US National Science Foundation Grants PHY–0354946 (TJG), and PHY-0321055 (KWM). One of us (JEF) wishes to acknowledge support from the University of Newcastle Outside Studies Program.

REFERENCES

1. K. Blum, *Density Matrix Theory and Applications,* 2nd ed. (Plenum, New York, 1996).
2. K.W. McLaughlin, O. Yenen, D.H. Jaecks, T.J. Gay, M.M Sant'Anna, D. Calabrese, and B. Jordan-Thaden, Phys. Rev. Lett. **88**, 123003 (2002).
3. J.W.McConkey, S.Trajmar, J.C.Nickel, and G.Csanak, J. Phys. B **19**, 2377-2392 (1986).
4. M.A.Khakoo and J.W.McConkey, J. Phys. B **20**, L175-L179 (1987).
5. J.A.Guest, K.H.Jackson, and R.N.Zare, Phys.Rev.A **28**, 2217-2228 (1983).
6. E.Flemming, O.Wilhelmi, H.Schmoranzer, and M.Glass-Maujean, J.Chem.Phys. **103**, 4090-4096 (1995).
7. T.P.Rakitzis and R.N.Zare, J.Chem.Phys. **110**, 3341-3350 (1999).
8. K.J.Goecke, J.Kessler, and G.F.Hanne, Phys. Rev. Lett. **59**, 1413-1415 (1987).
9. M.Eminyan and G.Lampel, Phys. Rev. Lett. **45**, 1171-1174 (1980).
10. T.J.Gay, J.E.Furst, K.W.Trantham, and W.M.K.P.Wijayaratna, Phys. Rev. A **53**, 1623-1629 (1996).
11. G.F.Hanne in *Novel Aspects of Electron-Molecule Collisions*,K.Becker ed. (World Scientific, Singapore, 1998).
12. A.S. Green, G.A. Gallup, M.A. Rosenberry, and T.J. Gay, Phys. Rev. Lett. **92**, 093201 (2004).
13. J.F.Williams and D.H. Yu, Phys. Rev. Lett. **93**, 073201 (2004).
14. W. Kedzierski, A. Abdellatif, J.W. McConkey, K. Bartschat, D.V. Fursa, and I. Bray, J. Phys. B **34**, 3367-3375 (2001).
15. C.P. Naβ, M. Eller, N. Ludwig, E. Reichert, and M.Webersinke, Z. Phys. D **11**, 71-80 (1989).

Spin-Resolved Collisions of Electrons with Rubidium Atoms: A Search for Relativistic Effects

W.E. Guinea[*], G.F. Hanne[*,†], M.R. Went[*,#], M.L. Daniell[*]
M.A. Stevenson[*], W.R. MacGillivray[*,&] and B. Lohmann[*]

[*]Centre for Quantum Dynamics, Griffith University, Nathan, Qld, Australia, 4111
[†]Physikalisches Institut, Universität Münster, Wilhelm-Klemm-Str. 10, 48149 Münster, Germany
[&]Faculty of Sciences, University of Southern Queensland, Toowoomba, Qld, Australia 4350
[#]Present address: Atomic and Molecular Physics Laboratories, Research School of Physical Sciences and Engineering, Australian National University, Canberra ACT, Australia, 0200

Abstract. The search for relativistic effects in electron-alkali scattering is currently a topic of considerable interest. The A_2 spin asymmetry parameter is a direct measure of relativistic effects in the electron-atom collision process, as it is entirely dependent on the spin-orbit effect. We present measurements of the A_2 spin asymmetry for the 5S → 5P transition in rubidium at incident energies of 15, 20, 30 and 50 eV and for elastic scattering at 15, 20, 30, 50 and 80eV. Our results indicate that under these collision conditions, relativistic effects are measurable, in qualitative agreement with the available theory.

INTRODUCTION

Recently there has been a search for relativistic effects in the scattering of electrons from heavy alkalis. This search was partly motivated by theoretical efforts which suggested that such spin-dependent relativistic effects should be measurable. Theoretical attempts to describe the role of the spin-orbit effect in spin polarised electron scattering from one electron atoms were first investigated by Burke and Mitchell [1]. Spin effects are relativistic in nature, as they arise from Dirac's application of relativity to quantum electro-dynamics. The spin-dependent interaction between the incident electron and a target atom is proportional to **L•S**, where **S** is the spin angular momentum vector of the incident electron and **L** is the total orbital angular momentum of the incident electron during the collision process. It was surprising that the effects of this interaction did not appear in measurements of the differential cross section or the Stokes parameters of such heavy elements as cesium [2]. Several theoretical treatments that ignore relativistic effects predicted the differential cross sections and Stokes parameters for these elements with a high degree of accuracy [2].

Andersen and Bartschat [3] demonstrated that these 'conventional' collision parameters are insensitive to spin dependent effects in the scattering process. Instead they identified several generalised Stokes parameters, three spin asymmetries and several so called circular dichroism parameters that are explicitly dependent on spin effects. The asymmetries are designated A_1, A_2 and A_{nn}. A_1 represents scattering of unpolarised electrons from polarised atoms, and is known as the interference asymmetry. A_{nn} represents scattering of polarised electrons from polarised atoms, and is called the exchange asymmetry. A_2 represents scattering of polarised electrons from unpolarised atoms, and is called the spin-orbit asymmetry. For elastic scattering it is identical to the Sherman function that describes the elastic scattering of polarised electrons from unpolarised targets [3]. The A_2 asymmetry is due to the spin-orbit effect on the continuum electron, and is thus a reliable indicator of the extent to which relativistic effects occur in the collision process. Experimentally one may build the spin asymmetries from the following equations:

$$NA_{nn} = \left[\left(N^{\uparrow\downarrow} + N^{\downarrow\uparrow}\right) - \left(N^{\uparrow\uparrow} + N^{\downarrow\downarrow}\right)\right]/P_a \bullet P_e \quad (1)$$

$$NA_1 = \left[\left(N^{\uparrow\downarrow} + N^{\uparrow\uparrow}\right) - \left(N^{\downarrow\downarrow} + N^{\downarrow\uparrow}\right)\right]/P_a$$

$$NA_2 = \left[\left(N^{\uparrow\uparrow} + N^{\downarrow\uparrow}\right) - \left(N^{\uparrow\downarrow} + N^{\downarrow\downarrow}\right)\right]/P_e$$

where $N = N^{\uparrow\downarrow} + N^{\downarrow\uparrow} + N^{\uparrow\uparrow} + N^{\downarrow\downarrow}$, and corresponds to the differential cross section, P_a is the target beam polarisation with respect to the scattering plane and P_e is the incident electron beam polarisation with respect to the scattering plane. The first superscript indicates the atomic spin with respect to the scattering plane, while the second superscript indicates the incident electron beam spin with respect to the scattering plane [2]. Measurements of these three spin asymmetries from the heavy alkali cesium clearly showed the existence of relativistic effects in the collision process [2, 4, 5]. Further measurements of these asymmetries in other heavy alkalis would be of enormous value in testing the validity of several relativistic theories. As such, measuring the A_2 parameter for rubidium would help fill a gap in our experimental knowledge and in the testing of theoretical predictions.

EXPERIMENTAL APPARATUS

The apparatus comprises four chambers: the source chamber, the differential pumping stage, the scattering chamber, and a chamber, containing a Mott polarimeter, which is used to determine the polarisation of the incident electron beam. The interaction region is in the centre of the scattering chamber. The differential pumping stage serves as a pressure buffer between the source chamber and the scattering chamber, the pressure ratio being about 1000. The source chamber houses the source of spin polarised electrons itself, and is maintained at a pressure of around 3×10^{-10} mbar.

A small wafer of GaAs produces the spin polarised electron beam. Deposition of small amounts of cesium and oxygen produce a negative electron affinity (NEA) surface, such that the conduction band minimum is higher in energy than the vacuum states. A near infrared diode laser, wavelength 820nm, is incident upon the surface of the crystal, and excites electrons from the valence to conduction band. The laser is incident via a linear polariser and a liquid crystal retarder (LCR). The incident laser light is thus circularly polarised, and the orientation, left hand circularly polarised (LHC) or right hand circularly polarised (RHC), can be changed manually or by computer control of the LCR applied voltage. For a discussion of GaAs spin-polarised sources, see [6]. As the interaction region is at ground, a negative potential applied to the crystal determines the energy of the ensuing electron beam.

A 90-degree electrostatic deflector guides the emitted electrons into an extensive electron optics train. The electron optical elements guide and focus the electron beam to the interaction region of the experiment. A Faraday cup is used to monitor the incident beam current. The Faraday cup is on a turntable, thus it can be moved to the side in order to conduct a measurement of the beam polarisation using the Mott chamber. An oven, containing rubidium, positioned at 120 degrees to the electron optics, is heated to provide the atomic beam of rubidium. A hemispherical electrostatic energy analyser mounted on the opposite side of the chamber is capable of detecting electrons scattered through angles 30-110 degrees in the scattering plane.

For the spin asymmetry measurements presented here, the number of counts for a particular spin orientation at a particular angle is recorded in the angular range 30-110 degrees in five-degree steps. The spin orientation at each angle is varied (around once every 10 seconds), so as to average over noise and any possible instrumental asymmetries. Normally one hundred of these spin flips is conducted at each angle. Up to 70 scans over the whole angular range may be taken in order to gain favourable statistics for a single asymmetry measurement.

RESULTS

Results for the A_2 asymmetry for elastic and inelastic electron scattering at an incident energy of 15eV are presented in Fig. 1. Results at incident energies of 20, 30 50 and 80eV are presented in figure 2. Only elastic results are reported for the 80eV measurement, as the inelastic differential cross section, at this energy, was too low to conduct a measurement. Importantly, the asymmetry is non-zero for both elastic and inelastic scattering at all energies. This indicates that spin-dependent effects are measurable at these intermediate energies. At lower angles the asymmetry is essentially zero for all results. This is not surprising as large asymmetries generally only occur near minima in the differential cross section [7]. The magnitude of the asymmetries overall are around one third of those seen for cesium [2, 4, 5]. Given the considerably different nuclear charge of rubidium and cesium, and the dependence of A_2 on the spin-orbit effect, this result is understandable.

Comparing the inelastic results, there appear to be similar trends and magnitudes across all energies except the 50eV result. The 15, 20 and 30eV results all show generally similar magnitude asymmetries (around 4% maximum values), though the 30eV result exhibits an apparently large asymmetry of 11.5% at 80 degrees. Assuming that the large asymmetry at 30eV is due to scatter in the data, the 50eV result exhibits the largest asymmetry (around 7%) of the inelastic results. On making a similar comparison as a function of energy for the elastic results, one finds that there the magnitude of the respective asymmetries is similar across all incident energies. Indeed the largest asymmetries for each particular energy vary from 4% (80eV) to 6% (20eV). Thus there is little variation in the magnitude of the asymmetries over this energy range for elastic scattering. There appears to be little difference in the magnitude of the asymmetries upon comparison of the elastic and inelastic results.

In each case, the results are presented with the applicable theory. Two theoretical approaches are presented here: a semi-relativistic Breit-Pauli R-matrix approach, with and without pseudostates, and a fully relativistic Dirac-Fock calculation. The first approach is a calculation by Bartschat; for details see Guinea et al [8]. The second is that of Stauffer [9].

The Dirac-Fock calculation has only been applied in this case to inelastic scattering at energies of 20eV to 50eV. The R-matrix approach has been applied at all energies with a variety of states in the expansion. Qualitatively, there is good agreement between theory and experiment, in that the theory predicts non-zero asymmetries at all energies. The magnitude of the asymmetries, as predicted by the BP 37 calculation is generally good. The exceptions to this are for both the 15eV results and the 20eV inelastic result. Of notable interest is the apparent angular shift that exists between the R-matrix theory and the results at all energies. The angular discrepancy between theory and experiment appears to be improving with the 37 state calculation, though the predicted magnitude in the cases noted above is still rather poor. The 37 state calculation generally appears to perform well in comparison to the 5 state calculation, which does not include pseudostates. A more complete calculation with a larger number of pseudostates would be useful in determining if the angular distribution and magnitude of the R-matrix prediction improves in comparison to the experimental results. Both qualitatively and quantitatively the RDW calculation is very good across those results it is compared with.

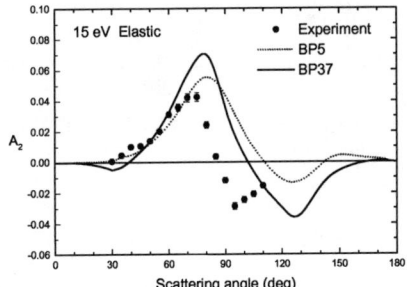 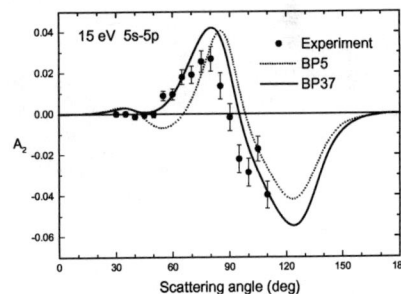

FIGURE 1. Results for the A_2 asymmetry for elastic and inelastic electron scattering from rubidium at an incident energy of 15eV.

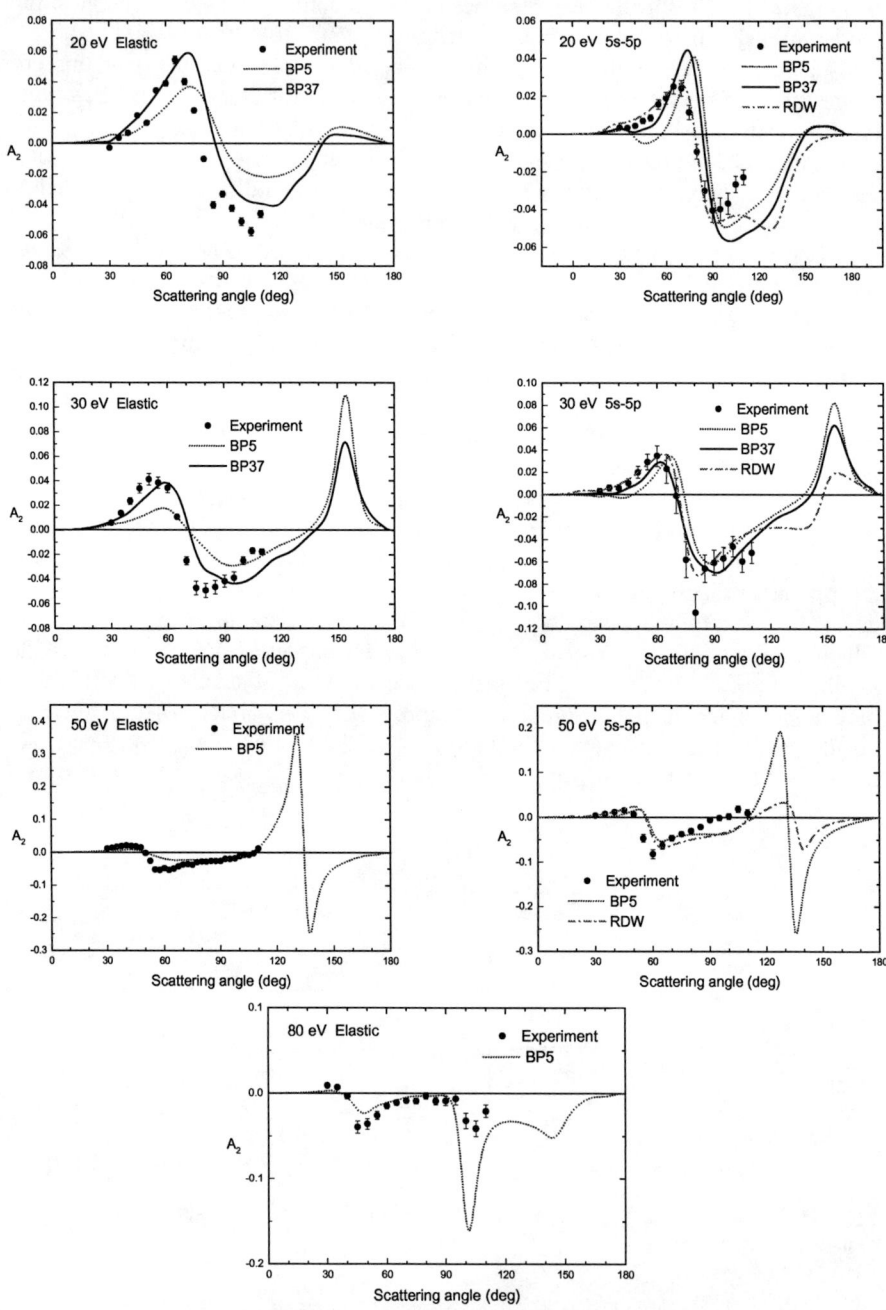

FIGURE 2. The A_2 asymmetry for elastic and inelastic electron scattering from rubidium at 20, 30 and 50eV incident energy, and for elastic scattering at 80eV incident energy.

ACKNOWLEDGMENTS

This work was supported by the Australian Research Council. WEG would like to acknowledge the assistance of an Australian Postgraduate Award, and GFH would like to acknowledge the support of the Volkswagen Stiftung.

REFERENCES

1. P.G. Burke and J.F.B. Mitchell, *J. Phys. B* **2**, 214 (1974)
2. G. Baum, N Pavlovic, B. Roth, K. Bartschat, Y. Fang and I. Bray, *Phys. Rev. A* **66**, 022705 (2002).
3. N. Andersen and K. Bartschat, *J. Phys. B* **35**, 4507 (2002)
4. G. Baum, S. Förster, N. Pavlović, B. Roth, K. Bartschat. and I. Bray, *Phys. Rev. A.* **70**, 012707 (2004)
5. G. Baum, W. Raith, B. Roth, M. Tondera, K. Bartschat, I. Bray, S. Ait-Tahar, I.P. Grant,. and P.H. Norrington, *Phys. Rev. Lett.* **82**, 1128-1131 (1999)
6. D.T. Pierce, R.J. Celotta, G.-C. Wang, W.N. Unertl, A. Galejs, C.E. Kuyatt, and S.R. Mielczarek, *Rev. Sci. Instrum.* **51**, 478 (1980)
7. Joachim Kessler, *Polarized Electrons*, 2nd Edition, Springer-Verlag, Berlin, (1985)
8. W.E. Guinea, G.F. Hanne, M.R. Went, M.L. Daniell, M. A. Stevenson, K. Bartschat, D. Payne, W.R. MacGillivray and B. Lohmann, *in press, J. Phys. B.* (2005)
9. A.D. Stauffer, *private communication*, (2003)

MULTI-AUGER DECAY IN NEGATIVE ION PHOTODETACHMENT

R. C. Bilodeau*,†, J. D. Bozek†, G. D. Ackerman†, N. D. Gibson**, C. W. Walter**, A. Aguilar†,‡, G. Turri*,†, I. Dumitriu* and N. Berrah*

*Western Michigan University, Physics Department, Kalamazoo, Michigan 49008-5151
†Lawrence Berkeley National Laboratory, Advanced Light Source, Berkeley, California 94720
**Department of Physics and Astronomy, Denison University, Granville, Ohio 43023
‡University of Nevada, Department of Physics, Reno, Nevada 89557-0058

Abstract. Inner-shell photoexcitation and detachment experiments on He$^-$ and S$^-$ ions are discussed. In both systems, negative ion Feshbach resonances produced by excitation into the partially filled valence p orbital are observed to decay via many-electron processes, such as multi-Auger decay. Absolute cross sections, near-threshold behavior, and relative product channel strengths (branching ratios) are also discussed.

Keywords: negative ions, Auger, double Auger, inner-shell, core, triply excited, hollow ion, Feshbach resonance, photodetachment, photoionization.
PACS: 32.80.Hd, 32.80.Gc

INTRODUCTION

The highly correlated nature of electrons in core-excited and hollow-ion states provides a valuable opportunity to study electron-correlation effects, and can stringently test and distinguish between detailed theoretical models. In negative ions, core-photoexcitation also involves a highly correlated ground state, proving a further challenge to theoretical studies. Decay pathways of core excited states typically involve multi-electron processes, e.g. Auger decay. States located above the double ionization limit can decay via 2-electron emission in two sequential Auger processes or in a single step with the demotion of one electron and the simultaneous emission of two others, i.e. double-Auger (DA) decay. For states located above higher ionization limits, sequential and simultaneous multi-Auger (one down, many out) processes become possible.

This article discusses recent experiments conducted on He$^-$ and S$^-$ with the focus on multiple electron emission in these systems. In He$^-$ these studies allowed for the observation of the $2s2p^2$ $^4P^e$ state, the lowest triply excited quartet state in He$^-$ [1]. While sequential Auger decay generally dominates 2-electron loss from core-excited states, the $2s2p^2$ $^4P^e$ state lies below the $2s^2$ threshold thus requiring a simultaneous DA process. In S$^-$, photodetachment or excitation of the $2p$ and $2s$ electrons resulted in the loss of up to 5 electrons, producing positive ions with charges up to +4 [2]. In at least some areas of the spectrum the observed signal must be due to simultaneous multi-Auger processes and other multi-electron processes combined with sequential (or "cascade") multi-Auger decay.

EXPERIMENTAL METHOD

The experiments were performed at the Advanced Light Source (ALS) using the Ion-photon Beamline [3]. A He⁻ beam was produced with a charge-exchange ion source (by NEC), while S⁻ ions were produced with a cesium-sputter source (SNICS II by NEC) [4]. The ions were mass-selected with a sector magnet and merged with a counter-propagating photon beam from ALS Beamline 10.0.1. The photon-ion interaction in the merged region leads to the loss of two or more electrons by photo multi-detachment, photodetachment followed by Auger decay, photoexcitation followed by autodetachment/ionization, or other higher-order processes such as double- or multi-Auger decay. A magnetic demerger directs the ion beam into a Faraday cup, where the ion beam current is monitored, while the produced positive ions (the signal) are deflected in the opposite direction by the magnetic demerger and counted by an MCP-based detection and counting system.

HOLLOW-ION RESONANCES IN HELIUM NEGATIVE IONS

Triply excited (hollow-ion) states of He⁻ were first observed in the attenuation of electron beams scattered on He [5], and later with the formation of He⁺ [6] in electron-scattering experiments. However, since the formation of quartet states is forbidden in electron scattering [7], the lowest triply excited quartet state in He⁻ (the $2s2p^2\ ^4P^e$ state) [6, 8] eluded observation until very recently [1], although the state had been predicted 25 years earlier [9]. This state has similarly not been observed in photoexcitation of any 3-electron system [10] (although the state was inferred in Li from Li⁺ collisions with gas targets [11]).

He⁻ forms the exotic $1s2s2p\ ^4P^o$ ground state bound by only 77.5116(6) meV [12] below the first excited state of He (the $1s2s$ state, which lies 19.819 eV above the $1s^2$ ground state [13]). Photoexcitation from the He⁻ ground state therefore opens the way to investigations of the He⁻ quartet manifold. In fact, calculations predicted a very large peak cross section of ≈600 Mb [14] for the fully allowed electric dipole transition, He⁻ $1s2s2p\ ^4P^o \xrightarrow{h\nu} 2s2p^2\ ^4P^e$. However, due to the absence of He states between the He⁺ $1s$ and He⁻ $2s2p^2\ ^4P^e$ states (see level diagram in Fig. 1), sequential Auger decay of the $2s2p^2\ ^4P^e$ state is impossible. As a result, the primary decay pathway of the He⁻ $2s2p^2\ ^4P^e$ state was expected to be single Auger decay to the He $1s2p\ ^3P^o$ state [15], with little probability for He⁺ production. Since present ion-beam experiments cannot measure neutral products, the weak He⁺ production channel and narrow width of the state (≈10 meV [1, 14, 15]) makes detection of the transition very challenging, and initial experiments [16] failed to observe the state.

A recent study by our group [1] obtained sufficient signal and resolution to observe He⁺ produced by the decay of the He⁻ $2s2p^2\ ^4P^e$ state following photoexcitation from He⁻ $1s2s2p\ ^4P^o$ (see Fig. 1). This represents the first observation of a simultaneous DA decay in a negative ion. Processes in other light systems indicate that DA decay is, in general, expected to be very weak (≪1 Mb [17]), and the measured 67 Mb peak cross section (after correction for the finite spectral bandwidth) is unprecedented. While this large cross section is mainly due to the large photoexcitation cross section, the observed

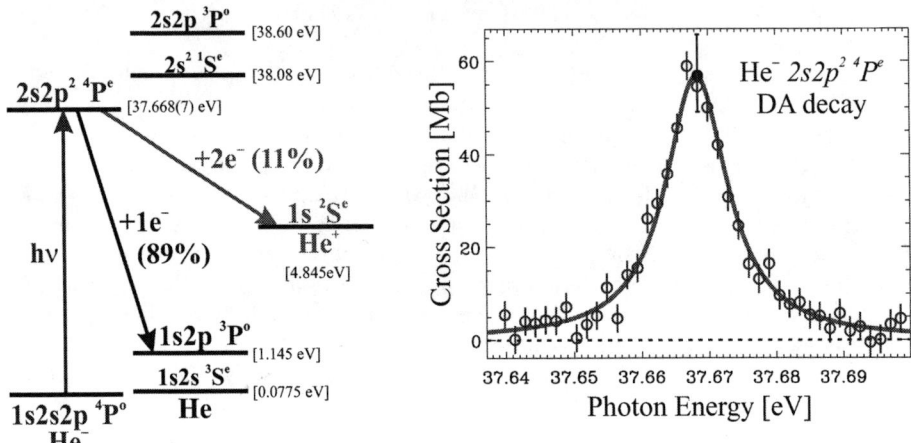

FIGURE 1. Partial energy level diagram (left panel) depicting the He$^-$ $2s2p^2$ $^4P^e$ decay pathways. The absence of He and He$^+$ states between the He$^+$ ground state and the He$^-$ $2s2p^2$ $^4P^e$ state requires that all He$^+$ production result from simultaneous DA decay to He$^+$ $1s\,^2S^e_{1/2}$. The measured He$^+$ signal is shown in the right panel [1]. The filled circle is an absolute cross section measurement. The curve is a Lorentzian convoluted with the spectral bandwidth (6 meV).

11% double-to-single Auger rate is also uncharacteristically large, given that values in other light systems are less than about 5% [17].

In addition to offering a tangible target for future theory on DA decay, the study also provides the first measurement of the line center and width of the $2s2p^2$ $^4P^e$ state. High resolution and high statistics measurements of the $2p3s3p$ $^4P^e$ and $2p3s3p$ $^4D^e$ were also obtained, and allowed for the determination of the linecenters, widths, and excitation cross sections for these states. In all cases, theory was found to be in qualitative agreement, but some significant deviations of the absolute cross sections and line positions were found [1].

CORE-EXCITED RESONANCE IN SULFUR NEGATIVE IONS

The signal observed in the region near the $2s$ detachment threshold of S$^-$ is shown in Fig. 2. Excitation of the $2s$ electron into the $3p$ shell (producing the $2s^{-1}3s^23p^6$ $^2S_{1/2}$ state) results in a clear resonance bound just below the $2s$ threshold, and indicates that filling the $3p$ shell stabilizes the orbital. Similar stabilization of valence p-shells has also been observed in He$^-$ [1] with $1s \rightarrow 2p$ excitation (see above) and in Te$^-$ [18] with $4d \rightarrow 5p$ excitation, and suggests that this is likely a general phenomenon. However, in these previous cases, the resonance formed was only observed in the single-charged product (total of 2 electrons removed) and did not produce any detectable higher charge-states. In the case of S$^-$, the resonance clearly couples strongly in all the observed charge states (up to 4 electrons removed in total). Fano profiles [19] fit to the S$^+$, S^{2+}, and S^{3+} production data yield a linecenter of 222.27(9) eV and a width of 1.44(4) eV (weighted

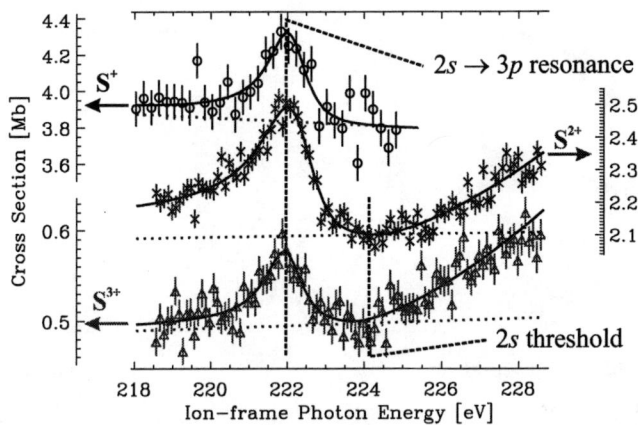

FIGURE 2. S$^+$ (○, top left scale), S^{2+} (×, right scale), and S^{3+} (△, bottom left scale) production following photodetachment and excitation at photon energies near the 2s threshold. The $2s \to 3p$ Feshbach resonance is observed near 222 eV, just below the 2s threshold. The solid curves are the Fano profile plus p-wave Wigner threshold law fits. The dotted lines indicate the (non-interfering) 2p detachment level.

average of the fits to the three channels).

It is important to note that the energy difference between the 3s and 3p valence shells is only ≈9 eV, while the ionization potential of S$^+$ is 23.34 eV [13]. Therefore, while two (sequential) Auger decay processes from $2s^{-1}3s^23p^6$ $^2S_{1/2}$ (e.g., to form S$^+$ $3s^03p^6$) could easily account for S$^+$ production, further Auger decay to higher charge states is not possible. The decay must instead proceed via higher order processes such as simultaneous multi-Auger decay and/or Auger plus shake-up processes. Since the larger charged products would presumably involve a larger number of electrons, the very similar cross sections observed for the resonance in the three channels (≈0.5, 0.4, and 0.08 Mb respectively), suggests that the highly correlated nature of the S$^-$ $2s^{-1}3s^23p^6$ $^2S_{1/2}$ state results in little preference over the number of electrons ejected during its decay.

NEAR-THRESHOLD PHOTODETACHMENT CROSS SECTIONS

The short-range potential that binds negative ions ($\propto r^{-4}$) gives rise to a near-threshold photodetachment cross section behavior significantly different than the behavior for photoionization of atoms and positive ions, which are bound in the long-range Coulomb potential ($\propto r^{-1}$). In negative ions, the near-threshold cross section follows the Wigner threshold law [20]: $\sigma = \sigma_\circ(h\nu - \varepsilon_t)^{\ell+1/2}$, with amplitude $\sigma_\circ$, threshold energy ε_t, and photoelectron angular momentum ℓ. This threshold law has been verified in both single- [21] and multi-photon [22] processes for the removal of valence electrons, and very recently has been shown to also apply to inner-shell photodetachment [23] despite the significant post-collision interaction (PCI) effects resulting from the rapid Auger decay.

Detachment of a 2s electron in S$^-$ produces the p-wave Wigner law ($\ell = 1$) seen in Fig. 2, with a threshold energy of 224.6(5) eV. Subtracting the $3s^23p^5$ $^2P_{3/2}$ binding

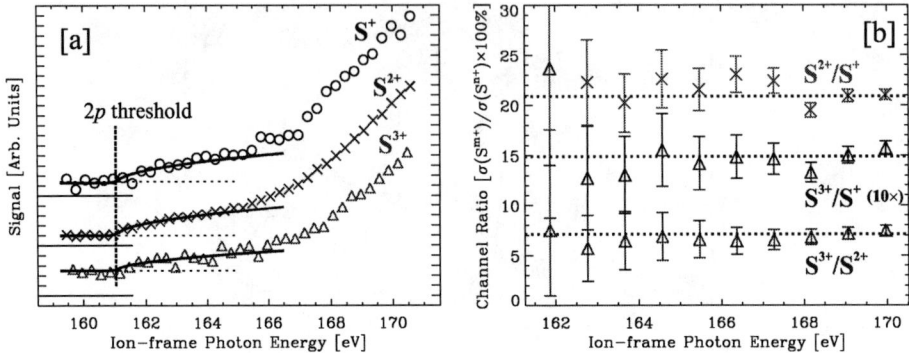

FIGURE 3. The left panel [a] shows the signal observed from S^+ ($\circ$), S^{2+} ($\times$), and S^{3+} ($\triangle$) near the $2p$ threshold. The curves are Wigner s-wave threshold laws ($\varepsilon_t = 161.04(6)$ eV [23]). Dotted lines indicate the below-threshold signal levels. The signals have been arbitrary scaled (and shifted by an amount indicated by the solid horizontal lines) in order to show that, over this range, the signals from all channels have nearly identical shapes. The right panel [b] shows the ratio of the $2p$ photodetachment strength between the channels for above-threshold photon energies. The $\sigma(S^{3+})/\sigma(S^+)$ ratio has been magnified by a factor of 10. The horizontal dotted lines are the weighted averages of data points. Note that for presentation purposes, 3 (for [a]) and 9 (for [b]) raw data points have been binned for each presented point.

energy (2.0771029(10) eV [24]), we obtain the S $2s^{-1}3s^23p^5$ core-excited state energy of 222.5(6) eV. Similarly, an s-wave law ($\ell = 0$) results from $2p$ detachment (Fig. 3a), the threshold energy (161.04(6) eV) of which allows us to obtain the S $2p^{-1}3s^23p^5$ core-excited state (158.76(6) eV) [23]. Due to the extreme difficulty in producing free S atoms, these basic core-excited states of neutral S had not been measured previously.

The S^+, S^{2+}, and S^{3+} production cross sections near the $2p$ photodetachment threshold is observed to have essentially identical shapes for each channel. This can be seen more quantitatively in Fig. 3b, which shows the ratio of the $2p$ photodetachment channel strengths $[\sigma(S^{m+})/\sigma(S^{n+})]$ for photon energies above-threshold. (Note that the non-$2p$ background signal, estimated using the below-threshold signal levels, was removed from each channel). The ratios are very nearly constant over this entire range and suggest that all charge states produced near the threshold result from the decay of the initial $2p$ photodetachment process. However, subsequent to the photodetachment, sequential Auger processes required to form S^{2+} or S^{3+} from S $2p^{-1}3s^22p^5$ are not energetically possible, indicating that here also, the decay involves the simultaneous ejection of more than one electron such as a simultaneous multi-Auger process.

SUMMARY

We have observed a strong Feshbach resonance with the excitation of an inner-shell s electron into the partly-filled valence p orbital of both He^- and S^-. Decay of the $2s2p^2\ ^4P^e$ state in He^- proceeds via a simultaneous DA process. Higher-order Auger decay processes are also likely to be important in the decay of the $S^-\ 2s^{-1}3s^23p^6\ ^2S_{1/2}$ state, which is observed in the S^+, S^{2+}, and S^{3+} product channels. We note that recent investigations in Si$^-$ [25] show a similar resonance with the formation of the $2s^{-1}3s^23p^4$ state, and suggest that stabilization of the valence p orbital in this fashion may occur in many negative ions [18]. Finally, from the inner-shell thresholds of S$^-$ it was possible to obtain core-excited $2s^{-1}3s^23p^5$ and $2p^{-1}3s^23p^5$ state energies of S. The production of high charge states (up to +3 was observed) indicates that many-electron processes, such as multi-Auger decay, are also important in the decay of these states.

ACKNOWLEDGMENTS

We thank B.S. Rude for his assistance during the experiments. This work was supported by DoE, Office of Science, BES, Chemical, Geoscience and Biological Divisions. The ALS is funded by DoE, Scientific User Facilities Division. This material is based in part on work supported by the National Science Foundation under Grant No. 0140233.

REFERENCES

1. R.C. Bilodeau, J.D. Bozek, A. Aguilar, G.D. Ackerman, G. Turri, and N. Berrah, Phys. Rev. Lett. **93**, 193001 (2004).
2. R.C. Bilodeau *et al.*, (to be published).
3. A.M. Covington *et al.*, Phys. Rev. A **66**, 062710 (2002).
4. J.A. Billen, IEEE Trans. Nucl. Sci. **28**, 1535 (1981).
5. C.E. Kuyatt, J.A. Simpson, and S.R. Mielczarek, Phys. Rev. **138**, A385 (1965); U. Fano and J.W. Cooper, Phys. Rev. **138**, A400 (1965).
6. J.J. Quéméner, C. Paquet, and P. Marmet, Phys. Rev. A **4**, 494 (1971).
7. D.S. Kim, H.L. Zhou, and S.T. Manson, J. Phys. B **30**, L1 (1997); *ibid.*, Phys. Rev. A **55**, 414 (1997).
8. K.T. Chung, Phys. Rev. A **51**, 844 (1995).
9. K.T. Chung, Phys. Rev. A **20**, 724 (1979).
10. D. Cubaynes *et al.*, Phys. Rev. Lett. **77**, 2194 (1996).
11. K.T. Chung, Phys. Rev. A **25**, 1596 (1982); M. Rødbro, R. Bruch, and P. Bisgaard, J. Phys. B **12**, 2413 (1979).
12. P. Kristensen, U.V. Pedersen, V.V. Petrunin, T. Andersen, and K.T. Chung, Phys. Rev. A **55**, 978 (1997).
13. Yu. Ralchenko *et al.*, NIST Atomic Spectra Database [Online: http://physics.nist.gov/asd3] (2005).
14. J. Xi and C.F. Fischer, Phys. Rev. A **59**, 307 (1999); J.L. Sanz-Vicario, E. Lindroth, and N. Brandefelt, Phys. Rev. A **66**, 052713 (2002); and references therein.
15. O. Zatsarinny, T.W. Gorczyca, and C. Froese Fischer, J. Phys. B **35**, 4161 (2002).
16. N. Berrah *et al.*, Phys. Rev. Lett. **88**, 093001 (2002).
17. M.S. Pindzola and D.C. Griffin, Phys. Rev. A **36**, 2628 (1987); R.L. Simons and H.P. Kelly, Phys. Rev. A **22**, 625 (1980); R. Wehlitz, M.T. Huang, K.A. Berrington, S. Nakazaki, and Y. Azuma, Phys. Rev. A **60**, R17 (1999).
18. H. Kjeldsen, F. Folkmann, T.S. Jacobsen, J.B. West, Phys. Rev. A **69**, 050501(R) (2004).
19. U. Fano and J.W. Cooper, Phys. Rev. **137**, A1364 (1965).
20. E.P. Wigner, Phys. Rev. **73**, 1002 (1948).
21. T. Andersen, Phys. Rep. **394**, 157 (2004); H.R. Sadeghpour *et al.*, J. Phys. B **33**, R93 (2000).
22. R.C. Bilodeau, M. Scheer, and H.K. Haugen, Phys. Rev. Lett. **87**, 143001 (2001).
23. R.C. Bilodeau *et al.*, Phys. Rev. Lett. (in press).
24. T. Andersen, H.K. Haugen, and H. Hotop, J. Phys. Chem. Ref. Data **28**, 1511 (1999).
25. R.C. Bilodeau, J.D. Bozek, G.D. Ackerman, G. Turri, and N. Berrah, (to be published).

Multiple direct and sequential Auger effect in the rare gases

F. Penent[1], P. Lablanquie[1], J. Palaudoux[1], L. Andric[1], T. Aoto[2], K. Ito[2], Y. Hikosaka[3], R. Feifel[4] and J.H.D. Eland[4]

1) LCP-MR, CNRS et UPMC, 11, rue P. & M. Curie, 75231 Paris, France
2) Photon Factory, IMSS, KEK, Tsukuba 305-0801, Japan
3) IMS, Okazaki 444-8585, Japan
4) PTCL, Oxford University, Oxford, United Kingdom

Abstract. The use of a magnetic bottle spectrometer [1] with synchrotron radiation allows multi dimensional electron spectroscopy to be performed by detecting in coincidence all electrons (2, 3, 4) ejected in multiple ionization events. Multiple Auger effect following inner-shell ionization can be investigated in this way. Application of the technique to rare gases (Xe $4d$ and Kr $3d$) double Auger decay reveals all the energy pathways involved. The dominant decay path proceeds by Auger cascade through autoionizing states of the doubly charged ion. Processes where 3 electrons are involved are also observed as direct double Auger and as involving precursor Rydberg series.

Keywords: photoionization, inner-shell ionization, Auger electron spectra, rare gases, magnetic bottle, electron time of flight spectra.
PACS: 32.80.Hd, 33.60.Cv, 39.90.+d

INTRODUCTION

Photoelectron spectroscopy (PES) and Auger Electron Spectroscopy (AES) are powerful analysis tools in many fields of science. Following inner-shell ionization, relaxation by Auger electron emission is an important process which gives a clear signature of the system.

As far as deep inner shells are concerned, cascade Auger decay leading to multiply charged ions is a common process in competition with X-ray radiative decay. However such a process is also possible for shallow inner-shells since the energy of the excited ion (with inner-shell vacancy) can be higher than triple (or even higher) ionization limits. This is, for instance, the case of Kr(3d) and Xe(4d) ionization thresholds which lie higher than triple ionization thresholds of Kr and Xe. It is hence possible to observe emission of two electrons instead of only one in "normal" Auger decay. This can proceed in two ways: 1- a sequential one, where the first Auger emission leaves the intermediate doubly charged ion in an autoionizing state which releases a second electron, 2- a direct process [1] where two electrons are ejected simultaneously and can share continuously the available energy. Such processes, although less important than "normal" Auger decay, are however not negligible but are difficult to study with conventional electron spectroscopy (using electrostatic analyzers) since they give electron peaks at lower energies but also continuous energy distributions difficult to separate from background. The proper way to study such processes is to detect the electrons in coincidence [2][3]. However, conventional coincidence techniques are mainly limited to the detection of two electrons out of three (or more) with low efficiencies due to limited angular acceptance, and a complete picture of the process is difficult to obtain. Complementary results are provided by electron/ion coincidences with analysis of the final charge of the ion [4] but the identification of decay processes remains difficult. Only a complete coincidence experiment where **all** electrons are detected can provide the necessary basis to fully disentangle all the processes following inner-shell ionization but also (direct) multiple ionization.

EXPERIMENT

In order to obtain a complete image of multiple ionization events, we have developed a synchrotron based experiment using a long time of flight (TOF) magnetic bottle electron spectrometer that allows coincident detection of the photoelectron with **all** (1, 2, 3...) subsequent Auger electrons and, more generally, of all electrons in multiple photoionization events. Such an apparatus was initially developed by J.H.D. Eland [5] with pulsed VUV lamps and has been implemented on synchrotron radiation sources 12 years after pioneering experiments with limited resolution [6]. Two independent (but similar) set-ups have been used: a first one in France at the Super ACO (SACO), (Orsay) and a second in Japan at the Photon Factory (PF), (Tsukuba). The single bunch mode, with a pulse period T = 120 ns at SACO and T = 624 ns at PF, was necessary for electron TOF measurements. The length of the TOF magnetic bottle was about 2.5 m to fit into the synchrotron environment while allowing good energy resolution. The strongly divergent magnetic field forming the bottleneck of the magnetic bottle is produced by a conical shaped strong permanent magnet (FeNdB) giving a magnetic field of about 0.5 T at 1 mm from the tip of the magnet. Electron trajectories are parallelized [7] after a few centimeters and the electrons are then guided by the field (0.5 to 1 mT) of the long solenoid towards a multi-hit detector (micro-channel plates (MCP) with a delay line position encoding detector or phosphor screen). The electrons' times of flight are measured with respect to the light pulse by a time to digital converter (TDC) with 250-500 ps time resolution. This experimental set-up provides almost 4π collection efficiency (>90%), very good energy resolution (down to 10~20 meV below 1 eV energy) and constant transmission from zero to at least 70 eV. The dead time of the detector only forbids detection of electrons with close energies. The only lost information concerns the initial angular distribution of the electrons which cannot be reconstructed from simulation of electron trajectories. Because of the repetition period of light pulses, the electron time of flight is only defined modulo [T] for the first electron. However, once a first electron is detected the time of flight of following electrons from the same multiple ionization event is not restricted to one period and a complete time coincidence map can be constructed from which energy correlations diagrams are built. A repelling potential is applied to the magnet in order to allow low energy electrons (down to zero eV) to arrive in a finite time (<8 µs). The potential of the needle, through which the target gas is injected to intersect the photon beam, is adjusted for the best energy resolution. Time to energy conversion is performed after calibration using photoelectrons of known energy. Although time of flight is only defined modulo [T], the absolute TOF of a photoelectron of known energy is easily derived by varying the energy step by step. Time to energy conversion is given by $T - T_0 = A/(E - E_0)^{1/2}$ but corrective terms (resulting from uncontrolled electric potentials) may be added to improve the fit quality. Additional calibration points are given by known Auger transitions in the rare gases allowing a self-consistent time to energy calibration.

EXPERIMENTAL RESULTS

First experimental results have been obtained for shallow inner-shell ionization of rare gas atoms: Xe(4d) and Kr(3d). Coincident detection of photoelectrons with all subsequent Auger electrons offers unprecedented possibilities.

The first interest of the method is to provide full disentanglement of complex Auger spectra resulting from the overlap of Auger lines that originate from different inner-shells ($d_{5/2, 3/2}$). Determination of the branching ratio follows immediately. For the sake of simplicity in data analysis, the photon energy was chosen (110 eV for Xe) in order that 4d photoelectrons are faster than any following Auger electrons. The absolute electron detection efficiency (about 40%) is derived from the ratio between the count rates for photoelectrons detected in coincidence with Auger electrons (taking also into account double Auger processes) and without coincidence (aborted coincidences). We verified that the transmission of the spectrometer is, within 5% uncertainty, independent of electron energy from 0 to at least 70 eV by recording successively He(2p) photoelectrons and He^+ ions signals (the ion detector is located close to the interaction region and the magnet was moved back to avoid trapping of ions in B-field).

A two dimensional (T_2 vs T_1) time coincidence map is published elsewhere for Xe(4d) photoionization [3], and the Auger electrons in coincidence with the photoelectron are easily and clearly observed. By filtering the photoelectron time of flight for $4d_{5/2, 3/2}$ sub-shells, the separation of N_4-OO, and N_5-OO lines is straightforward and it becomes very easy to give the branching ratio of some overlapping structures hidden in normal Auger spectra [9]. Such filtered spectra are presented in figure 1 with peak numbering from ref. [9] for the prominent lines.

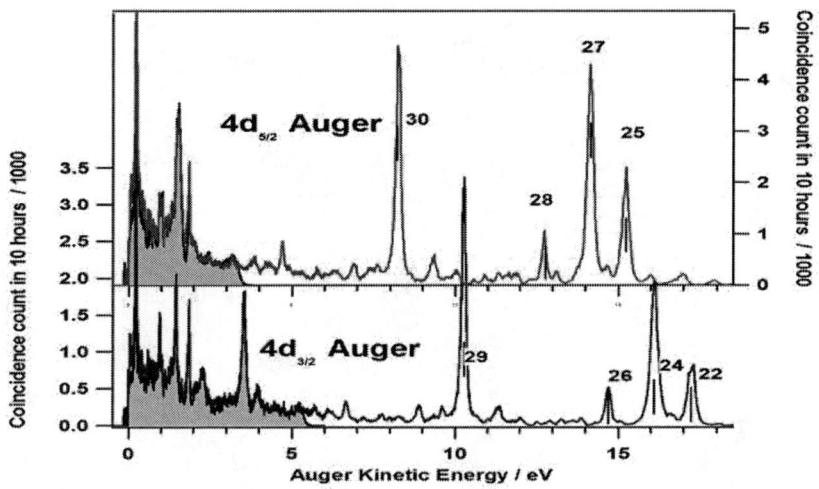

FIGURE 1: N-OO Auger spectra filtered against N_4 and N_5 sub-shells

Up to now, no clear observation has been made for the low energy region of the spectra (E<5.4 eV for N_4-OO, E<3.4 eV for N_5-OO transitions). Because we can detect more than 2 electrons in coincidence with a high efficiency, we have a clear interpretation of this low energy region. By selecting the events where three electrons are detected in coincidence, we can plot the energy of the second or third electron following the selected $4d_{3/2}$ or $4d_{5/2}$ photoelectron and we can scale this signal (taking into account the electron detection efficiency) to the signal where only two electrons are detected. The result is the grey area on figure 1, which corresponds to the emission of two electrons following the photoelectron. The threshold for this process corresponds to the Xe triple ionization. It shows that once an intermediate state of Xe^{++} is formed above triple ionization threshold it decays by emitting a second Auger electron. Radiative decay of Xe^{++*} could only be a very minor process. Direct double Auger could also contribute to this process. The grey area in fig.1 is hence due to N-OOO transitions.

A much deeper analysis can be done on 2D energy maps by plotting the energy of the second and third electron in coincidence after filtering the time of flight of the (faster) photoelectron to select the $4d_{3/2}$ or $4d_{5/2}$ sub-shell. A detailed analysis of this process is published elsewhere [8]. Another way to plot such data is given in figure 2. The vertical axis gives the sum of the energies ($E_1 + E_2$) of the two Auger electrons, while projection on the diagonal gives the energy of the faster Auger electron.

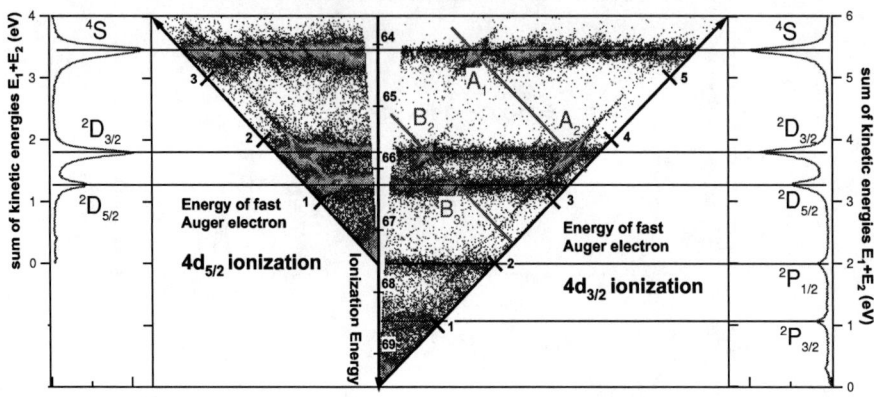

FIGURE 2: Energy map of the double Auger decay following Xe(4d) inner shell ionization. The horizontal lines correspond to the final Xe^{3+} states ($^4S_{1/2}$, $^2D_{5/2}$, $^2D_{3/2}$, $^2P_{3/2}$, $^2P_{1/2}$). The $^2P_{3/2}$, $^2P_{1/2}$ states cannot be accessed after 4d$_{5/2}$ ionization. The projection on E_1+E_2 gives the energy of Xe^{3+} states.

The sum of the energies of the two Auger electrons gives the energy of the final state of Xe^{3+} below the respective 4d$_{3/2,5/2}$ ionization thresholds. Each horizontal line corresponds to a given state of Xe^{3+}, so the data give the spectrum of triply charged Xe directly. The threshold for Xe^{3+} $^4S_{1/2}$ is found at 64.09±04 eV. The branching ratio towards Xe^{3+} final states is not statistical [8].

Much more information is given by the energy distribution along the horizontal lines for a given final state. This distribution is not smooth, and prominent peaks are visible which are characteristic of cascade decays. Although events are characterized by electron pairs (E_1, E_2) that leave two possibilities for the position of the intermediate state, it remains easy to distinguish a given state as it appears on different lines in figure 2 due to two different inner shells and up to five final states. Two intermediate states at 66.00 and 67.27 eV from the Xe ground state are most heavily involved in the cascade Auger process, contributing about a third of the triple ionization signal; they appear as different spots (as for instance A and B) on the diagram depending on the initial hole and on the final state. Although there is a lack of calculations for such high energy states and processes, they are probably high lying 5s^{-2} correlation satellites [8] and result mainly from a two electron process where a 5s electron fills the inner vacancy while another is ejected. These states autoionize to the final Xe^{3+} states. The electron peak corresponding to this last step has a natural width (resulting from lifetime) smaller than 28meV (which can be measured when the instrumental energy resolution becomes good enough, i.e. for low energy electrons) much lower than the natural width of the first step of about 100meV due to the Xe(4d^{-1}) lifetime. This gives also a simple way to disentangle the different steps in the Auger cascade. In addition to these two prominent states a Rydberg series 5s^{2}5p^{3}nl converging to $^2D_{3/2}$ threshold and decaying to $^4S_{1/2}$, is found. Such states are precursors of the direct double Auger process and suggest three electron processes. No clear structure is observed for the decay to the highest Xe^{3+} $^2P_{1/2,3/2}$ states from the 4d$_{3/2}$ hole and a direct double Auger decay with any sharing of the available energy between the two electrons seems possible. The study of angular correlation between the two Auger electrons with an adequate experimental set-up should clarify this [10].

Similar observations have been made for Kr(3d) inner-shell ionization and will be published in a forthcoming paper; they mostly reveal cascade Auger decay. Since the available energy for the double Auger decay is in this case about 20eV, it is also possible to form Kr^{3+} $4s4p^4$ final states in addition to the five $4p^3$ states (as observed in Xe). One important observation is that the Kr triple ionization potential is found to be more than one eV lower than the value listed in atomic data tables [11]. This result alone demonstrates the utility of our method for spectroscopy of multiply charged ions. It was also possible to observe quadruple ionization in Xenon following "4p" inner-shell ionization by detecting 4 electrons in coincidence and summing up their energies. In this case, the cascade process is more difficult to follow since for a triplet of energies (E_1, E_2, E_3) six combinations of energy levels are possible. Nevertheless, if the same Xe^{3+} intermediate state decays towards different final $5p^2$ Xe^{4+} states its position can be determined. It should be possible in this way to find the main cascade decay channels from the three dimensional energy maps. This analysis is in progress.

ACKNOWLEDGMENTS

Our acknowledgements go to LURE and Photon Factory synchrotron laboratories for providing single bunch mode to perform these experiments.

REFERENCES

1 T. A. Carlson and M.O. Krause, *Phys. Rev. Lett.* **14**, 390 (1965).
2 J.Viefhaus et al, *Phys. Rev. Lett.*, **92**, 083001, (2004); Lablanquie et al, *Phys. Rev. Lett.*, **87**, 053001, (2001)
3 F. Penent et al, *J. El. Spect. Rel. Phen.*, **144–147**, 7 (2005)
4 B. Kammerling, B. Krassig and V. Schmidt, *J. Phys. B*, **25**, 3621 (1992)
5 J. H. D Eland et al,. *Phys. Rev. Lett.* **90**, 053003-1 (2003); J.H.D. Eland et al. *Chem. Phys.* **290**, 27 (2003)
6 K. Okuyama et al, *Phys.Rev. A* **41**, 4930 (1991)
7 P. Kruit and F. H. Read, *J. Phys. E: Sci. Instrum.* **16**, 313 (1983).
8 F. Penent et al, *Phys. Rev. Lett.*, in press [2005]
9 T.X. Carroll et al, *J. El. Spect. Rel. Phen.*, **125**, 127 (2002)
10 J. Viefhaus et al, *private communication* (2005)
11 J. Sugar and .A Musgrove, *J. Phys. Chem. Ref. Data* **20**, 859 (1991).

First principles calculations of the double photoionization of atoms and molecules using B-splines and Exterior Complex Scaling

F. Martín[*], D. A. Horner[†], W. Vanroose[**], T. N. Rescigno[†] and C.W. McCurdy[†,‡]

[*]*Departamento de Química C-9, Universidad Autónoma de Madrid, 28049 Madrid, Spain*
[†]*Lawrence Berkeley National Laboratory, Chemical Sciences, Berkeley, CA 94720*
[**]*Departement Computerwetenschappen, Katholieke Universiteit Leuven, Belgium*
[‡]*Departments of Applied Science and Chemistry, University of California, Davis, CA 95616*

Abstract. We report a fully ab initio implementation of exterior complex scaling in B-splines to evaluate total, singly and triply differential cross sections in double photoionization problems. Results for He and H_2 double photoionization are presented and compared with experiment.

Keywords: Double photoionization, B-splines, Exterior Complex scaling
PACS: 33.80.Eh, 32.80.Fb, 34.10.+x

INTRODUCTION

Exterior complex scaling (ECS) has been shown to be a powerful method to solve atomic and molecular scattering problems, because it allows the correct imposition of continuum boundary conditions without their explicit analytic application [1]. In ECS, the coordinates are scaled outside a fixed radius,

$$r \to \begin{cases} r & r \leq R_0 \\ R_0 + (r-R_0)e^{i\eta} & r > R_0 \end{cases} \quad (1)$$

where R_0 defines the radius within which the wave function will be the usual function of real-valued coordinates, and η is a scaling angle. In an exact or converged calculation the solutions of the Schrödinger equation for $r < R_0$ do not depend on η, because exterior complex scaling provides the exact solution with the coordinates taken on a particular complex contour [2]. Setting $\eta \neq 0$ while imposing the boundary condition that *on the ECS contour* the wave function vanishes as $r \to \infty$ effectively imposes outgoing scattering boundary conditions on the exact solution, and that is why the ECS approach provides a path to compute collision amplitudes. It is this property that makes ECS particularly useful in cases where the asymptotic boundary conditions make traditional methods difficult to apply.

The B-spline method has been extensively applied to atomic and molecular photoionization problems [3, 4, 5]. An important property of B-splines is that they are able to span a large volume to any degree of accuracy without encountering the numerical problems that prevent the use of exponentially decreasing basis functions. This is crucial for the description of continuum states, especially when the asymptotic region is needed. In

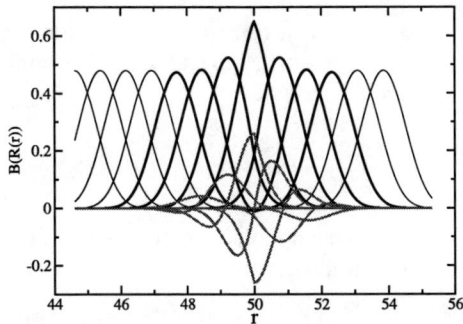

FIGURE 1. 8th order B-splines on the complex exterior scaling contour with $R_0 = 50$ and $\eta = 40^0$. Heavy black lines are the real parts of the only complex splines. Grey lines are the imaginary parts.

addition, B-spline basis sets are effectively complete, which is an ideal property in those problems where the entire spectrum is needed [3]. The double ionization continuum lies in this category.

In this paper we report an implementation of ECS with B-splines [6] to evaluate total, singly and triply differential cross sections in double photoionization problems. The method has been successfully applied to study double photoionization of He and H_2 [7, 8, 9]. A few results are presented and compared with experiment.

THEORY

A complex B-spline basis. B-splines that scale according to the ECS transformation are defined by setting a series of knots $t_i \leq t_{i+1}$ on the complex contour and by using the usual recursion relation [10] for B-splines of order k,

$$B_i^k(r) = \frac{r - t_i}{t_{i+k-1} - t_i} B_i^{k-1}(r) + \frac{t_{i+k} - r}{t_{i+k} - t_{i+1}} B_{i+1}^{k-1}(r) \quad (2)$$

together with the definition of B-splines of order $k = 1$

$$B_i^1(r) = \begin{cases} 1 & \text{for } t_i \leq r < t_{i+1} \\ 0 & \text{otherwise} \end{cases} \quad (3)$$

A basis of B-splines is defined by a grid of breakpoints, ξ_i, coinciding with the knots, t_i (which may be multiple), that appear in the recursion relation above. The breakpoints can be placed arbitrarily on this contour but one of them and its corresponding knot must be placed at $t_i = R_0$. In this way, B_i^k has a discontinuous first derivative with respect to r at $r = R_0$, because the derivative of the contour itself is discontinuous at that point. The discontinuity in the first derivative of all the B-splines that span the point R_0 is essential to reproduce that of the exact wave function. Figure 1 shows a typical B-spline basis of order $k = 8$ and the discontinuities of the first derivatives at $r = R_0$. Only B-splines that straddle the point R_0 have both real and imaginary components. All other B-splines are real, whether they are on the complex part of the contour or not.

Transition amplitudes. The triply differential cross section (TDCS) for an atom or molecule to absorb one photon, of frequency ω and for two electrons, one having energy E_1, to emerge into solid angles $d\Omega_1$ and $d\Omega_2$ is

$$\frac{d\sigma}{dE_1 d\Omega_1 d\Omega_2} = \frac{4\pi^2}{\omega c} k_1 k_2 |f(\mathbf{k}_1, \mathbf{k}_2)|^2. \tag{4}$$

The amplitude, $f(\mathbf{k}_1, \mathbf{k}_2)$, is associated with the purely outgoing wave function Ψ_{sc}^+ that is the solution of the driven Schrödinger equation for the "first order wave function," which we can write in the velocity form as,

$$(E_0 + \omega - H)|\Psi_{sc}^+\rangle = \varepsilon \cdot (\nabla_1 + \nabla_2)|\Psi_0\rangle, \tag{5}$$

where ε is the polarization unit vector, ∇_1 and ∇_2 are the gradient operators for the electronic coordinates, and Ψ_0 is the initial bound state of the system. The amplitude for double ionization corresponding to Ψ_{sc}^+ can be evaluated, aside from an irrelevant overall phase discussed elsewhere [1, 7], from the integral expression,

$$f(\mathbf{k}_1, \mathbf{k}_2) = \langle \Phi^{(-)}(\mathbf{k}_1, \mathbf{r}_1) \Phi^{(-)}(\mathbf{k}_2, \mathbf{r}_2)| [E - T - v(r_1) - v(r_2)] |\Psi_{sc}^+(\mathbf{r}_1, \mathbf{r}_2)\rangle, \tag{6}$$

where E is the excess energy above the double ionization threshold, T is the two-electron kinetic energy operator, and $v(r)$ is the nuclear attraction potential seen by one electron in the field of the bare nuclei.

For a He atom, the functions $\Phi^{(-)}(\mathbf{k}, \mathbf{r})$ would be the standard atomic Coulomb wave functions, but in the case of H_2, in the Born-Oppenheimer approximation where the two electrons leave behind two bare protons, positioned at $\pm \mathbf{A}$, they are the continuum states of the H_2^+ ion. In the latter case, they are the solutions of

$$\left[\frac{k^2}{2} + \frac{\nabla^2}{2} + \frac{1}{|\mathbf{r} - \mathbf{A}|} + \frac{1}{|\mathbf{r} + \mathbf{A}|} \right] \Phi^{(-)}(\mathbf{k}, \mathbf{r}) = 0, \tag{7}$$

and satisfy the usual relation, $\Phi^{(-)}(\mathbf{k}, \mathbf{r}) = (\Phi^{(+)}(-\mathbf{k}, \mathbf{r}))^*$.

Wave functions. To solve Eq. (5) with the proper outgoing scattering boundary conditions, we use the ECS version of the B-spline basis. The two-electron continuum function, Ψ_{sc}^+, in Eq. (5) is written, for a fixed value of L (atomic case) or M (molecular case) as a sum of products of two-dimensional radial wave functions and spherical harmonics. The radial functions are made of B-splines. The ground state is also constructed using B-splines in the standard fashion [5]. This transforms Eq. (5) in a set of linear equations.

DOUBLE PHOTOIONIZATION OF HE

The components of Ψ_{sc}^+ reveal much of the dynamics of the photoionization process at a glance. The first two of them, the *kskp* and *kpkd* contributions, are plotted in Fig. 2 for a photon energy 20eV above the double ionization threshold. These plots show only the direct contribution and are thus not symmetric under interchange of r_1 and

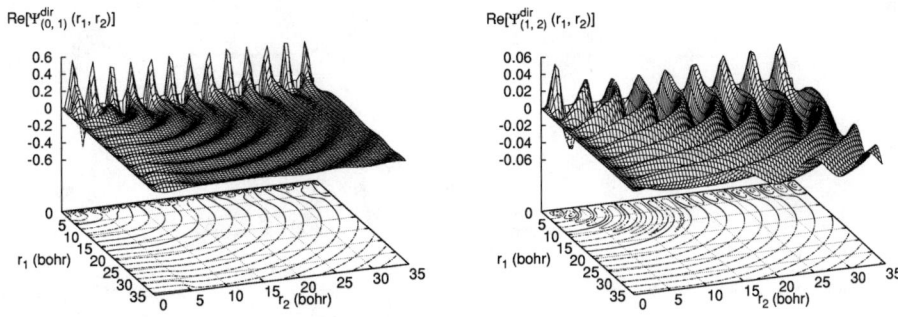

FIGURE 2. Real part of direct contribution to the He wave function at 20 eV. The panels from left to right show the contributions from the *kskp* and *kpkd* partial waves.

r_2. In the first of them we see the single ionization contribution as an outgoing wave parallel to the r_2 axis and confined to small r_1. The *kskp* contribution also displays the outgoing waves for double ionization as wave fronts at constant hyperradius. For the higher angular component the relative importance of single ionization decreases since it proceeds through higher ionization thresholds; thus the outgoing double ionization wave fronts are more apparent. As l_1 and l_2 increase the wave function components rapidly decrease in magnitude as can be seen in the *kpkd* contribution.

Braüning et al. [11] have measured absolute TDCS's for a photon energy of 20 eV above threshold. These experiments and essentially all others on this system were performed in "coplanar geometry", that is, with the polarization vector and both momenta $\mathbf{k}_1$ and $\mathbf{k}_2$ lying in the same plane. The measurements provide a rigorous test of the theoretical description of the double photoionization process. In Fig. 3 we compare the ECS results with the experimental ones for the cases in which the first electron exits parallel with an angle of $\theta_1 = 60°$ and $\theta_1 = 90°$ with respect to the polarization axis.

DOUBLE PHOTOIONIZATION OF H_2

In the case of a molecular target, the theoretical challenge is greater and potentially more interesting. Previous treatments have generally made use of a correlated initial state in combination with a final state that is simply an uncorrelated product of Coulomb wave functions or treated only in united-atom limit [12, 13]. Here, we treat both initial and final states on equal footing and take a completely non perturbative approach.

Fig. 4 compares length and velocity forms of the calculated integral cross section, which should be identical in a completely converged calculation, as they are in ECS calculations on double photoionization of helium [7, 8]. The remaining discrepancy here is likely due to an insufficient number of angular momentum values in the final state, Ψ_{sc}^+, which we treat with fewer partial waves than the initial state. Including higher angular momenta in a molecular calculation increases the size of the calculation faster

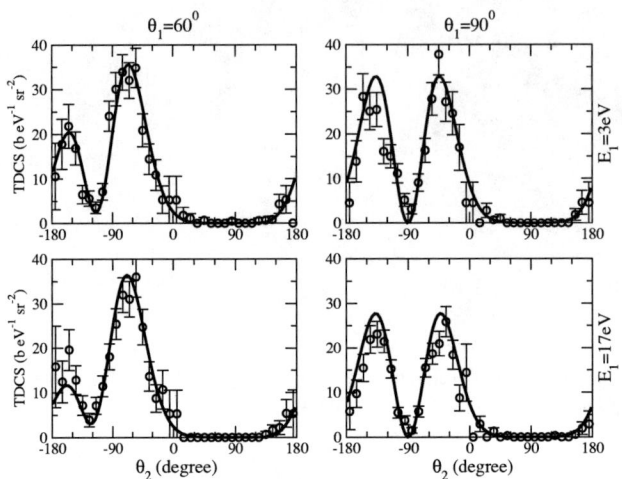

FIGURE 3. He TDCS for photon energy 20 eV above threshold, at energy sharings with $E_1 = 3$ eV (upper panels) and 17 eV (lower panels) for $\theta_1 = 60^0$ and $\theta_1 = 90^0$. Circles: experiment by Bräuning et al. [11]. Thick solid curve: Present result.

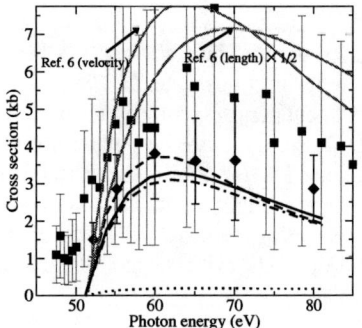

FIGURE 4. Integral cross section of H_2. Solid curve: velocity form. Dashed: length form. Dotted: $^1\Sigma_u^+$ contribution to velocity form. Dot-dashed: $^1\Pi_u$ contribution to velocity form. Grey lines, theory of ref. [12], length form (lower) times 1/2 and velocity form (upper). Experiments: squares, Dujardin et al. of 1987 [14] and diamonds, Kossmann et al. of 1989 [15].

than in an atomic calculation. The lower symmetry generates a larger number of distinct "double continua" for each pair of l_1, l_2 angular momenta since all m_1, m_2 adding up to M label distinct double continua. Fig. 4 shows a dramatic difference between double photoionization and either single photoionization of H_2 or double photoionization of He. The $^1\Pi_u$ contribution dominates by about a factor of 10 over that of the $^1\Sigma_u^+$ symmetry. Recent COLTRIMS experiments [13, 16] noted the dominance of the Π_u contribution and therefore of polarization perpendicular to the molecular axis.

In the same COLTRIMS experiments, the TDCS has been measured for *oriented* H_2 molecules. In Fig. 5, we present a comparison of calculated and measured TDCS

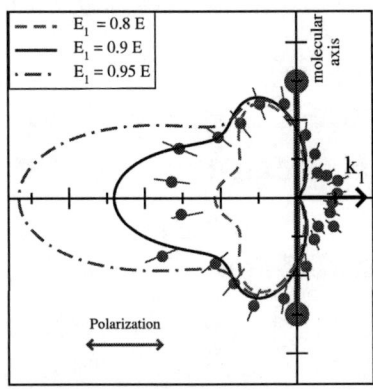

FIGURE 5. Measured and calculated H_2 TDCS for photon energy 75.5 eV an "in-plane" geometry. Experiment averages all $E_2 > 0.8E$. Circles: experiment by Weber *et al.* [13].

for the case in which two electrons exit in the plane defined by the axis of the H_2 molecule and the polarization direction. Three different energy sharings compatible with the experimental measurements are shown. The TDCS is very sensitive to the energy of each electron as well as to the description of correlation in the ground state of the molecule and the final continuum state. All in all, the figure shows that results of the calculations are compatible with the experimental measurements.

REFERENCES

1. C. W. McCurdy, M. Baertschy, and T. N. Rescigno, *J. Phys. B*, **37**, R137 (2004).
2. B. Simon, *Phys. Lett. A*, **71**, 211 (1979).
3. H. Bachau, E. Cormier, P. Decleva, J. E. Hansen, and F. Martín, *Rep. Prog. Phys.*, **64**, 1815–1942 (2001).
4. F. Martín, *J. Phys. B*, **32**, R197–R231 (1999).
5. I. Sánchez, and F. Martín, *J. Phys. B*, **30**, 679–692 (1997).
6. C. W. McCurdy, and F. Martín, *J. Phys. B*, **37**, 917 (2004).
7. C. W. McCurdy, D. A. Horner, T. N. Rescigno, and F. Martín, *Phys. Rev. A*, **69**, 032707 (2004).
8. D. A. Horner, J. Colgan, F. Martín, C. W. McCurdy, M. S. Pindzola, and T. N. Rescigno, *Phys. Rev. A*, **70**, 064701 (2004).
9. W. Vanroose, F. Martín, T. N. Rescigno, and C. W. McCurdy, *Phys. Rev. A*, **70**, 050703R (2004).
10. C. de Boor, *A Practical Guide to Splines*, Springer, New York, 1978.
11. M. Bräuning, R. Dörner, C. L. Cocke, M. H. Prior, B. Kriässig, A. S. Kheifets, I. Bray, A. Bräuning-Demian, K. Carnes, S. Dreuil, V. Mergel, P. Richard, J. Ulrich, and H. Schmidt-Böcking, *J. Phys. B: At. Mol. Opt. Phys.*, **31**, 5149 (1988).
12. H. L. Rouzo, *J. Phys. B*, **19**, L677 (1986).
13. T. Weber, A. O. Czasch, O. Jagutzki, A. K. Müller, V. Mergel, A. Kheifets, E. Rotenberg, G. Meigs, M. H. P. MH, S. Daveau, A. Landers, C. L. Cocke, T. Osipov, R. D. M. no, H. Schmidt-Böcking, and R. Dörner, *Nature*, **431**, 437 (2004).
14. G. Dujardin, M. J. Besnard, L. Hellner, and Y. Malinovitch, *Phys. Rev. A*, **35**, 5012 (1987).
15. H. Kossmann, O. Schwarzkopf, B. Kämmerling, and V. Schmidt, *Phys. Rev. Lett.*, **63**, 2040 (1989).
16. T. Weber, A. Czasch, O. Jagutzki, A. Müller, V. Mergel, A. Kheifets, J. Feagin, E. Rotenberg, G. Meigs, M. H. Prior, S. Daveau, A. L. Landers, C. L. Cocke, T. Osipov, H. Schmidt-Böcking, and R. Dörner, *Phys. Rev. Lett.*, **92**, 163001 (2004).

Auger Electron- Photoelectron Coincidence Experiments in Ar

P. Bolognesi[1], M. Coreno[1,2], A. De Fanis[3], V. Feyer[1,4], S. Turchini[5], N. Zema[5], T. Prosperi[5] and L. Avaldi[1,2]

[1] *CNR-Istituto di Metodologie Inorganiche e dei Plasmi, Area della Ricerca di Roma 1, CP10-00016 Monterotondo Scalo, Italy*
[2] *INFM-TASC, Gas Phase Beamline at Elettra, Area Science Park, Basovizza, Italy*
[3] *JASRI / SPring-8, Sayo-gun Hyogo 679-5198 Japan*
[4] *Institute of Electron Physics, National Academy of Sciences, Uzhgorod, Ukraine*
[5] *CNR-Istituto di Struttura della Materia, Via Del Fosso del Cavaliere 100, 10133 Roma, Italy*

Abstract. The photoionisation of the Ar 2p has been studied by Auger electron-photoelectron coincidence experiments using both linearly and circularly polarised radiation. The results show that the Ar 2p photoionisation can be described in LS coupling and that these coincidence experiments allow to obtain the basic quantities that define the photoionisation process.

Keywords: complete experiment, photoelectron-Auger electron coincidence, indirect photodouble ionisation, Argon $2p_{3/2}$ shell.
PACS: 32.80.Fb, 32.80.Hd

1. INTRODUCTION

In the Auger electron-photoelectron coincidence experiments

$$h\nu + A \rightarrow A^+(\text{inner-shell}^{-1}) + e_{ph} \rightarrow A^{2+} + e_{Auger} + e_{ph}$$

the inner shell photoionisation process and its non-radiative decay to the double continuum can be investigated in detail. The process can be regarded as, and indeed is, an indirect photodouble ionisation. When the intermediate state can be treated as a well-defined isolated resonance in the double ionisation continuum and the competing direct double ionisation neglected then the process can be considered a sequential one, consisting of two incoherent steps. In such a case these coincidence experiments are a very suitable tool to provide i) a better and unambiguous spectroscopic characterization of the emitted electron spectra [1] and ii) a deep insight into the two constituent processes [2]. Indeed they have been proposed as "complete experiments" [3] where the transition amplitudes and phases either of photoionisation or of Auger decay can be obtained experimentally. However, so far Auger electron-photoelectron coincidence data have been always combined [2, 4] to non coincidence data, like the measurement of asymmetry parameter of the angular distribution of the photoelectron and Auger electron to approach a "quantum mechanical complete experiment". Here we show that the amplitudes and phases for Ar 2p photoionisation can be obtained using only data from coincidence experiments.

2. EXPERIMENTAL

The Ar 2p photoionisation has been studied with both linearly and circularly polarized radiation at the Gas Phase and Circular Polarised beamlines, respectively, of the Elettra storage ring, Trieste using the multicoincidence end-station. A detailed description of the beamlines [5, 6], the experimental station and experimental procedures [7] have been given elsewhere, and will not be repeated here. The experiment has been performed in the perpendicular plane, at the photon energy $h\nu$=253.6 eV, i.e., 5 eV above the Ar L_3 threshold. The energy resolution was about 80 and 200 meV for the detection of the photoelectron and the Auger electron respectively. All the data are internormalised and therefore each set of the experimental coincidence angular correlations is reported on a common scale of counts.

The polarisation of the incident radiation was complete in the case of the measurements with linearly polarised radiation (S_1=1), while only partial in the case of measurements with circularly polarised radiation (S_3=0.80±0.05) [6].

3. RESULTS

The Ar $2p_{3/2}$ photoionisation and its following decay

$$h\nu + Ar \rightarrow Ar^+ 2p_{3/2}^5(^2P^o) + e_{ph}(\ell = 0,2)$$
$$\rightarrow Ar^{2+}3p^4(^3P^e) + e_{Auger}(\ell = 1) \quad (1.1)$$
$$\rightarrow Ar^{2+}3p^4(^1D^e) + e_{Auger}(\ell = 1,3) \quad (1.2)$$
$$\rightarrow Ar^{2+}3p^4(^1S^e) + e_{Auger}(\ell = 1) \quad (1.3)$$

have been studied via the measurement of the angular distribution of the $L_3M_{23}M_{23}$ Auger electrons in coincidence with a photoelectron detected at fixed positions (namely at 0, 30 and 60°) with respect of the direction of the polarization of the incident radiation. In case of the $Ar^{2+}(^1S^e)$ state we have also performed the complementary experiment, in which the angular distribution of the low energy photoelectron is measured in coincidence with the Auger electron detected at a fixed position.

The basic quantities to describe the Ar $2p_{3/2}$ photoionisation in the dipole approximation are the three dipole matrix elements $D_j = \langle 2p^6 {}^1S \| \vec{d} \| 2p^5 {}^2P_{3/2} e_{lj} \rangle = |D_j| e^{-i\Delta_j}$ with l=0, j=1/2, and l=2 j=3/2, 5/2. An absolute phase is irrelevant, so we shall consider only two independent phase differences Δ_{ij}=Δ_i−Δ_j. Thus the quantities that fully characterise the process are five. In the two step approximation the coincidence angular pattern, which is represented by the triple differential cross section (TDCS) $d^3\sigma/d\Omega_1 d\Omega_2 dE_1$, is given by a general expression [3], which disentangles the properties of the light polarization, the geometry of the two electron emission and the dynamical parameters

$$TDCS \propto \rho_{00}(J,\vartheta_1,\varphi_1)\left[1+\sum_{k_2>0,even}^{2\ell_{max}}\alpha_{k_2}\sum_{q_2}A_{k_2 q_2}(J,\vartheta_1,\varphi_1)\sqrt{\frac{4\pi}{2k_2+1}}Y_{k_2 q_2}(\vartheta_2,\varphi_2)\right] \quad (2)$$

The labels 1 and 2 refer to the photoelectron and Auger electron respectively. ρ_{00} describes the angle-dependent intensity for the non-coincident observation of the photoelectron, α_{k_2} are the Auger anisotropy coefficients, $A_{k_2 q_2}$ the angle-dependent alignment tensor and $Y_{k_2 q_2}$ the spherical harmonics. In the case of linear polarization and perpendicular plane geometry ($\varphi_1=\varphi_2=0°$), for the $Ar^{2+}(^1S^e)$ final state where $k_2=2$ and $\alpha_2=-1$ the TDCS becomes

$$TDCS \propto \rho_{00}(\frac{3}{2},\vartheta_1)\left[1-\sqrt{\frac{4\pi}{5}}\sum_{q_2}A_{2q_2}(\frac{3}{2},\vartheta_1)Y_{2q_2}(\vartheta_2,)\right] \quad (3)$$

The magnitudes $|D_j|$ of the dipole matrix elements and their relative phases enter in the expressions of the A_{2q_2} statistical tensors [3]. The Δ_{ij} appear only via the cosine function, thus the sign of these phases remains undetermined.

In order to determine also this sign a set of measurements with circularly polarised radiation has been performed. Kabachnik and Schmidt [8] have demonstrated theoretically that the angular distribution of the Auger electrons is different for the left and right circular polarization, provided the Auger electron is detected in coincidence with the photoelectron. In this formulation the circular dichroism, Δ, defined as the difference between the TDCS for right, σ^+, and left, σ^-, circularly polarised radiation, can be expressed as a function of the matrix elements and relative phases of the two steps. The expression of Δ_{ij} [8 and formula (4) in section 3.2] displays a direct dependence on the sine of the Δ_{ij}'s, therefore a measurement of the Δ_{ij}'s allows the determination of their signs, too.

3.1 Results with linearly polarised radiation

The experimental TDCS are shown in figure 1. Only the angular distribution measured at $\vartheta_1=0°$, i.e. with the photoelectron detected along the direction of polarisation of the incident radiation, displays a cylindrical symmetry about the direction of polarisation. For the other two angular distributions neither the direction of the photoelectron nor the axis of polarisation of the incident radiation represents an axis of symmetry.

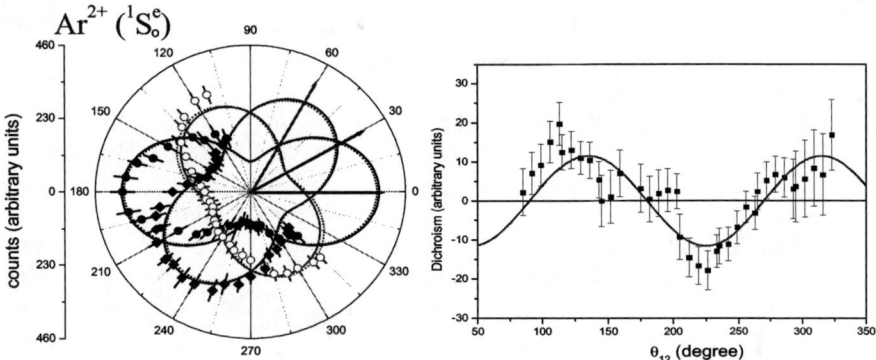

FIGURE 1. (a) TDCS for the $Ar^{2+}(^1S^e)$ state. The photoelectron was detected at $\vartheta_1=0$ (dots), 30 (lozanges) and 60°(circle), respectively. The curves are the results of a simultaneous fit of formula (3) to the three sets of data. (b) $Ar^+(2p_{3/2})$ circular dichroism versus the mutual angle θ_{12} between the fixed $2p_{3/2}$ photoelectron and the $L_3MM(^1S^e)$ Auger electron.

Formula (3) has been used in a simultaneous fit of the three sets of data at different θ_1. In the fitting we have written formula (3) in terms of the amplitude ratios $|D_{1/2}/D_{3/2}|$ and $|D_{5/2}/D_{3/2}|$ (because in this experiment no absolute values are measured), phase differences $\Delta_{1/2,3/2}$ and $\Delta_{5/2,3/2}$, and a scaling factor. In this way the number of free parameters is reduced to five. First we worked in the *LSJ* approximation and neglected the spin–orbit interaction in the continuum, i.e., $|D_{5/2}/D_{3/2}|=3$ and $\Delta_{5/2,3/2}=0$ [9]. Next we removed the above mentioned constraints and repeated the fit. In this latter case it should be noted that, depending on the initial guess of the parameters, it is possible to achieve almost equivalent fits of the experimental data, according to statistical tests, with values of the phases which differ more than their uncertainties. We selected the set of parameters with the lowest χ^2 value. The TDCSs calculated with the LSJ (solid line in Figure 1)and jj (dashed line) parameters are practically undistinguishable and both give a reasonable representation of the data. The result of this analysis shows that the *LSJ* approximation is sufficient for the description of the Ar $2p_{3/2}$ photoionisation and therefore it can be described only by three parameters $D_s = \sqrt{3/2} \cdot D_{1/2}$, $D_d = \sqrt{15} \cdot D_{3/2}$ and $\Delta_{sd} = \Delta_{1/2,3/2}$.

TABLE 1. Amplitudes and relative phases from the fit with and without the constraints of the LSJ approximation. The values in the last raw are the reduced χ^2.

	LSJ	jj
$|D_{1/2}/D_{3/2}|$	1.85 ±0.10	1.77 ±0.13
$|D_{5/2}/D_{3/2}|$	3	2.60 ±0.25
$\Delta_{1/2,3/2}$	62° ±2°	62° ±13
$\Delta_{3/2,5/2}$	0	0° ±13
χ^2	1.2	1.3

3.2 Results with circularly polarised radiation

In the measurements with circularly polarised radiation the direction of the photoelectron does not introduce a preferential axis and the relevant quantity is the relative angle, θ_{12}, between the two electrons. Thus the data from the three angular distributions measured simultaneously can be added together to extract the experimental dichroism (figure 1.b). In calculating Δ we have assumed the helicity as positive when the radiation emerging from the beamline has the electric vector rotating clockwise as observed from the experimental chamber. Following Kabachnik and Schmidt [8] in the case of Ar $2p_{3/2}$, Δ can be written as

$$\Delta = TDCS(\sigma^+) - TDCS(\sigma-) = \sigma_{3/2}(E)c(E_A)\frac{\omega_a}{4\pi^2}\frac{9}{2\sqrt{2}}\frac{D_s D_d \sin(\Delta_{sd})}{D_s^2 + D_d^2}\sin 2\theta_{12}$$

$$\propto \sin(\Delta_{sd})\sin 2\theta_{12} \qquad (4)$$

where $\sigma_{3/2}$ is the partial ionisation cross section of the Ar $2p_{3/2}$ state, $c(E_A)$ the Lorenzt factor [8] and ω_A the Auger partial branching ratio. A fit of this expression to the experimental data results in a negative value of the term $\sin\Delta_{sd}$.

4. DISCUSSION AND CONCLUSIONS

The values of the amplitudes and phase can now be used to predict other observables of the process. Let's consider the other two decay channels of the $2p_{3/2}$ vacancy (reactions (1.1) and (1.2)). They are characterized by the same matrix elements of the photoionisation, but different α_2 depending on the final Ar^{2+} ion state. Thus the TDCSs for channels (1.1) and (1.2) have been fitted with formula 2, using the obtained Dj values and only α_2 as free parameter. The results shown in figure 2.a prove that a satisfactory representation of the data is obtained in both reactions. The α_2 values obtained, 0.38 ± 0.02 and -0.44 ± 0.02 for the $^3P^e$ and $^1D^e$ final states, respectively, are in good agreement with previous experimental determinations (0.37 ± 0.1 and -0.48 ± 0.1) [10]. Then the Dj values have also been used to calculate the complementary TDCS, where the photoelectron angular distribution is measured in coincidence with the $L_3M_{23}M_{23}$ ($^1S^e$) Auger electron detected at 30° (figure 2.b). The shape of the TDCS is very different from the noncoincidence angular distribution of the photoelectrons. The TDCS, calculated using the values from Table 1 and just a scaling factor as a free parameter, is in reasonable agreement with the experiment. The other observables in a photoionisation experiment are the asymmetry parameter β of the photoelectrons, the alignment A_{20}, and the orientation A_{10} of the photoion, and the ξ, η, and ζ spin polarization parameters of the photoelectron [11]. All these quantities have been calculated and collected in Table II where they are compared with previous experimental determinations and theoretical predictions.

In summary it has been shown that a complete description of Ar 2p photoionisation can be achieved via an Auger electron-photoelectron coincidence experiment.

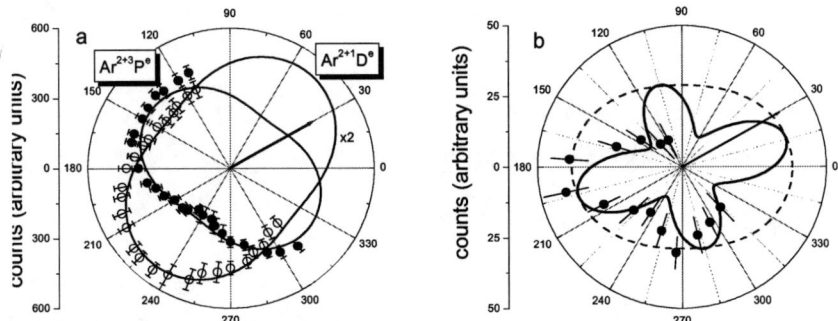

FIGURE 2. a) TDCS for the $^3P^e$ and $^1D^e$ states of Ar^{2+}; b) angular correlation between the photoelectron and an Auger electron detected at 30°. The dotted line represents the non coincidence angular distribution of the photoelectron.

TABLE 2. Values of the observables of Ar $2p_{3/2}$ photoionisation calculated with the amplitudes obtained in LSJ approximation.

	This work	Previous experiments	Theory
β_{ph}	0.17±0.06	0.24±0.06[13] 0.19±0.03[14]	0.32[13] 0.6[17]
A_{20}	0.32±0.02	0.27±0.1[15]	0.15[15] 0.12[16]
A_{10}	-0.13±0.05		0.53[16]
ξ	-0.53±0.02		
η	-0.36 ±0.01		
ζ	-0.72 ±0.02		

This procedure has the advantage that only one experimental measurement is needed and overcomes the need to measure several quantities, which sometimes result not independent [12], in different experiments. It is also shown that the photoionisation of Ar $2p_{3/2}$ can be described in the LSJ approximation.

REFERENCES

1. von Raven, E., Meyer, M., Pahler, M., and Sonntag, B., *J. Electron Spectrosc. Rel. Phenom.* **52**, 677 (1990)
2. Kämmerling, B. and Schmidt, V., *Phys. Rev. Lett.* **67**, 1848 (1991) and *Phys. Rev. Lett.* **69**, 1145 (1992)
3. Kabachnik, N. M., *J. Phys. B:At. Mol. Opt. Phys.* **25**, L389 (1992)
4. De Fanis, A. et al *J. Phys. B:At. Mol. Opt. Phys.* **32**, 5739 (1999)
5. Blyth, R.R., et al. *J. Electron Spectrosc. Relat. Phenom.* **101-103**, 959 (1999)
6. Desiderio, D. et al *Synchrtron radiat. News* **12**, 34 (1999)
7. Bolognesi, P. et al *J. Elect. Spect. Rel. Phenom.* **141**, 105 (2004)
8. Kabachnik, N. M., and Schmidt, V.,*J. Phys. B:At. Mol. Opt. Phys.* **28**, 233 (1995)
9. Heinzmann, U., *J. Phys. B:At. Mol. Opt. Phys.* **13**, 4353 (1980)
10. Sarkadi, L., Vajnai, T., Palinkas, J., Kover, A., Vegh, J., and Mukoyama, T., *J. Phys. B* **23**, 3643 (1990)
11. Huang, Li K.-N., Phys. Rev. A **22**, 223 (1980).
12. Kabachnik ,N.M.. J. Electron Spectrosc. Relat. Phenom. **137-140**, 305 (2004)
13. Lindle, D.W. et al *Phys. Rev.* **A31**, 714 (1985)
14. Avaldi,L. et al. *J. Phys.B:At. Mol. Opt. Phys.* **27**, 3953 (1994)
15. U. Becker and B. Langer, Phys. Scr., T **78**, 13 (1998);
16. U. Kleiman and B. Lohmann, J. Electron Spectrosc. Relat. Phenom. **131-132**, 29 (2003)
17. M. Kutzner, Q. Shamblin, S. E. Vance, and D. Winn, Phys. Rev. A **55**, 248 (1997).

Analysis of the resonant Auger decay during ultrafast fragmentation of CH₃F

G. Prümper*, V. Carravetta†, Y. Muramatsu*, X.J. Liu*, K. Ueda*,
Y. Tamenori**, M. Kitajima‡, M. Hoshino‡ and H. Tanaka‡

*Institute of Multidisciplinary Research for Advanced Materials, Tohoku University, Sendai 980-8577, Japan
†Institute for Chemical Physical Processes CNR, via Moruzzi 1, 56124 Pisa, Italy
**Japan Synchrotron Radiation Research Institute, Sayo-gun, Hyogo, Japan
‡Department of Physics, Sophia University, Chiyoda-ku, Tokyo 102-8554, Japan

Abstract. After inner shell excitation the ultrafast fragmentation of small molecules can proceed on a time scale similar to that of the Auger decay. In this case the nuclear motion affects the observed electron spectrum in various ways. The best known example is the Doppler splitting of atomiclike Auger lines, but non-parallel potential surfaces are equally important. Using an electron energy - ion momentum coincidence experiment and an ab initio calculation of potential curves, we discuss the main features seen in the non-coincident and coincident resonant Auger spectrum by decay of the $F(1s) \rightarrow 6a^*$ core excited state of CH_3F.

Keywords: inner shell, Doppler, ultrafast, fragmentation, Auger
PACS: 34.50.Lf, 07.81.+a, 29.30.-h

INTRODUCTION

The promotion of a core electron in a molecule into an unoccupied anti bonding valence orbital causes relaxation of the nuclear frame and electronic shell. The time scale of a possible competition of the two processes is of the order of a few femtoseconds and the term "ultrafast dissociation" is often used for describing this dissociation [1, 2, 3]. Recently inner shell excitation leading to "ultrafast dissociation" of small molecules has been used as a tool to study the details of femtosecond nuclear dynamics [4, 5, 6] by analyzing the electron spectra. In this paper we try to model the observed Auger electron spectra that occur after the F(1s) excitation of CH_3F. The reaction that can be described most easily is the atomiclike Auger decay happening inside the F atom.

$$CH_3F + h\nu \rightarrow CH_3F^*(F1s^{-1}6a_1^*),$$
$$CH_3F^*(F1s^{-1}6a_1^*) \rightarrow CH_3 + F^*(1s2p^6\ ^2S),$$
$$F^*(1s2p^6\ ^2S) \rightarrow F^+(2p^4\ ^1D) + e^-. \quad (1)$$

During the dissociation, the F* atom emits an Auger electron with a kinetic energy close to 656.5 eV [7]. The lifetime broadening of 0.2 eV corresponds to a lifetime of about 3.3 fs. A detailed peak shape analysis making use of the Doppler structure of this atomiclike Auger line was performed for the corresponding reaction in SF_6 and CF_4 [11] for different values of the detuning from the resonance. Because of the clear visibility of this line it could be separated from the other Auger decay channels, by treating the rest of the spectrum as background. However the understanding of the complete Auger spectrum is desirable.

In general no simple reaction formula can be written down. Instead the Auger decay happens during the molecular deformation or fragmentation. In this work we restrict the analysis to decays following the 1s-excitation to the first unoccupied molecular orbital, written as the first step in equation 1. It turns out in the theoretical analysis that the Auger final state consists of many different molecular electronic states. These states will be labeled according to their symmetry A_1, A_2 and E. Further labels are used to describe the main electronic configurations of the different states. The calculation assumes a rigid CH_3 fragment and treats the C-F bond length as the only free parameter. This crude approximation is obviously violated as many different charged fragments such as F^+, CH_2^+, CH^+, C^+ and H^+ are observed as reaction products. Nevertheless we use this approach to calculate the corresponding potential curves and assume the early states of the nuclear motion before the Auger decay can be approximated using this approach. This is possible because the F(1s) Auger lifetime is only 3.3 fs. Following the Auger decay diabatic nuclear motion hopping from one diabatic potential curve to another is likely. Also deviations from the computed potential curves especially due to vibrational excitation and emission of hydrogen atoms or protons have to be expected. Nevertheless we consider the preliminary model calculation presented here as an important step towards a more complete understanding of ultra fast fragmentation phenomena.

EXPERIMENT

The experimental setup, the data acquisition system and the data evaluation technique are described in detail elsewhere [8, 9, 10]. Here we give only a brief account. The setup consists of a hemispherical electron spectrometer (Gammadata-Scienta SES-2002) and an ion time of flight (TOF) spectrometer mounted inside a vacuum chamber. An effusive CH_3F beam is crossed by the synchrotron radiation. Electrons pass the pusher electrode of the ion spectrometer and enter the electron spectrometer. Ions are extracted using pulsed fields triggered by the electron detection. The experiment has been carried out on the c branch of the high resolution photochemistry figure-8 undulator beamline 27SU [13, 14, 15, 16]. The photon energy bandwidth does not contribute to the observed electron energy width and was estimated as 50 meV. Additionally, conventional high resolution electron spectra were recorded using a gas cell. The result is shown in figure 3. Using a vertical polarization of the photon beam reveals a sharp feature around 656 eV that also appears in coincidence with the F^+ ion.

THE CALCULATION

The aim of this work is to get a first impression of the rich structure one can expect from the Auger decay happening in the dissociating molecule and to establish a crude correspondence between the Auger final states and the observed ionic fragments. The first step of the calculation is to generate a set of Auger profiles. Each profile belongs to

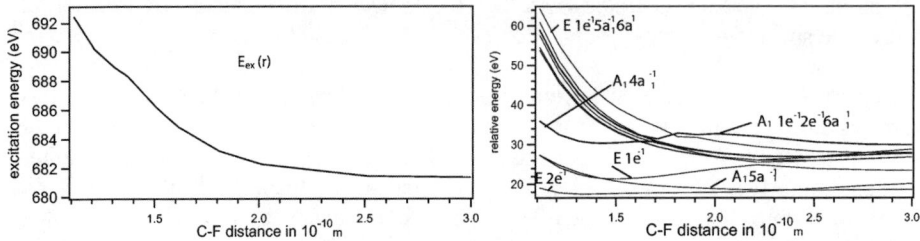

FIGURE 1. Left: Adiabatic potential curve for the core excited state $E_{ex}(r)$. The zero energy is defined by the ground state energy at the equilibrium geometry r=1.413 10^{-10}m. Right: Adiabatic potential curves for the final states.

a specific final fragmentation or bound state. The shape of the peak depends essentially on whether of the slope of the potential curve of the core excited state or that of the Auger final state is steeper. The potential curves of the ground state, the core excited state and the Auger final states were calculated using the configuration interaction method (CI) with a 5ζ quality basis set assuming a fixed geometry of the CH_3 part. This geometry is the result of a self consistent field optimization for CH_3F. While the ground state and the core excited state have been computed using an extended set of determinants, a limited CI calculation including only the main contributions determining the participator and some of the lower spectator states was performed for the final Auger states. The results of this calculation are shown in figure 1. The calculated equilibrium distance of 1.413 10^{-10} m is used as the starting point of the dissociation reaction. This corresponds roughly to the resonant excitation with zero detuning. The potential curve $E_{ex}(r)$ of the core excited state shown in figure 1 determines the fragmentation dynamics. Equation 2 describes the classical motion of the two particle system CH_3 and F. The reduced mass of the two fragments CH_3 and F is m_{eff}=8.38 amu.

$$\frac{d}{dt^2}r(t) = -1/m_{eff}\frac{d}{dr}(E_{ex}(r(t))) \quad (2)$$

Numerical integration of equation 2 with the initial conditions: $r(t=0)$ = 1.413 10^{-10} m and $\frac{d}{dt}r(t=0) = 0$ yields the internuclear distance as a function of time. The decay rate of the excited state is modeled by an exponential decay given by the experimental Auger lifetime of 3.3 fs. For each molecular final state the Auger energy is given by the difference of the two potential curves at the current decay distance r(t). Thus the resulting spectrum is given by this energy difference weighted with the decay number at the given distance. The lifetime broadening effect was modeled by a subsequent convolution with a gaussian profile with 0.2 eV FWHM. All computed spectra are shifted +7eV in kinetic energy for better agreement with the experimental data. The reason for this shift is probably the different levels of accuracy used in the computation of the core excited state and the Auger final states. The resulting Auger spectra are shown in figure 2.

The main approximation of our method is, that in the early state of the dissociation process the nuclear dynamics and the Auger decay can be described by the potential

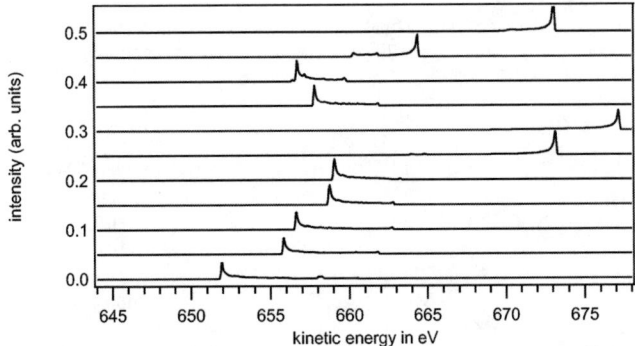

FIGURE 2. Calculated Auger peak profiles for the different final state potential curves. Top to bottom: $A_1\, 5a_1^{-1}\, ...\, E\, 1e^{-1}\, 5a_1^{-1}\, 6a_1^1$ same order as in table 1.

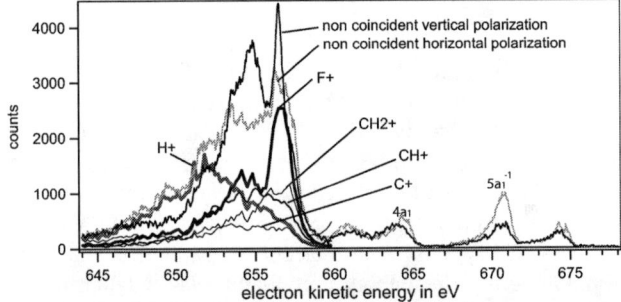

FIGURE 3. Electron spectra on top of the resonance (zero detuning). Top two graphs: Measured high resolution electron spectra of CH_3F. Sharp features only become visible for the vertical polarization. Other graphs: Electron spectra coincident with different ionic fragments for horizontal light polarization. Sharp features around 656 eV appear in coincidence with the F^+ and the CH_2^+ fragments. In the coincidence experiment the kinetic energy range was limited to the interval 644 eV to 660 eV. The intensity of the non coincident measurements was scaled to fit the plot.

curves that belong to the frozen geometry of the CH_3 remainder. Subsequent processes lead to further vibrational excitation and fragmentation.

ASSIGNMENT OF THE STATES

We assign the potential curves and peaks based on comparison with other calculations and experimental findings. Because of the dispersion of the peaks, with kinetic energy above 662eV we assign them to the participator Auger decays. The corresponding spectra for the detuning of the photon energy from the resonance are not shown here. The best agreement between calculation and experiment was obtained by a shift of +7eV for the calculated Auger kinetic energies.

The Auger peaks with the high kinetic energy are assigned to molecular ionic states

TABLE 1. List of Auger final states. The second column describes the main contribution at the equilibrium geometry. (*)=linear combination of $2e^{-2}\,6a_1^1$, $1e^{-1}\,2e^{-1}\,6a_1^1$ and $1e^{-2}\,6a_1^1$

symmetry	main conf.	Peak energy	type of final state
A_1	$5a_1^{-1}$	673.1 eV	participator Auger
A_1	$4a_1^{-1}$	664.4 eV	participator Auger
A_1	$1e^{-1}\,2e^{-1}\,6a_1^1$	656.7 eV	spectator Auger decay
A_2	(*)	657.8 eV	spectator Auger decay
E	$2e^{-1}$	677.1 eV	participator Auger
E	$1e^{-1}$	673.1 eV	participator Auger
E	$1e^{-1}\,2e^{-1}\,6a_1^1$	659.0 eV	spectator Auger decay
E	$2e^{-2}\,6a_1^1$	658.7 eV	spectator Auger decay
E	$1e^{-1}\,2e^{-1}\,6a_1^1$	656.6 eV	spectator Auger decay
E	$2e^{-1}\,5a_1^{-1}\,6a_1^1$	655.8 eV	spectator Auger decay
E	$1e^{-1}\,5a_1^{-1}\,6a_1^1$	651.9 eV	spectator Auger decay

resulting from participator Auger decay, leading to CH_3^+ ions or decay products. The isolated line at 656.5 eV that belongs to the atomiclike Auger decay described by equation 1 can not be assigned uniquely to a specific spectator Auger decay. However the broad structure between 645 eV and 655 eV coincides in energy with the various spectator Auger decay channels. The higher lying potential curves correspond to spectator Auger decay states with similar energies. As the Auger electron carries away less energy the higher excess energy must lead to electronic excitation within the fragments or to nuclear motion. Obviously one energy loss channel is the emission of protons. Thus in this kinetic energy region the main ion yield is H^+. Due to the large number of the electronic states and vibrational excitations no pronounced structures can be resolved there.

CONCLUSION

We have shown that the overall structure of the resonant $F(1s \rightarrow 6a^*)$ Auger spectrum of CH_3F can be understood qualitatively by using adiabatic potential curves varying only the C-F bond length. The different Auger final states can partially be resolved using the electron ion coincidence technique. The deviations between the observed spectra and the calculations are assigned to the lack of nuclear degrees of freedom for vibrational excitations and fragmentation. Open questions are the evolution of the electron charge distribution after the Auger decay i.e. the connection between the potential curves and the ionic fragments observed experimentally and the influence of the other nuclear degrees of freedom.

ACKNOWLEDGMENTS

The experiment was carried out with the approval of the SPring-8 program review committee and was partly supported by the Japan Society of the Promotion of Science (JSPS) in the form of Grants-in-Aid for Scientific Research. The staff of Spring-8 are greatly acknowledged for providing an excellent experimental facility.

REFERENCES

1. P. Morin and I. Nenner, Phys. Rev. Lett. **56**, 1913 (1986).
2. O. Björneholm, S. Sundin, S. Svensson, R.R.T. Marinho, A. Naves de Brito, F. Gel'mukhanov, and H. Ågren, Phys. Rev. Lett. **79**, 3150 (1997).
3. O. Björneholm, M. Bässler, A. Ausmees, I. Hjelte, R. Feifel, H. Wang, C. Miron, M.N. Piancastelli, S. Svensson, S.L. Sorencen, F. Gel'mukhanov, and H. Ågren, Phys. Rev. Lett. **84**, 2826 (2000).
4. F. Gel'mukhanov and H. Ågren, Phys. Rev.A, **54**, 379 (1996).
5. E. Pahl, L. S. Cederbaum, H.-D. Meyer, and F. Tarantelli, Phys. Rev. Lett. **80**, 1865 (1998).
6. F. Gelmukhanov, V. Kimberg, and H. Ågren, Phys. Rev. A **69**, 020501(R) (2004).
7. S. Svensson, L. Carlsson L, N. Mårtensson, P. Baltzer, and B. Wannberg, J. Electr. Spectrosc. Relat. Phenom. **50**, C1 (1990).
8. G. Prümper, K. Ueda, Y. Tamenori, M. Kitajima, N. Kuze, H. Tanaka, C. Makochekanwa M. Hoshino and M. Oura, Phys. Rev. A, **71**, 052704 (2005).
9. G. Prümper, Y. Tamenori, A. De Fanis, U. Hergenhahn, M. Kitajima, M. Hoshino, H. Tanaka and K. Ueda, J. Phys. B: At. Mol. Opt. Phys. **38**, 1 (2005).
10. G. Prümper, K. Ueda, U. Hergenhahn, A. De Fanis, Y. Tamenori, M. Kitajima, M. Hoshino, and H. Tanaka, J. Electr. Spectrosc. Relat. Phenom. **144**, 227, (2005).
11. M. Kitajima, K. Ueda, A. De Fanis, T. Furuta, H. Shindo, H. Tanaka, K. Okada, R. Feifel, S.L. Sorensen, F. Gel'mukhanov, A. Baev, H. Ågren, Phys. Rev. Lett. **91**, 213003 (2003).
12. O. Kugeler, G. Prümper, R. Hentges, J. Viefhaus, D. Rolles, U. Becker, S. Marburger, and U. Hergenhahn, Phys. Rev. Lett. **93**, 033002 (2004).
13. H. Ohashi, E. Ishiguro, Y. Tamenori, H. Kishimoto, M. Tanaka, M. Irie, T. Tanaka, and T. Ishikawa, Nucl. Instrum. Methods A **467-468**, 529 (2001).
14. H. Ohashi, E. Ishiguro, Y. Tamenori, H. Okumura, A. Hiraya, H. Yoshida, Y. Senba, K. Okada, N. Saito, I.H. Suzuki, K. Ueda, T. Ibuki, S. Nagaoka, I. Koyano, and T. Ishikawa, Nucl. Instrum. Methods A **467-468**, 533 (2001).
15. Y. Tamenori, H. Ohashi, E. Ishiguro, and T. Ishikawa, Rev. Sci. Instrum. **73**, 1588 (2002).
16. T. Tanaka and H. Kitamura, J. Synchrotron Radiation **3**, 47 (1996).

Correlation Effects in Electron-Diatomic Molecule Inelastic Collisions

B. Joulakian[*], V. Serov[§] and N. Lahmidi[*]

[*]Laboratoire de Physique Moléculaire et des Collisions, Institut de Physique, 1 Bld Arago, Technopôle 2000, 57078 Metz Cedex 3 France
Fax: 33 (0)387315801, Phone: 33 (0)387315858
e-mail: joulak@univ-metz.fr
[§]Chair of Theoretical and Nuclear Physics, Saratov State University, 83 Astrakhanskaya Saratov 410026 Russia

Abstract. The dissociative (e,2e) ionisation of diatomic hydrogen by fast electrons is studied theoretically as a vertical transition from the lowest vibrational and rotational level of the fundamental electronic state $^1\Sigma_g^+$ of H_2 to the first dissociative $^2\Sigma_u$ state of H_2^+, using prolate spheroidal solutions. After verification of the perturbative procedure in the non-dissociative case, for which experimental and theoretical results exist, the variation of the multiply differential cross section of the dissociative ionization is studied to show the importance of the left-right correlation. The results show the anisotropy of the proton production and the break down of the dynamically well-understood behaviors in the case of simple (e,2e) ionization.

INTRODUCTION

Double excitation of two electron atomic or molecular systems is one of the most challenging problems of atomic physics [1]. Understanding the nature of the electron-electron correlation beyond the independent particle model has been the subject of a large number of publications.

Ionization-excitation and double ionization of the two electron systems can be considered as a particular kind of double excitation. In the case of ionization excitation of atomic species [2], like helium, spectroscopic and electron detection techniques are necessary for the measurement of the corresponding cross sections, while for dissociative diatomic species, like diatomic hydrogen or lithium, spectroscopic detection is not necessary if one can detect, in coincidence, the ejected electron and the emerging proton [3].

In this paper we report on our theoretically investigation of the dissociative ionization of H_2 as a vertical transition from the fundamental electronic $^1\Sigma_g^+$ state of H_2 to the first dissociative $^2\Sigma_u$ state of H_2^+ and try to show the particularities of this process with respect to the simple non dissociative ionization [4, 5]. We present our approach to the determination of the corresponding transition matrix elements using prolate spheroidal solutions, which by their nature contain the symmetry of the problem and are well adapted to this type of problems. These solutions are very rarely employed in collision problems, as they need particular attention in the construction of the corresponding partial waves [6], which we master now.

THEORY

Following E.S.Chang and U. Fano (1972) [7] we express the multiply differential cross section (5DCS) of the (e,2e) ionization from a degenerate initial state n_i to the degenerate final state n_f in the following form:

$$\sigma^{(5)}(n_f n_i) = \frac{d^5\sigma_{(n_f n_i)}}{d\Omega_e d\Omega_s d k_s^2/2} = (2\pi)^4 \frac{k_e k_s}{k_i} \int \frac{d\Omega_{Euler}}{8\pi^2} \sum_{m_i}^{\ell_i} \frac{1}{2\ell_i + 1} \sum_{m_f}^{\ell_f} \left| T_{n_f n_i}^{m_f m_i}(\alpha,\beta,\gamma) \right|^2 \quad (1)$$

where ℓ_f and ℓ_i represent respectively de final and initial degrees of degeneracy, (α, β, γ), the Euler angles of the target and (k_i, k_s and k_e), the moduli of the incident, scattered and the ejected electrons respectively. Using the closure relation over the final vibrational and rotational states and using the axial symmetry of the diatomic target, we can reduce this expression to the following form:

$$\sigma^{(5)} = 2\pi \frac{k_e k_s}{k_i K^2} \int\int \left| T_{fi}(\theta_\rho, \varphi_\rho) \right|^2 \sin\theta_\rho d\theta_\rho d\varphi_\rho \quad (2)$$

where the transition matrix element $T_{fi}(\theta_\rho, \varphi_\rho)$ for a fixed internuclear distance ρ and a fixed orientation $(\theta_\rho, \varphi_\rho)$ of the target in the laboratory can be reduced in the cases of high values of the incident electron energy to the following expression:

$$T_{fi} = \left\langle \chi(\vec{k}_e, \vec{r}_1) \Phi_{2\Sigma_u}(\vec{r}_2, \rho) \left| -2\cos(\frac{\vec{K}.\vec{\rho}}{2}) + e^{i\vec{K}.\vec{r}_1} + e^{i\vec{K}.\vec{r}_2} \right| \Phi_{1\Sigma_g}(\vec{r}_1, \vec{r}_2, \rho) \right\rangle \quad (3)$$

where $\vec{K}$ represents the momentum transfer. Here the exchange between the ejected and bound residual electron is neglected. These electrons are described by the solutions of the two-center Schrödinger equation in prolate spheroidal coordinates. Here

$$\chi^-(\vec{k},\vec{r}) = \frac{(4\pi)}{(2\pi)^{3/2}} \sum_{\ell=0}^{\infty} \sum_{m=-\ell}^{\ell} i^\ell e^{-i\Delta_\ell} T_{\ell m}(k,\lambda) S^*_{\ell,m}(k,\mu) S_{\ell,m}(k,\cos\chi) \quad (4)$$

represents the partial wave expansion of the wave representing the ejected electron in the field of the residual diatomic ion where $T_{\ell m}(k,\lambda)$ and $S_{\ell m}(k,\mu)$ are the solutions of the following equations [8]

$$\left[\frac{d}{d\mu}\left\{(1-\mu^2)\frac{d}{d\mu}\right\} - \left\{ A + \frac{k^2\rho^2\mu^2}{4} + \frac{m^2}{(1-\mu^2)} - (z_a - z_b)\rho\mu \right\} \right] S_l^m(\mu) = 0 \quad (5)$$

$$\left[\frac{d}{d\lambda}\left\{(\lambda^2-1)\frac{d}{d\lambda}\right\}+\left\{A+\frac{k^2\rho^2\lambda^2}{4}-\frac{m^2}{(\lambda^2-1)}+(z_a+z_b)\rho\lambda\right\}\right]T_l^m(\lambda)=0 \qquad (6)$$

with $\lambda=\dfrac{r_a+r_b}{\rho}$, $\mu=\dfrac{r_a-r_b}{\rho}$, φ representing the prolate spheroidal co-ordinates.

For the initial state wave function of the two bound electrons, four different wave functions were used, which are given in reference [9], these are:

a) Coulson type wave functions

$$\Phi_{target}^{^1\Sigma_g^+}(\rho,\vec{r}_1,\vec{r}_2)=N(\rho)\{1s\sigma_g(1)1s\sigma_g(2)\} \qquad (7)$$

with $1s\sigma_g = \sum\limits_{i=1}^{N} c_i\varphi_i = c_1(e^{-\alpha_1 r_a}+e^{-\alpha_1 r_b})+c_2(e^{-\alpha_2 r_a}+e^{-\alpha_2 r_b})+c_3(r_a e^{-\alpha_2 r_a}+r_b e^{-\alpha_2 r_b})+...$

b) Nordsieck wave function

$$\Phi_{target}^{^1\Sigma_g^+}(\rho,\vec{r}_1,\vec{r}_2)=N(\rho)\{1s\sigma_g(1)1s\sigma_g(2)-2p\sigma_u(1)2p\sigma_u(2)\} \qquad (8)$$

with $1s\sigma_g = e^{-\alpha\lambda}(e^{\beta\mu}+e^{-\beta\mu})$ and $2p\sigma_u = e^{-\alpha\lambda}(e^{\beta\mu}-e^{-\beta\mu})$

c) Weinbaum or Mueller and Eyring wave functions

$$\Phi_{target}^{^1\Sigma_g^+}(\rho,\vec{r}_1,\vec{r}_2)=N_1(\rho)\{1s\sigma_g(1)1s\sigma_g(2)\}+N_2(\rho)\{2p\sigma_u(1)2p\sigma_u(2)\} \qquad (9)$$

One can choose more sophisticated multi-configuration correlated wave functions to calculate the 5DCS, but this would add complications, not necessary at the present stage of our study, when it is still necessary to test the numerically calculated spheroidal continuum solutions [10].

RESULTS

After the verification of our procedure by reproducing the confirmed results of the simple nondissociative (e,2e) ionization of H_2 [4, 5], we have calculated the 5DCS of the dissociative ionization for the energy value of the incident electron E_i=4087eV and for that of the ejected electron E_e=20 eV.

In figure (1a) we show the variation of the 5DCS of the dissociative ionization of H_2 with the ejection angle θ_e for the fixed scattering angle θ_s=1°. One can see that the three wave functions, that take into account the left-right correlation, yield similar behaviors, while the Coulson-type wave function constructed by σ_g orbitals only, which in the non dissociative case gives the expected behavior, yields in the dissociative case a completely different shape. This can be explained by the fact that the electron-electron correlation has less influence on the simple nondissociative ionization, since it is produced by direct interaction of the incident electron with the bound electrons, while in the case of dissociative ionization the second bound electron of the residual ion is excited to a higher-energy electronic state. This excitation is due to the sudden ejection of the first electron followed by the influence of the suddenly produced "hole" on the residual electron.

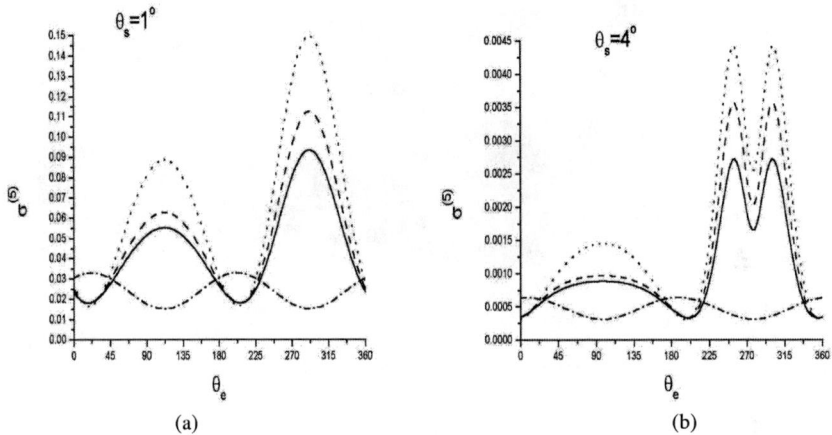

FIGURE 1. The variation of the 5 fold differential cross section of the dissociative ionization of H_2 in terms of the ejection angle for a given scattering angle θ_s. The energy of the incident electron E_i=4087eV and that of the ejected electron 20eV. The solid line shows the results obtained by Mueller's wave function, the dashed line, by Weinbaum's function, the dotted line, by Nordsieck's function and the dashed-dotted line, by Coulson's wave function.

A Similar observation can be made at a larger scattering angle θ_s=4° on figure (1b). This case reveals yet another peculiarity of the dissociative ionization, namely, the pronounced minimum at θ_e=277° that corresponds to the direction of the momentum transfer vector $\vec{K}$. For this particular scattering angle we have $\vec{K}=\vec{k}_e$ and the conservation of the momentum implies that the ionization is the result of a head on collision in which only the incident and the ejected electron take part. In other words, in this situation the residual ion does not participate to the reaction, so that the ionization is possible, but without excitation.

A global view of the 5DCS versus both the scattering and the ejection angles is shown on figure (2). In the region of small scattering angles θ_s we see that the diatomic target behaves as an atomic one, producing the usual lobes of the 5DCS. This is due to the fact that the incident electron has very large impact parameter and the molecular nature of the target is not very important. The specificity of the ionization-excitation process appears most evidently in the "forbidden zone" around θ_s =4° and θ_e =278°.

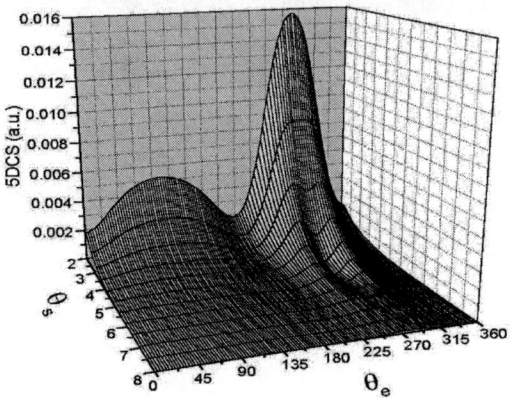

FIGURE 2. The variation of the 5 fold differential cross section of the dissociative ionization of H_2 in terms of the ejection and the scattering angles, for the same energy values as in figures (1a and 1b) and using Mueler's function

In the case of coincidence detection of the scattered electron and the emerging proton, one must integrate around all possible ejection directions in space.

$$\sigma^{(5)}(\theta_p,\varphi_p) = 2\pi \frac{k_e k_a}{k_i K^2} \int |T_{fi}(\theta_e,\varphi_e)|^2 \sin\theta_e d\theta_e d\varphi_e \qquad (10)$$

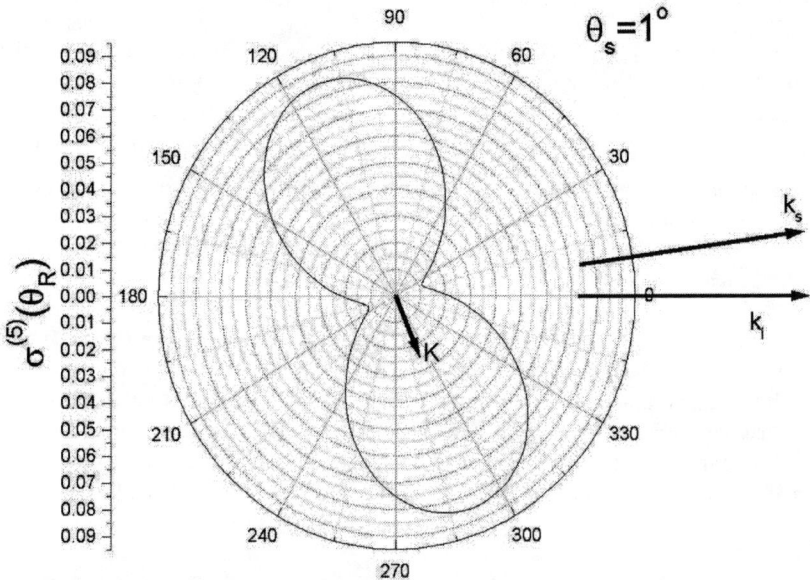

FIGURE 3. The variation of the 5 fold differential cross section of the dissociative ionization of H_2 in terms of the polar angle θ_p for the same energy values as in figure 1

On figure (3) we present the variation of the corresponding multiply differential cross section in terms of the polar angle θ_p in the plane of incidence ($\varphi_p=0$) defined by the incidence and scattering directions. This shows that the proton emerges in the direction of the momentum transfer.

CONCLUSION

We have shown that the ionization excitation of diatomic systems is a probe to study electron-electron correlation, and that left-right correlation must be included in the calculation. We have also shown that performing the cumbersome calculations
using partial wave developments with prolate spheroidal solutions is now realizable. We predict the existence of a forbidden zone in the region of the Bethe ridge, which appears in the non-dissociative case. We predict also that in the case of a coincidence detection of the proton with the scattered electron, the favorable direction of the proton is that of the momentum transfer.

ACKNOWLEDGMENTS

We want to thank the CINES (Centre Informatique National de l'Enseignement Supérieur) for the attribution of computational facilities

REFERENCES

1. Tanner G., Richter K. and Rost J. M., Rev. Mod. Phys. **72** 498-540 (2000).
2. Dogan M.and Crowe A., J. Phys. B: At. Mol. Opt. Phys. **35** 2773-2781 (2002).
3. Ulrich J., Moshammer R., Dorn A, Dörner R., Schmidt L. Ph. H. and Schmidt-Böcking H., Rep. Prog. Phys. **66** 1463 (2003).
4. Chérid M., Lahmam-Bennani A., Zurales R.W., Lucchese R.R., Duguet A., Dal Cappello M.C. and Dal Cappello C., *J. Phys. B: At. Mol. Opt. Phys* **22**, 3483-3499 (1989).
5. Weck P. Joulakian B Hanssen J. Fojon O.A. Rivarola R.D. Phys.Rev. A **62** 0114701 (2000)
6. Serov, V Joulakian B. B., Pavlov D. V., Puzynin I. V. and Vinitsky S. I., Phys. Rev. A **65** 062708 (2002).
7. Chang E.S.and Fano U. Phys. Rev. A <u>6</u>, page 173 (1972).
8. Morse PM, and Feshbach H *Methods of theoretical Physics (NewYork: Mcgrow –Hill) 1508 (1953)*.
9. McLean D, Weiss A. and Yoshimine M. Rev. Mod. Phys. **32** 211-218 (1960).
10. Serov, V Joulakian B. B., Derbov V. L. and Vinitsky S. I., J. *J. Phys. B: At. Mol. Opt. Phys* **38**, 2765-2773 (2005).

$(\gamma, 2\gamma)$ studies on multiply excited states of H_2 and N_2 in the vacuum ultraviolet range

Takeshi Odagiri

Department of Chemistry, Tokyo Institute of Technology, O-okayama 2-12-1, Meguro-ku, Tokyo 152-8551, Japan

Abstract. The doubly differential cross sections for the emission of two photons from excited fragments in photoexcitation of molecules have been measured as a function of incident photon energy by using a new method, $(\gamma, 2\gamma)$ *method*. The cross section curves that are free from ionization have been obtained for the first time. It has been shown that the $(\gamma, 2\gamma)$ method is a powerful tool for investigating the spectroscopy and dynamics of the multiply excited molecules

Keywords: multiply excited molecules, neutral dissociation, synchrotron radiation, photoexcitation
PACS: PAC NO. 33.80.-b

INTRODUCTION

Doubly and multiply excited molecular states in the vacuum ultraviolet range are highly correlated systems in which the independent electron model and Born-Oppenheimer approximation are less valid than for ground and lower excited states. We measured the cross sections for the emission of the fluorescences from excited hydrogen atoms and radicals produced in the photoexcitation of CH_4, NH_3 and H_2O as a function of incident photon energy and found much larger cross sections due to the doubly excited states than expected within the independent electron model [1, 2, 3]. From these investigations as well as electron impact ones [4, 5, 6, 7, 8, 9] the fluorescences from excited neutral fragments are found to be sensitive probes for the precursor doubly excited molecules. However, in the energy range higher than the thresholds for the dissociative ionization with excitation, $AB + h\nu \rightarrow A^* + B^+ + e^- \rightarrow A + h\nu' + B^+ + e^-$, the contribution from such dissociative direct-ionization dominates the fluorescence cross section curves and thus prevents the resonance peaks attributed to the higher-lying doubly or multiply excited states from being observed. Recently we have developed a new method, $(\gamma, 2\gamma)$ *method*, to overcome the difficulty [10, 11, 12]: two fluorescence photons from neutral fragments are detected in coincidence to observe the following process (1),

$$AB + h\nu \rightarrow AB^{**} \rightarrow A^* + B^* \\ \rightarrow A + B + h\nu' + h\nu''. \quad (1)$$

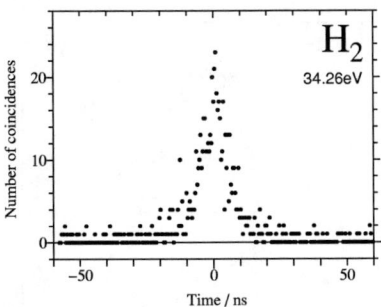

FIGURE 1. An example of the coincidence time spectra measured at the incident photon energy of 34.26 eV for H_2. The origin of time is taken at the peak.

The ionization-free fluorescence cross section curves highlight the structures due to the doubly or multiply excited states that are unresolved in the usual fluorescence cross section curves by (γ, γ) method. In this report, experimental results on H_2 [10] and N_2 [11, 12] are presented.

EXPERIMENTAL

The $(\gamma, 2\gamma)$ experiments were performed at the BL20A of the Photon Factory, KEK. Details of the experiments are described elsewhere [10, 11, 12]. In Brief, the linearly-polarized synchrotron radiation from the 2.5 GeV storage ring in the multi-bunch mode was monochromatized by a 3 m normal incidence monochromator and entered a gas cell filled with molecular hydrogen or nitrogen. Two fluorescence photons emitted parallel to the electric vector of the incident light and opposite to each other were detected in coincidence by two VUV-photon detectors, each of which is composed of an MgF_2 window and a microchannel plate. Since the filter range of the VUV-photon detector is 115-150 nm only the Lyman-α photons are detected in H_2 experiment, while several fluorescence photons from neutral N^* atoms are detected in N_2 experiment [13, 14]. Pulses from both detectors are fed into a standard delayed-coincidence system. An example of the coincidence time spectra is shown in figure 1. The bandpass of the wavelength of the incident light was 0.28 nm (the energy width of 280 meV at the incident photon energy of 35 eV) for H_2 and 0.12 nm (the energy width of 160 meV at 40 eV) for N_2.

True coincidence rate is normalized for the gas pressure in the gas cell, flux of the incident photons in the interaction region and geometrical factor, giving the relative value of doubly differential cross section of process (1), $d^2\sigma_2/d\Omega_i d\Omega_j$, where Ω_i and Ω_j are solid angles for the emission of two fluorescence photons. The normalization procedure is described elsewhere [10, 11, 12]. The relative cross sections for the emission of the fluorescence differential with respect to the solid angle for the photon, $d\sigma_1/d\Omega_i$, were also measured by using the same apparatus.

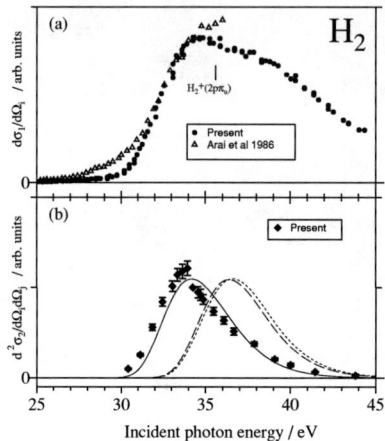

FIGURE 2. The singly differential cross sections for the emission of the Lyman-α fluorescence (a) and doubly differential cross sections for the emission of two Lyman-α photons (b) in photoexcitation of H_2. The symbols are experimental cross sections, closed circles(a) and closed diamonds(b): our results [10], open triangles(a): Arai et al.(1986) [15]. The curves are calculated cross sections (see text), solid curve: $Q_2{}^1\Pi_u(1)$, dashed curve: $Q_2{}^1\Pi_u(2)$, dotted curve: $Q_2{}^1\Sigma_u^+(1)$. The vertical bar in (a) indicates the threshold energy of the dissociative ionization resulting in the emission of a Lyman-α photon, process (2) (see text).

RESULTS AND DISCUSSION

H_2 [10]

Figure 2(a) shows the singly differential cross sections for the emission of the Lyman-α fluorescence, $d\sigma_1/d\Omega_i$, in photoexcitation of H_2 as a function of incident photon energy. The vertical line at 35.6 eV in figure 2(a) indicates the threshold energy for the dissociative ionization resulting in the emission of a Lyman-α photon,

$$H_2 + h\nu \rightarrow H_2^+({}^2\Pi_u(2p\pi_u)) \rightarrow H(2p) + H^+ + e^-. \qquad (2)$$

In the energy range below the threshold of process (2), the singly differential cross section, $d\sigma_1/d\Omega_i$, is attributed to neutral dissociation of the doubly excited $Q_1{}^1\Sigma_u^+(1)$, $Q_1{}^1\Pi_u(1)$ and $Q_2{}^1\Pi_u(1)$ states of H_2 [15, 16], while above the threshold the cross section involves the contribution from the dissociative direct-ionization, process (2). On the other hand, in the doubly differential cross sections for the emission of two Lyman-α photons, $d^2\sigma_2/d\Omega_i d\Omega_j$, shown in figure 2(b), no contribution from ionization is included and hence we obtain the cross section curve attributed entirely to the doubly excited states of H_2. Figures 2(a) and (b) clearly show that the cross section curve free from ionization obtained by the $(\gamma, 2\gamma)$ method provides us with much more advantage in observing the doubly or multiply excited states of molecules embedded in the ionization continuum in comparison with the fluorescence cross section curve measured with the (γ, γ) method.

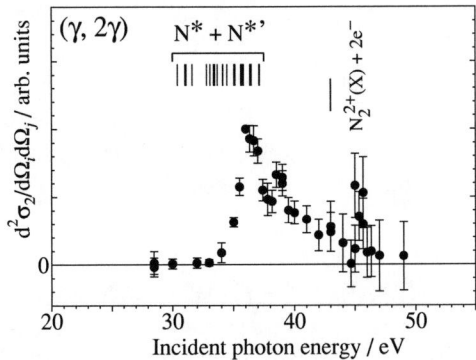

FIGURE 3. The doubly differential cross sections for the emission of two photons of N fluorescence in photoexcitation of N_2. The short vertical bars and long vertical bar at 43 eV represent dissociation limits leading to the emission of two photons of N fluorescence and the double ionization threshold of N_2 [18], respectively.

It is expected that the states responsible for the formation of two H(2p) atoms are the Q_2 states of H_2 built on the $(2p\pi_u)$ core leading to H(2p)+H$^+$. Among the molecular states resulting from two H(2p) atoms only a $^1\Pi_u$ state is optically allowed from the ground electronic state of H_2 [17]. Thus a $^1\Pi_u$ state in the Q_2 series of H_2 is expected to make a large contribution to the doubly differential cross sections, $d^2\sigma_2/d\Omega_i d\Omega_j$. The expectation is confirmed by a theoretical calculation based on the reflection approximation and semiclassical treatment for the decay dynamics. Curves in figure 2(b) represent the calculated cross section curves for the $Q_2{}^1\Pi_u(1)$, $Q_2{}^1\Pi_u(2)$ and $Q_2{}^1\Sigma_u^+(1)$ states of H_2. All the calculated curves are normalized to the same height and the calculated curve for the $Q_2{}^1\Pi_u(1)$ state is normalized to the experimental data around 34 eV. It has turned out that the doubly differential cross sections, $d^2\sigma_2/d\Omega_i d\Omega_j$, are well reproduced by the calculated cross section curve for the $Q_2{}^1\Pi_u(1)$ state. We conclude that (i) precursor doubly excited state for the production of two H(2p) atoms is the $Q_2{}^1\Pi_u(1)$ state of H_2 and (ii) the decay dynamics of the $Q_2{}^1\Pi_u(1)$ state is well approximated by the semiclassical picture.

N_2 [11, 12]

Figure 3 shows the doubly differential cross sections for the emission of two photons of N fluorescence (process (1)), $d^2\sigma_2/d\Omega_i d\Omega_j$, in photoexcitation of N_2, which reveals unknown resonance structures in the range 35 - 46 eV. They seem to be due to singly, doubly, or triply excited states of N_2. It should be noted that the resonance peak at 45 eV is above the double ionization threshold (43 eV [18]) and the width of this peak is narrower than those of other peaks. It seems that this peak originates from a double Rydberg molecular state built on $N_2{}^{2+}$ state.

SUMMARY

We have developed a new experimental method, $(\gamma, 2\gamma)$ method, for investigating the multiple photoexcitation in molecules. The $(\gamma, 2\gamma)$ cross sections as a function of incident photon energy without any background from the direct ionization provided us with new information on the spectroscopy and dynamics of multiply excited molecules.

ACKNOWLEDGMENTS

The author wishes to thank Prof. Noriyuki Kouchi and all the members of the laboratory for their excellent collaboration. The author also wishes to thank Prof. Masashi Kitajima for his helpful discussion. The experiments were carried out under the approval of Photon Factory Program Advisory Committee for proposal no 2003G006 and 2004G196. This research was partially supported by Japan Society for the Promotion of Science, Grant-in-Aid for Scientific Research (B), no 17350006 and Grant-in-Aid for Young Scientists (B), no 17750007.

REFERENCES

1. M. Kato, K. Kameta, T. Odagiri, N. Kouchi and Y. Hatano, *J. Phys.*, B **35**, 4383–4400 (2002)
2. M. Kato, T. Odagiri, K. Kameta, N. Kouchi and Y. Hatano, *J. Phys.*, B **36**, 3541–3554 (2003)
3. M. Kato, T. Odagiri, K. Kodama, M. Murata, K. Kameta and N. Kouchi, *J. Phys.*, B **37**, 3127–3148 (2004)
4. T. Odagiri, N. Uemura, K. Koyama, M. Ukai, N. Kouchi and Y. Hatano, *J. Phys.*, B **29**, 1829–1839 (1996)
5. N. Uemura, T. Odagiri, Y. Hirano, Y. Makino, N. Kouchi and Y. Hatano, *J. Phys.*, B **31**, 5183–5196 (1998)
6. T. Odagiri, K. Takahashi, K. Yoshikawa, N. Kouchi and Y. Hatano, *J. Phys.*, B **34**, 4889–4900 (2001)
7. T. Odagiri, H. Fukuzawa, K. Takahashi, N. Kouchi and Y. Hatano, *Nukleonika*, **48**, 95–102 (2003)
8. T. Odagiri and N. Kouchi, *Physica Scripta*, **T110**, 183–187 (2004)
9. H. Fukuzawa, T. Odagiri, T. Nakazato, M. Murata, H. Miyagi and N. Kouchi, *J. Phys.*, B **38**, 565–578 (2005)
10. T. Odagiri, M. Murata, M. Kato and N. Kouchi, *J. Phys.*, B **37**, 3909–3917 (2004)
11. M. Murata, T. Odagiri and N. Kouchi, *J. Electron Spectrosc. Relat. Phenom.*, **144-147**, 147–149 (2005)
12. M. Murata, T. Odagiri and N. Kouchi, submitted to *J. Phys.* B
13. M. Ukai, *J. Electron Spectrosc. Relat. Phenom.*, **79**, 423–428 (1996)
14. A. Ehresmann, S. Machida, M. Kitajima, M. Ukai, K. Kameta, N. Kouchi, Y. Hatano, E. Shigemasa and T. Hayaishi, *J. Phys.*, B **33**, 473–490 (2000)
15. S. Arai, T. Yoshimi, M. Morita, K. Hironaka, T. Yoshida, H. Koizumi, K. Shinsaka, Y. Hatano, A. Yagishita and K. Ito, *Z. Phys.*, D **4**, 65–71 (1986)
16. N. Kouchi, M. Ukai and Y. Hatano, *J. Phys.*, B **30**, 2319–2344 (1997)
17. G. Herzberg, *Molecular Spectra and Molecular Structure I. Spectra of Diatomic Molecules*, Princeton, Van Nostrand, 1967
18. G. Dawber, A. G. McConkey, L. Avaldi, M. A. MacDonald, G. C. King and R. I. Hall, *J. Phys.*, B **27**, 2191–2209 (1994)

…

Fragmentation Mechanism of Highly Excited C_{70} Cations in the Extreme Ultraviolet

Koichiro Mitsuke*[†], Hideki Katayanagi*[†], Junkei Kou*,
Takanori Mori*, and Yoshihiro Kubozono[¶]

*The Institute for Molecular Science, Myodaiji, Okazaki 444-8585, Japan
[†]Graduate University for Advanced Studies, Myodaiji, Okazaki 444-8585, Japan
[¶]Department of Chemistry, Faculty of Science, Okayama University, Okayama 700-8530, Japan

Abstract. The ion yield curves for C_{70-2n}^{z+} (n = 1-8, z = 2 and 3) produced by photoionization of C_{70} were measured in the photon energy ($h\nu$) range of 25 – 150 eV. The appearance $h\nu$ values were higher by ca. 34 eV than the thermochemical thresholds for dissociative ionization of C_{70} leading to C_{70-2n}^{z+}. Evaluation was made on the upper limits of the internal energies of the primary C_{70}^{z+} above which $C_{70-2n+2}^{z+}$ fragments cannot escape from further dissociating into C_{70-2n}^{z+}+C_2. These critical internal energies agreed well with appearance internal energies of C_{70}^{z+} theoretically obtained corresponding to the threshold for the formation of C_{70-2n}^{z+}. The photofragmentation of the parent C_{70}^{z+} ions is considered to be governed by the mechanism of internal conversion of their electronically excited states, statistical redistribution of the excess energy among a number of vibrational modes, and sequential ejection of the C_2 units.

Keywords: Fullerene ions; Photofragmentation; C_{70}; Reaction rate; Binding energies
PACS: 33.20.Ni; 33.80.Eh; 36.40.Wa

INTRODUCTION

Geometrical structures and electronic properties of fullerenes have attracted widespread attention because of their novel architecture, novel reactivity, and novel catalytic behaviors as typical nanometer-size materials. However, spectroscopic information is very limited in the extreme UV region, which was due to difficulties in acquiring a sufficient quantity of the sample. This situation has changed in recent years, since the techniques of synthesis, isolation, and purification have advanced so rapidly that appreciable amounts of fullerenes can be readily obtained.

Our group have measured the yields of parent ions produced from C_{60} and C_{70} by mass spectrometry, determining relative partial cross sections for production of singly and doubly charged ions [1]. Figure 1 shows typical yield curves of C_{60}^{+} and C_{60}^{2+} in the $h\nu$ range from 25 to 180 eV. We have demonstrated that the metastable parent ions can accommodate an internal energy of more than ~ 40 eV [2,3]. Moreover, the ratios between the yields of doubly and singly charged ions from C_{60} and C_{70} increase with increasing photon energy and asymptotically reach 0.5 – 0.6 and 2 – 3 for C_{60} and C_{70}, respectively, at $h\nu$ > 80 eV [4]. These ratios are one order of magnitude higher than those documented by many references dealing with photoionization of ordinary

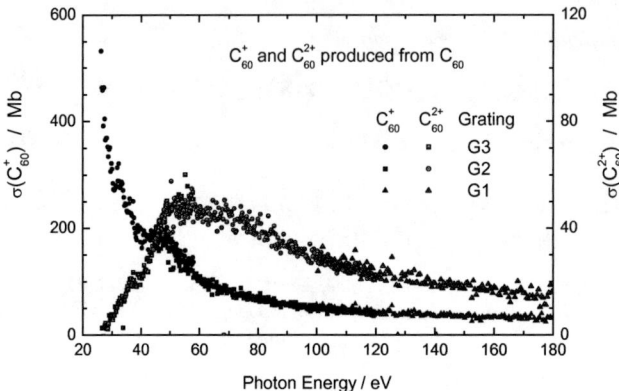

FIGURE 1. Partial cross sections σ for formation of C_{60}^+ and C_{60}^{2+} from C_{60}. Each σ curve includes the contribution of not only the parent but also fragment ions produced by the C_2-loss processes, so that it is essentially equal to σ involving all the ionic species in a particular charge state.

molecules. The above observations are well understood as that the excess internal energy is transmitted so quickly among enormous vibrational degrees of freedom that unimolecular reactions, such as autoionization, direct dissociation, electronic predissociation or coulomb explosion, are substantially suppressed and the lifetime of the parent ions is extended beyond the order of a millisecond.

Nevertheless, fragmentation occurs to some extent when fullerenes gain much excess energy in electronic excitation and ionization processes, as revealed by mass spectrometric studies of C_{60} and C_{70} in combination with laser multiphoton ionization [5], heavy-ion excitation [6], or electron impact ionization [7,8]. Decomposition of C_{60}^{z+} or C_{70}^{z+} ($z \geq 1$) primarily formed is known to lead to fragment ions with even-numbered carbon atoms (C_{60-2n}^{z+} or C_{70-2n}^{z+}, $n=1, 2,...$) via loss of multiple C_2 units from parent ions in high-vibrational states. In contrast, experimental studies of single photon excitation are very limited [1-3,9-13]. Recent results on the relative cross section of the fullerene fragments from C_{60} show that the appearance photon energies of C_{60-2n}^+ and C_{60-2n}^{2+} are higher by 30 - 40 eV than the thermochemical thresholds for dissociative photoionization of C_{60} [2,3,12]. Assuming a scheme of stepwise photofragmentation processes

$$C_{60} + h\nu \rightarrow C_{60}^{z+} + ze^- \rightarrow \cdots \rightarrow C_{60-2n}^{z+} + nC_2 + ze^- \qquad (1)$$

Kou et al. have interpreted the above large kinetic shift in terms of unimolecular decay modelled by quasiequilibrium theory [2,3]. They calculated the appearance internal energies of C_{60}^{z+} corresponding to the threshold for the formation of C_{60-2n}^{z+}. These theoretically obtained appearance energies agreed well with the critical internal energies of the primary C_{60}^{z+} required for the formation of C_{60-2n}^{z+}. This agreement suggests the following fragmentation mechanism: large excess energy is redistributed among a number of vibrational modes of C_{60}^{z+} and C_2 units are consecutively ejected.

Conversely there has been no effort to estimate the critical internal energies of C_{70}^{z+} for production of C_{70-2n}^{z+} from photoionization of C_{70}. They might differ from those for C_{60-2n}^{z+} from C_{60}, because the symmetry of C_{70} is much lowered and binding energies for component reactions

$$C_{70-2n+2}^{z+} (n \geq 1) \rightarrow C_{70-2n}^{z+} + C_2 \qquad (2)$$

are expected to depend strongly on the cluster size 70-2n. Obviously the binding energy should have a sudden increase at $n = 6$ due to a stable structure of a truncated icosahedron of C_{60}^{z+}. It is therefore less easy to tell whether the analogous statistical model is applicable to fragmentation starting from C_{70}^{z+}. To answer this question we determined the appearance photon energies of C_{70-2n}^{z+} in the present study, and made close comparison between the calculated and observed critical internal energies of the primary C_{70}^{z+} for the formation of C_{70-2n}^{z+}.

EXPERIMENTAL METHOD

All the measurements have been carried out at the bending magnet beamline BL2B constructed in the UVSOR synchrotron radiation facility in Okazaki, equipped with an 18 m spherical grating monochromator of Dragon type [14]. The Experimental set-up for photoionization mass spectrometry of the fullerene family has been described in detail elsewhere [1,15,16]. Briefly a molecular beam of C_{70} was produced by heating the sample powder to approximately 680 K. Monochromatized synchrotron radiation was focused onto the C_{70} beam. The produced photoions were extracted by a pulsed electric field, mass-separated by a time-of-flight (TOF) mass spectrometer, and detected with a microchannel plate electron multiplier. To normalize the ion counts the fluxes of the molecular and light beams were monitored throughout the measurement by a silicon photodiode and a crystal-oscillator surface thickness monitor, respectively.

RESULTS AND DISCUSSION

Ion Yield Curves for the Fragment Ions

Taking TOF mass spectra by scanning the monochromator allowed us to measure the yield curves for C_{70-2n}^{z+} from C_{70} as a function of $h\nu$. Figure 2 illustrates the ion yields of C_{70-2n}^{2+} and C_{70-2n}^{3+}, respectively, divided by the yields of the parent ion C_{70}^{2+} and C_{70}^{3+}. These curves are considered to provide fractional abundances of C_{70-2n}^{z+} within ~ 25 μs after ionization, at least around the onset region for each ion. Singly charged fragment ions were hardly detected in the whole energy range of the measurement. At $n \leq 4$ the appearance photon energies $AE(n,z)$ for a given z shift to higher $h\nu$ positions with decreasing size of 70-2n and the curves rise more gently towards a peak. Values of $AE(n,z)$ are found to be higher by ca. 34 eV than the thermochemical thresholds for dissociative photoionization of C_{70} leading to C_{70-2n}^{z+}. We have found similar large kinetic shifts for dissociative photoionization of C_{60} [2,3].

In Fig. 2 the yield curves for C_{70-2n}^{z+} with $n \geq 5$ behave quite differently from those with $n \leq 4$. This finding confirms that the cluster size of 60 is a magic number even in the course of fragmentation of ionized fullerenes, in connection with pronounced stability of C_{60} and C_{60}^{z+}.

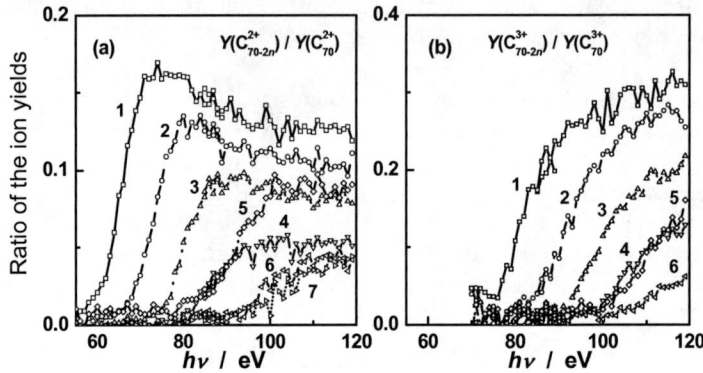

FIGURE 2. Relative ion yield curves of $C_{70-2n}{}^{z+}$ ions obtained from time-of-flight mass spectra. (a) $z = 2$, $n = 1\text{-}7$ and (b) $z = 3$, $n = 1\text{-}6$.

Critical internal energies of $C_{70}{}^{z+}$ for the formation of $C_{70-2n}{}^{z+}$

When C_{70} is photoionized at $h\nu = AE(n,z)$, we are able to write the critical internal energy $E_{max}(n,z)$ initially transmitted to $C_{70}{}^{z+}$ as

$$E_{max}(n,z) = AE(n,z) + E_v - IP(z) \qquad (3)$$

by assuming the kinetic energy of the emitted photoelectron to be zero. Here, E_v denotes the vibrational energy of C_{70}, and $IP(z)$ is the ionization potential of C_{70} for the formation of $C_{70}{}^{z+}$. In this study, $E_v = 3.5$ eV is assumed and $IP(2) = 18.8$ eV and $IP(3) \sim 35$ eV are taken from the literature [7,17]. We have evaluated $E_{max}(n,z)$ from eq. (3) and listed the resultant values in Table 1 for $C_{70-2n}{}^{z+}$. It is expected that $E_{max}(n,z)$ is nearly equal to the upper limit of the internal energies of the primary $C_{70}{}^{z+}$ above which $C_{70-2n+2}{}^{z+}$ fragments cannot escape from further dissociating into $C_{70-2n}{}^{z+}$ + C_2. At $n \leq 4$ $E_{max}(n,z)$ increases steadily with n by 7 eV - 9 eV, which may reflect relatively weak n-dependences of the binding energies for reaction (2). Furthermore $E_{max}(n,z)$ for a given n depends very weakly on z. These results, together with the prominent kinetic shifts, suggest that internal conversion of the electronically excited states of $C_{70}{}^{z+}$ results in redistribution of the excess energy among the vibrational degrees of freedom followed by statistical ejection of the C_2 units.

Table 1. Critical internal energies $E_{max}(n,z)$ of the primary $C_{70}{}^{z+}$ ions above which $C_{70-2n+2}{}^{z+}$ fragments cannot escape from further dissociating into $C_{70-2n}{}^{z+} + C_2$. All values are in eV.

n	$C_{70-2n+2}{}^{z+}$	$C_{70-2n}{}^{z+}$	Observed,[a] $E_{max}(n,z)$		Calculated[b] $E_{RRKM}(n)$
			$z=2$	$z=3$	
1	$C_{70}{}^{z+}$	$C_{68}{}^{z+}$	44±2	44±2	44
2	$C_{68}{}^{z+}$	$C_{66}{}^{z+}$	51±2	50±2	52
3	$C_{66}{}^{z+}$	$C_{64}{}^{z+}$	59±2	60±2	57
4	$C_{64}{}^{z+}$	$C_{62}{}^{z+}$	66±2	66±2	63
5	$C_{62}{}^{z+}$	$C_{60}{}^{z+}$	65±2	67±2	68

[a] Obtained using Eq. (3) from the appearance photon energies for formation of $C_{70-2n}{}^{z+}$ from C_{70}.
[b] Appearance internal energies of $C_{70}{}^{z+}$ for formation of $C_{70-2n}{}^{z+}$ at 25 μs after photoionization of C_{70}. The RRKM model is employed to derive the rate constants for reaction (2) from which the fractional abundance curves are calculated.

Analysis of the critical internal energies of the primary C_{70}^{z+}

We have employed the RRKM theory to derive the curves of fractional abundance (breakdown graphs) for C_{70}^{z+} and C_{70-2n}^{z+} ions ($n = 1 - 5$) as a function of the internal energy of the primary C_{70}^{z+} [18]. The microcanonical rate constant $k_n(\varepsilon)$ for reaction (2) of $C_{70-2n+2}^{z+}$ having internal excitation energy ε can be given by [2,3,8,9,19]

$$k_n(\varepsilon) = \frac{\alpha \, G^*(\varepsilon - E_0^n)}{h \, N(\varepsilon)} \qquad (n = 1 - 6) \qquad (4)$$

Here, α is the reaction path degeneracy, E_0^n is the critical activation energy for reaction (2), $G^*(\varepsilon - E_0^n)$ is the number of states for the transition state (activated complex), and $N(\varepsilon)$ is the density of states of $C_{70-2n+2}^{z+}$. For $n \geq 2$ the ε value of $C_{70-2n+2}^{z+}$ is computed under the assumption that the energy available after dissociation of $C_{70-2n+4}^{z+} \rightarrow C_{70-2n+2}^{z+} + C_2$ is statistically partitioned between $C_{70-2n+2}^{z+}$ and C_2. We have used Haarhoff's approximation [19] to calculate $G^*(\varepsilon - E_0^n)$ and $N(\varepsilon)$, assuming vibrational frequencies of $C_{70-2n+2}^{z+}$ to be replaced by those of a neutral C_{70} reported by Wang et al. [20] and Brockner and Menzel [21].

The critical activation energies E_0^n in eq. (4) are taken from the binding energies of $C_{70-2n+2}^{z+}$ ($n = 1 - 6$) for reaction (2) in the literature. We adopted a set of the binding energies reported by Gluch et al. [22] which we modified by normalizing to the binding energy of C_{60}^+ (= 9.2 eV) proposed by Wörgötter et al. for the process $C_{60}^+ \rightarrow C_{58}^+ + C_2$ [8]. Wörgötter et al. obtained this value by assuming a totally loose transition state in which the C_{58}^+ and C_2 fragments tumble freely. This transition state, named TS3 in [8], was consistent with a large frequency factor in the Arrhenius relation which has been expected by a very large rotational partition function of C_2.

Figure 3 shows the fractional abundance K_n for C_{70-2n}^{z+} ions with the TS-3 model as a function of the internal energy of the primary C_{70}^{z+} at 25 μs after photoionization of C_{70}. We defined the appearance internal energy $E_{RRKM}(n)$ for the formation of C_{70-2n}^{z+} as the internal energy of C_{70}^{z+} corresponding to $K_n = 0.03 K_n^{max}$. The values of $E_{RRKM}(n)$ determined from Fig. 3 are listed in the sixth column of Table 1. For the

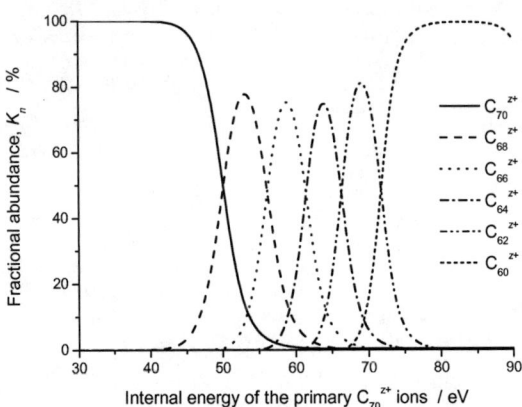

FIGURE 3. Fractional abundance curves of C_{70}^{z+} and C_{70-2n}^{z+} ions at 25 μs after photoionization of C_{70} obtained by using the RRKM model to calculate the rate constants for reaction (2).

formation of C_{68}^{z+}, C_{66}^{z+}, and C_{64}^{z+} $E_{RRKM}(n)$ are 44, 52, and 57 eV, respectively, in good agreement with the observed critical internal energies $E_{max}(1,z)$, $E_{max}(2,z)$, and $E_{max}(3,z)$, respectively. These results are the manifestation of the validity of the present statistical treatment at $n \leq 4$: large amounts of the internal energy of C_{70}^{z+} are equilibrated among the vibrational degrees of freedom, and subsequent fragmentation proceeds through reaction (2) via a transition state with the activation energy of E_0^n.

In future prospects, we are planning to perform photoelectron-photoion coincidence spectroscopy and photofragment momentum spectroscopy to elucidate statistical energy partitioning and consecutive C_2 ejection in highly excited fullerene cations.

ACKNOWLEDGMENTS

We are grateful to Mr. Y. Haruyama, Dr. Y. Takabayashi, and Mr. E. Kuwabara of Okayama University for valuable assistance during sample preparation. This work has been supported by a grant for scientific research from Research Foundation for Opto-Science and Technology.

REFERENCES

1 J. Kou, T. Mori, S. V. K. Kumar, Y. Haruyama, Y. Kubozono and K. Mitsuke, *J. Chem. Phys.* **120**, 6005-6009 (2004).
2 J. Kou, T. Mori, Y. Kubozono and K. Mitsuke, *Phys. Chem. Chem. Phys.* **7**, 119-123 (2005).
3 J. Kou, T. Mori, Y. Kubozono and K. Mitsuke, *J. Electron Spectrosc. Relat. Phenom.* **144-147**, 247-250 (2005).
4 H. Katayanagi, J. Kou, T. Mori, Y. Kubozono and K. Mitsuke, unpublished data.
5 P. Wurz and K. R. Lykke, *J. Phys. Chem.* **96**, 10129-10139 (1992); D. Ding, R. N. Compton, R. E. Haufler and C. E. Klots, *J. Phys. Chem.* **97**, 2500-2504 (1993); J. Laskin, B. Hadas, T. D. Märk and C. Lifshitz, *Int. J. Mass Spectrom.* **177**, L9-L13 (1998).
6 B. Walch, C. L. Cocke, R. Voelpel and E. Salzborn, *Phys. Rev. Letters* **72**, 1439-1442 (1994); T. LeBrun, H. G. Berry, S. Cheng, R. W. Dunford, H. Esbensen, D. S. Gemmell, E. P. Kanter and W. Bauer, *Phys. Rev. Letters* **72**, 3965-3968 (1994).
7 M. Foltin, M. Lezius, P. Scheier and T. D. Märk, *J. Chem. Phys.* **98**, 9624-9634 (1993); P. Scheier, B. Dünser, R. Wörgötter, M. Lezius, R. Robl and T. D. Märk, *Int. J. Mass Spectrom. Ion Proc.* **138**, 77-93 (1994).
8 R. Wörgötter, B. Dünser, P. Scheier, T. D. Märk, M. Foltin, C. E. Klots, J. Laskin and C. Lifshitz, *J. Chem. Phys.* **104**, 1225-1231 (1996).
9 R. K. Yoo, B. Ruscic and J. Berkowitz, *J. Chem. Phys,* **96**, 911-918 (1992).
10 T. Drewello, W. Krätschmer, M. Fieber-Erdmann and A. Ding, *Int. J. Mass Spectrom. Ion Process.* **124**, R1-R6 (1993).
11 M. Fieber-Erdmann, W. Krätschmer and A. Ding, Supplement to *Z. Phys. D* **26**, S308-310 (1993).
12 A. Reinköster, S. Korica, G. Prümper, J. Viefhaus, K. Godehusen, O. Schwarzkopf, M. Mast and U. Becker, *J. Phys. B* **37**, 2135-2144 (2004).
13 S. Aksela, E. Nõmmiste, J. Jauhiainen, E. Kukk, J. Karvonen, H. G. Berry, S. L. Sorensen and H. Aksela, *Phys. Rev. Letters* **75**, 2112-2115 (1995).
14 H. Yoshida and K. Mitsuke, *J. Synchrotron Rad.* **5**, 774-776 (1998); M. Ono, H. Yoshida, H. Hattori and K. Mitsuke, *Nucl. Instrum. Methods Phys. Res. A* **467-468**, 577-580 (2001).
15 J. Kou, T. Mori, M. Ono, Y. Haruyama, Y. Kubozono and K. Mitsuke, *Chem. Phys. Letters*, **374**, 1-6 (2003).
16 T. Mori, J. Kou, M. Ono, Y. Haruyama, Y. Kubozono and K. Mitsuke, *Rev. Sci. Instrum.* **74**, 3769-3773 (2003).
17 H. Steger, J. de Vries, B. Kamke, W. Kamke and T. Drewello, *Chem. Phys. Letters* **194**, 452-456 (1992).
18 Chemical reaction theory developed for evaluating the microcanonical rate constant of the unimolecular dissociation, by assuming equipartitioning of the available energy in the transition state of the energy-rich polyatomic molecule.
19 W. Forst, "Theory of Unimolecular Reactions", Chap. 6, Academic, New York, 1973.
20 X. Q. Wang, C. Z. Wang, and K. M. Ho, *Phys. Rev. B* **51**, 8656-8659 (1995).
21 W. Brockner and F. Menzel, *J. Molec. Structure* **378**, 147-163 (1996).
22 K. Gluch, S. Matt-Leubner, O. Echt, B. Concina, P. Scheier, and T. D. Märk, *J. Chem. Phys.* **121**, 2137-2143 (2004).

Direct measurement of spectral momentum densities of ordered and disordered semiconductors by high energy EMS

C. Bowles, M.R. Went, A.S. Kheifets and M. Vos

Atomic and Molecular Physics Laboratories, Research School of Physical Sciences and Engineering, Australian National University, Canberra ACT 0200, Australia

Abstract. High Energy solid state electron momentum spectroscopy (EMS) is capable of directly measuring spectral functions of ordered and disordered solid matter. In this paper we investigate the spectral functions for the group IV semiconductors Ge and Si. We attempt to resolve the electronic structure differences in amorphous, polycrystalline and crystalline atomic arrangements of the semiconductors. We examine the experimental differences in polycrystalline and amorphous Ge, and draw conclusions as to the similarities/differences between the two states of matter.

Keywords: Electron Momentum Spectroscopy, band structure, semiconductor, electronic structure
PACS: 61.43.Dq, 71.20.Mq, 71.23.Cq, 73.61.Cw

1. INTRODUCTION

Semiconductor samples can be prepared in an amorphous, polycrystalline or single-crystal form. Each form is expected to have a different electronic structure resulting from differences in the degree of short and long range order. In this paper we will present experimental spectral functions for these three states of order for semiconductors. Single crystals have both short and long range order. Polycrystalline samples consists of many small randomly oriented single crystals separated by grain boundaries. Except for the small number on atoms that are located at the grain boundaries each atom is in a very similar environment as in a single crystal. The measured electronic structure of polycrystalline samples is thus expected to resemble the spherically-averaged electronic structure of a single crystal. Amorphous semiconductors can be described by a continuous random network (CRN) [1]: each atom still has a co-ordination number of four but is positioned in a distorted tetrahedron with a distribution of bond lengths and bond angles. CRN structures are more ordered than a classic amorphous solid. With limited short range order in the CRN samples the question arises, how different is the polycrystalline and amorphous electronic structures of semiconductors?

From a theoretical point of view the amorphous phase presents a challenge as the electronic structure can not be described by Bloch functions due to the absence of a periodicity in the potential. One approach is to approximate the amorphous phase as a crystal with an extremely large, disordered, unit cell [2], another approach is based on Green's function techniques [3]. We will present experimentally measured spectral functions to try to resolve the validity of some of the theoretical assumptions.

In EMS a high energy electron impinges upon a solid target, scattering from a bound electron which is ejected. These two electrons are then measured in coincidence and via

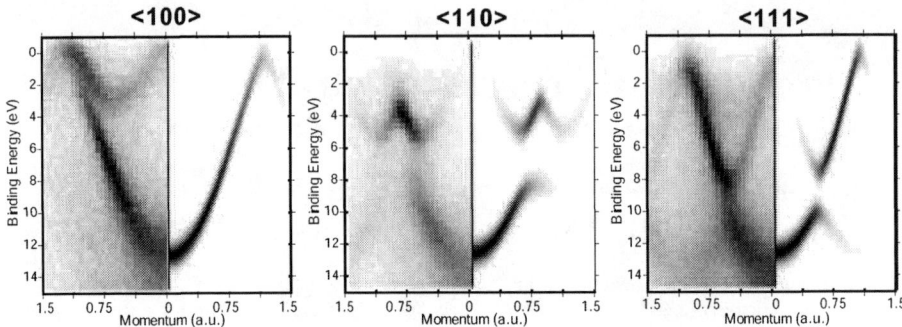

FIGURE 1. The spectral functions for a single crystal silicon sample measured for momenta along the a)< 100 >, b)< 110 >, c)< 111 > crystallographic direction. Experimental results are shown in the left half, theoretical calculations in the right half of each panel.

conservation laws we can directly determine the binding energy and the momentum of the bound electron an instant before the collision took place [4]. The EMS cross-section can be shown to be directly proportional to the target spectral function, ie the modulus squared of the momentum space one electron wavefunction [5]. In our spectrometer the detectors are positioned in such a way that only target electrons with momentum along the vertical direction contribute to the coincidence count rate. Thus for single crystal targets the anisotropy in the electronic structure can be resolved by rotating the sample. In EMS real momentum of the bound electron is measured, not the crystal momentum and thus polycrystalline and amorphous solids are also viable targets. By using EMS to measure the the spectral functions of ordered and disordered semiconductors we attempt to establish experimentally to what extent their electronic structures differ.

2. CRYSTALLINE SILICON

The experimental spectral function of crystalline Si has been measured by the solid state EMS group at the Australian National University for experimental information see Vos et al. [6, 7]. In Fig. 1 we show the experimental spectral functions for the three high symmetry directions of single crystal Si and compared them to calculations based on full potential linear muffin tin orbital (FP-LMTO) density functional theory. Anisotropy in the electronic structures of the Si crystal result in large variations of the EMS results for different crystal directions (Fig. 1). The FP-LMTO theory matches the measured band dispersion amazingly well. Theory fails to accurately predict the relative band intensities, with, for example, theory predicting the bottom of the band to be most intense, whereas in the experiment the top of the band has the largest intensity. The clear differences between the measured and calculated spectral functions for the Si <100> direction can be attributed to finite experiment momentum resolution [8]. These results have been analysed in more detail by Kheifets et. al. [9] and Bowles et. al. [10]. With the validity of the method established by the good agreement between experiment and theory for single crystal samples, we are now in a good position to study the less

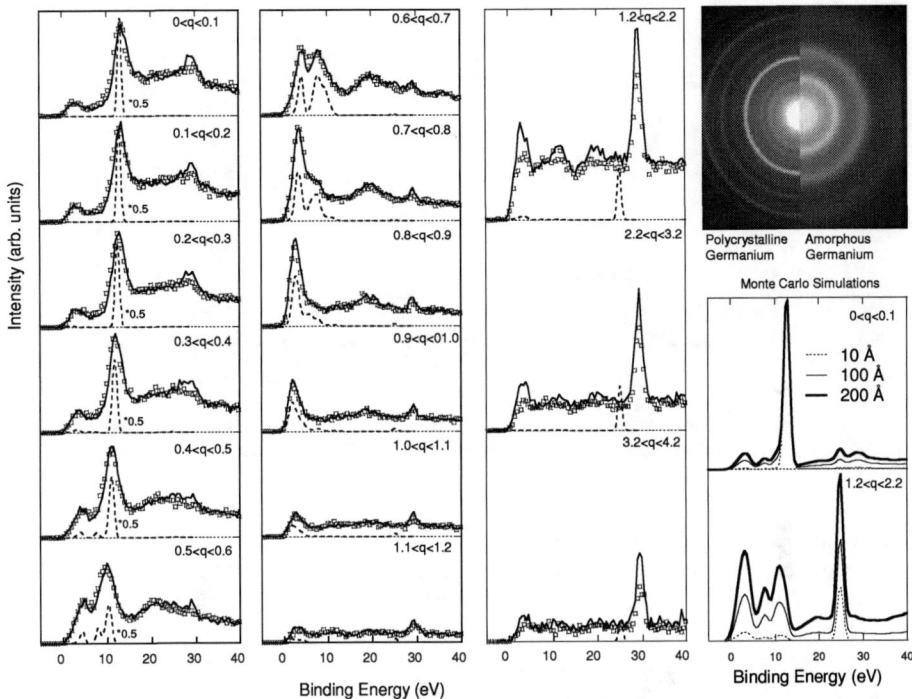

FIGURE 2. The experimental spectra of amorphous Germanium (squares), polycrystalline Germanium (full line) at momentum intervals as indicated. Also shown are calculated spectra based on the spherically averaged theory of single crystal. Intensity of theory in the left panel is divided by 2. The diffraction patterns shown at the right is measured *in situ* for the polycrystalline and amorphous film. The lower right panel shows the influence of the thickness on the 3d intensity as derived from Monte Carlo simulations.

understood amorphous form, and compare its electronic structure with the results for polycrystalline films.

3. POLYCRYSTALLINE VERSUS AMORPHOUS GERMANIUM

Amorphous semiconductor films, suitable for EMS can be made by evaporation on thin (30 Å) carbon films. Polycrystalline films can be made by annealing these films. However in the case of silicon a reaction at the Si/C interface produced silicon carbide [11]. This problem does not occur for germanium. The measured spectral function for amorphous and polycrystalline germanium as well as their electron diffraction patterns are shown in Fig. 2. The polycrystalline diffraction pattern is a 360° smearing of the single crystal diffraction pattern, resulting in sharp concentric circles. Amorphous Ge samples however with their large distribution of bond lengths and angles give much broader diffraction rings.

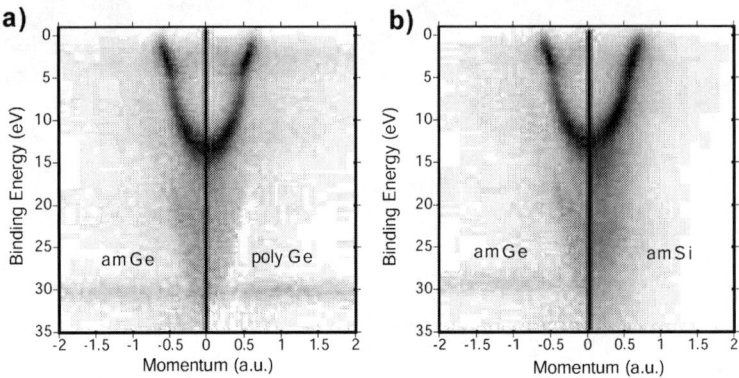

FIGURE 3. a) The measured spectral functions for disordered Ge, amorphous (left) and polycrystalline (right). (b) The measured spectral functions of amorphous Ge (left) and amorphous Si (right).

It is thus surprising that, in spite of the large differences in the electron diffraction patterns, the measured spectral function of polycrystalline and amorphous germanium are very similar indeed. The two sets of spectra were normalised using a single factor. Within the statistical limits the spectra near zero momentum are identical, with the exception that the plasmon loss peak (16 eV below the main peak) is slightly more pronounced in the polycrystalline case. This is in agreement with the findings of Zeppenfeld and Raether that the plasmon energy loss peak is more intense in electron energy loss spectra of polycrystalline samples compared to amorphous samples [12]. The second difference is in the spectrum for the 0.6-0.7 a.u. momentum range where there is a minimum in intensity near 5 eV for the polycrystalline case. In the amorphous case this minimum is less pronounced. This momentum range coincides for many crystalline orientations with the first Brillouin zone boundary, and hence we are sensitive for the splitting between the inner and outer valence band in these momentum range. In spite of the spherical averaging this splitting is still evident in the polycrystalline data. The fact that this is less evident for the amorphous case is expected, as the concept of the first Brillouin zone is not so well defined for amorphous semiconductors where the atomic bonds and angles vary by about 10% [13] relative to the crystalline case. Band gaps between the conduction band and valence band are found experimentally and in calculations for amorphous semiconductors [14, 15].

Most surprisingly the most pronounced differences are found in the high momentum spectra (above 1.2 a.u.) of Fig. 2. In the polycrystalline case the 3d electron appear systematically more intense compared to the amorphous case. The normalisation constant of the two measurements was chosen such that the valence band features had equal intensity. Normalisation of the 3d features to equal height would make the amorphous Ge valence band more intense than the polycrystalline one. The most likely cause of this phenomenon is that the variation in intensity is due to different elastic multiple scattering contributions due to different sample thicknesses. This possibility was investigated in Monte Carlo calculations, using the result of the FP-LMTO calculation as input [16], but assuming different sample thicknesses. These results are shown in Fig. 2 as well.

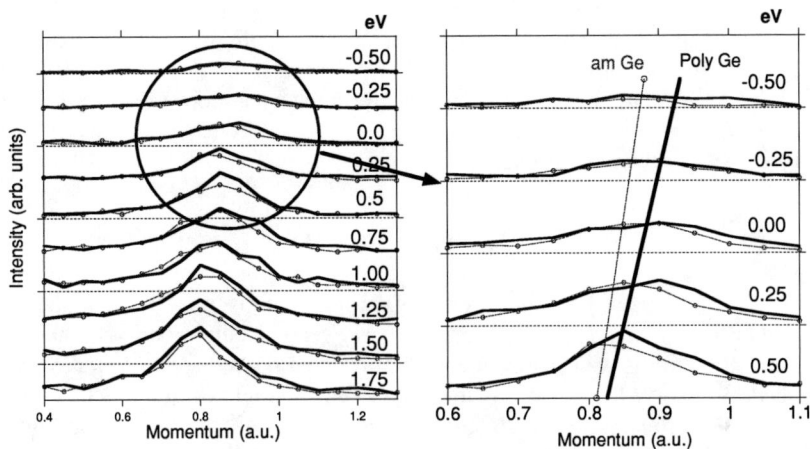

FIGURE 4. Left: Momentum profiles at different binding energies. The 'dispersion' appears weaker for the amorphous sample compared to the polycrystalline one. Right: a closer look at the differences in the slope of the sp-hybridised band of the amorphous and polycrystalline spectral functions.

Consider first the intensity at the bottom of the valence band. Elastic scattering will remove intensity away from zero momentum, and will cause intensity to be shifted from the peak at zero momentum to the background at the same energy, but larger momentum values. For the non-dispersing 3d feature elastic scattering will cause intensity still to contribute to the 3d peak, but now at the 'wrong' momentum. Hence it appears that the 3d intensity increases with thickness relative to the valence band intensity.

In Fig. 2 we show the results of the FP-LMTO calculations as well. In this calculation we treated the 3d electrons as valence electrons. In this way we obtain intensities for valence electrons and the 3d electrons in a uniform way. The calculated Ge 3d position is at somewhat smaller binding energy as the measured 3d position. Due to life time broadening (not included in the calculation) the maximum peak height near the bottom of the band is larger in the theory than in the experiment. For easy comparison we re-scaled theoretical intensity in the left panel of Fig. 2. The measured 3d intensity is significantly larger than the calculated one, another indication that elastic multiple scattering reduces the valence band intensity more than the 3d intensity.

We compare the measured spectral function of amorphous germanium as a grey-scale plot with that of polycrystalline germanium Fig.3(a) and with amorphous silicon in Fig. 3(b). Again the differences are minor. For Ge the 3d level can just be distinguished near 29.5 eV binding energy, and this feature is of course absent in Si. The silicon features are somewhat broader. The amorphous silicon spectra resemble the theoretical results of Hickey and Morgan calculations at least semi-quantitatively. [3]

Upon examining the spectral function of amorphous and polycrystalline Ge one more difference is noticeable. The slope at the top of the bands near the Fermi level is slightly different. This point is emphasised in Fig. 4. The band gap is due to a periodic potential that interacts strongly with electron states with **k** vectors near the Brillouin zone boundary. In the amorphous case this periodic potential is less well defined due to the lack of long range order and the dispersion for a more disordered sample would deviate less from a free electron behaviour. The polycrystalline slope near the top of the band (13.3 ±3.6 eV/a.u.) is much smaller then the amorphous slope (20.4 ±8.1 eV/a.u.). This electronic structure effect could be an experimental indication of the crystalline order differences of the two samples.

4. CONCLUSION

Single crystal Si spectral functions were shown in comparison to full potential linear muffin tin orbital calculations. Agreement was in general quite good. The anisotropy of the band structure is well resolved and the observed band dispersion was very well reproduced by theory. Based on this understanding of single crystal silicon result we want to compare the electronic structure of amorphous silicon and amorphous and polycrystalline germanium. For momenta near the edge of the first Brillouin zone the spectra are split in a lower and upper band contribution for polycrystalline Ge but somewhat less for amorphous Ge and Si. This can be attributed to the lack of long range order in the latter cases. A noticeable difference in the Ge 3d level to valence band intensity ratio between amorphous and polycrystalline Ge and amorphous Ge was found but is not understood. The position of the maxima in the Ge momentum profiles near the top of the band are more dependent on their binding energy than the amorphous ones. Besides these minor differences we find a surprisingly large similarity between the amorphous and polycrystalline Ge spectra.

This research was made possible by a research grant from the Australian Research Council.

REFERENCES

1. W. Zachariasen, *J. Am. Chem. Soc.*, **54**, 3841–3851 (1932).
2. S. Bose, K. Winer, and O. Andersen, *Phys. Rev. B*, **37**, 6262–6277 (1988).
3. B. Hickey, and G. Morgan, *J. Phys. C:Solid State Phys.*, **19**, 6195–6209 (1986).
4. I. McCarthy, and E. Weigold, *Rep. Prog. Phys.*, **54**, 789–879 (1991).
5. E. Weigold, and I. McCarthy, *Electron momentum spectroscopy*, Physics of Atoms and Molecules, Kluwer Academic, 1999.
6. M. Vos, G. Cornish, and E. Weigold, *Rev. Sci. Instrum.*, **71**, 3831–3840 (2000).
7. M. Vos, and E. Weigold, *J. Elec. Spec. Rel. Phenom.*, **112**, 93–106 (2000).
8. M. Vos, V. Sashin, C. Bowles, A. Kheifets, and E. Weigold, *J. Phys. Chem. of Sol.*, **65**, 2035–2039 (2004).
9. A. Kheifets, V. Sashin, M. Vos, E. Weigold, and F. Aryasetiawan, *Phys. Rev. B*, **68**, Art. No. 233205 (2003).
10. C. Bowles, A. Kheifets, V. Sashin, M. Vos, and E. Weigold, *J. Elec. Spec. Rel. Phen.*, **141**, 95–104 (2004).
11. Y. Q. Cai, M. Vos, P. Storer, A. S. Kheifets, I. E. McCarthy, and E. Weigold, *Solid State Communications*, **95**, 25–29 (1995).
12. K. Zeppenfeld, and H. Raether, *Zeitschrift für Physik*, **193**, 471 (1966).
13. S. Moss, and J. Gracyk, *Phys. Rev. Lett.*, **23**, 1167 (1969).
14. D. Weaire, *Phys. Rev. Lett.*, **26**, 1541–1543 (1971).
15. D. Weaire, and F. Thorpe, *Phys. Rev. B*, **4**, 2508–2520 (1971).
16. M. Vos, and M. Bottema, *Physical Review B*, **54**, 5946–5954 (1996).

Towards Electron Momentum Spectroscopy Studies of Clusters: A New Apparatus

KL Nixon, G Hewitt, B Gilbert, A Dunn, R Northeast, M Ellis, DS Slaughter, P Euripides, WD Lawrance and MJ Brunger

School of Chemistry, Physics and Earth Sciences, Flinders University, GPO Box 2100, Adelaide 5001, South Australia, Australia.

Abstract. This paper reports on our progress in attempting to realise an apparatus for studying atomic and molecular clusters. Specifically, we describe our design and present some preliminary results for a new triple coincidence electron momentum spectrometer. This apparatus is an (e,2e + ion) configuration where the two electrons from an electron impact ionisation event and the residual ion are detected in coincidence in order to probe the intermolecular bonding in van der Waals clusters.

Keywords: van der Waals clusters, electron momentum spectroscopy.
PACS: 34.80.Gs

INTRODUCTION

The intrinsic chemistry of gas phase reactions is now quite well understood. A comparable molecular level understanding for reactions in solutions is the driving force for many present spectroscopic studies, as intermolecular interactions involved in solvation play an important role in the reaction dynamics of solution chemistry. van der Waals clusters, which are two or more atoms or molecules held together by the same intermolecular forces controlling condensed phase behaviour, provide a means to study these crucial intermolecular interactions in the gas phase.

Electron momentum spectroscopy (EMS) is a direct probe of the molecular electronic wavefunction. It has successfully been applied to atomic and molecular targets in the gas phase as well as solid-state targets [1]. The targets, applications and apparatus of EMS experiments have been continually developed since the establishment of the technique. Recent EMS studies of large organic molecules, and the corresponding theoretical calculations [2], demonstrate the success of applying EMS to increasingly complex targets in order to investigate the nature of the intramolecular bonding and chemically important properties of the molecules. Applying EMS to van der Waals clusters is thus a natural evolution for this technique, but it provides a significant technical challenge in terms of its practical realisation. Nonetheless, EMS is in principle an ideal technique to provide detailed information on the intermolecular bonding in van der Waals clusters, due to its high sensitivity to electrons with small momenta, ie those far from the nucleus. It is precisely these electrons that are involved in the intermolecular bonding within the clusters. The experimental results will therefore provide data on the electronic wavefunctions of the clusters and enable us to evaluate the quality of theoretical calculations of non-covalent interactions.

A new EMS apparatus is being developed at Flinders University to investigate intermolecular interactions by studying van der Waals clusters in the gas phase. The remainder of this paper outlines the experimental design and our progress in apparatus construction and testing, which makes these experiments possible.

EXPERIMENTAL DESIGN CONSIDERATIONS AND APPARATUS

EMS experiments use kinematically complete electron impact ionisations, ie (e,2e) events, where both outgoing electrons, the scattered incident electron and the ejected target electron, are detected and analysed in coincidence, to probe a target. When studying molecules, the conditions for the kinematic regime are chosen such that a plane wave impulse approximation description for the reaction mechanism should be valid [1].

van der Waals clusters are formed in an ultra cold gas using a supersonic free jet expansion. A supersonic expansion creates a range of cluster sizes, therefore the ability to determine the cluster size associated with the detected (e,2e) event is crucial. To this end, a time of flight mass spectrometer (ToFMS) needs to be used to analyse the ions produced from (e,2e) events for not only their mass, but also to determine if fragmentation occurred. Employing a supersonic expansion requires the experiment to be pulsed which reduces the sampling rate for data collection. Additionally, the supersonic expansion jet consists of only a fraction of the desired clusters seeded in a carrier gas, reducing the target density and therefore the count rate. Thus, the EMS spectrometer must be designed for maximum count rates to compensate for these effects. To this end, relatively low incident energies will be used as the (e,2e) cross section will be larger, an asymmetric geometry will be employed, again as the cross section is relatively higher, optics capable of 2-D imaging coupled to large active area detectors will be used, as will a carrier gas with a high ionisation energy to avoid interference with the cluster ionisation. Based on their experience for typical number densities obtained in cluster formation in pulsed supersonic expansions, (e,2e) count rates obtained with a 1-D system at a less favourable geometry and the improvement in (e,2e) count rates in going from a 1-D to a 2-D system as found by Storer [3], Lawrance and Brunger estimated that the cluster count rates would be comparable to those found in the current effusive source 1-D Flinders EMS apparatus [4].

The vacuum system of our new apparatus consists of two stainless steel chambers, the expansion chamber in which the clusters are produced, and the collision chamber which houses the spectrometer. The chambers are 450 mm diameter, 410 mm high pumped with two VHS 10 diffusion pumps and 650 mm diameter, 520 mm high pumped with two VHS 6 diffusion pumps, respectively. The cold traps on each of the four pumps are cooled with recirculated methanol at ~ -70°C. Typical pressures of 3×10^{-8} and 7×10^{-8} torr are achieved before baking.

van der Waals clusters will be produced in a supersonic expansion using a piezoelectric pulsed nozzle operating at ~ 100 Hz. The vertically propagating beam is skimmed (Au plated Cu skimmer with an orifice diameter of 1 mm) before it enters the interaction region.

The incident beam of electrons over an energy range of 800 – 2kV is produced by an electron gun designed and fabricated in-house. A Pierce element is used to extract the electrons emitted by a thoriated tungsten filament. The electrons are then accelerated and focussed by two three-element cylindrical lens stacks. Three sets of X and Y deflectors and three collimating apertures are incorporated throughout the gun. The aim of the gun is to produce a stable, high-current and well-collimated electron beam. All elements were fabricated from titanium with molybdenum apertures.

An asymmetric geometry is used to detect the two outgoing electrons of interest. The scattered analyser remains fixed at a polar angle (θ) of ~ 20° with respect to the incoming electron beam, while the ejected electron analyser is nominally at $\theta = 68°$ and rotated ± ~ 40° to sample a range of momenta. Currently, the two outgoing electrons are decelerated and focussed with in-house designed cylindrical lens stacks. Each stack consists of two three-element lenses, with two sets of deflectors and a collimating aperture incorporated throughout each of the stacks. All the lens elements were fabricated from titanium with molybdenum apertures. Hemispherical deflectors with mean radius of 90 mm are coupled to each lens stack to disperse the electrons in terms of their energy and the azimuthal angle (ϕ) with which they leave the interaction region. These electrons are subsequently detected with 40 mm active area MCP/RAE 2-D detectors [5]. The large radius hemispherical deflectors should provide reasonable energy resolution, which is essential for selecting non dissociative states of the cluster ion. The inner and outer hemispheres were hydroformed from non-magnetic stainless steel.

The ToFMS is mounted above the scattering plane to detect the resulting cluster ions leaving the interaction region. The ToFMS is based on the Wiley and McLaren design [6] with a two-stage acceleration region and 2nd order space focussing [7] to minimise the flight time broadening due to the spatial distributions of the ion beam and to give better mass resolution. Geometry constraints restrict the length of the drift tube and hence impact on the mass resolution attainable. As the ToFMS is used to distinguish between cluster sizes, of say H_2O or benzene as an example, the mass resolution requirements are quite relaxed and should be easily attainable in the length available. Short flight times to maximise the sampling rate, at the cost of mass resolution, are also achievable considering the required mass resolution. In addition to cluster size identification, the ToFMS can be used to give information about the dissociation energy of the detected fragments when operated in an ion imaging mode and therefore can distinguish between detected ions resulting from a simple ionisation event or from an ionisation and fragment process. All elements in the current ToFMS were fabricated from non-magnetic stainless steel.

Helmoltz coils and CO-NETIC shielding have been implemented to give a field free interaction region and minimal magnetic fields throughout the electron's trajectory. The Helmholtz coils are tuned to give zero magnetic field in the interaction region which leaves residual magnetic fields of less than 20 mGauss at the extremities of the collision chamber.

The diagnostic signals from the experiment are sampled by two multifunction simultaneous sampling boards (United Electronic Industries: PD2-MFS-8-1M-12). The analog-input subsystem of each board consists of eight simultaneous sample and hold buffers and a multiplexer. The multiplexer is configured to read these buffers in single-ended mode. This mode is adequate, because the input signal levels are at least 200 mV, which is an order of magnitude greater than the ambient noise. The order in which the buffers are read is determined by the board's channel list FIFO buffer. The

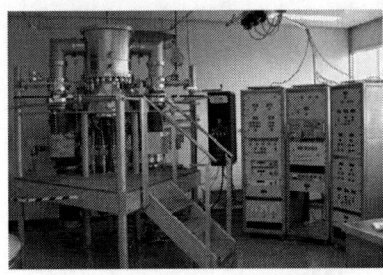 

FIGURE 1. Digital photographs of the new triple coincidence spectrometer at Flinders University. a) external vacuum chambers and control electronics of the apparatus. b) Internal elements of the spectrometer showing the electron gun, two hemispherical deflectors and the ToFMS

multiplexed buffer signals are coupled to a 12-bit ADC through a programmable gain amplifier. The converted values are transferred from the on-board 16 KB FIFO to a 300 KB buffer in the memory space of the data acquisition application. The board's driver uses programmed I/O to transfer the data into the 300 KB buffer at a rate of up to 1 million samples per second. The digital-analog subsystem of the board consists of two 12-bit D/A converters. The output range of these converters is ± 10 V. One of the 12-bit D/A converters is used to control the ramp supply of the electron gun. During a single experimental scan, the ramp supply is ramped from 20 – 50 V. This corresponds to a 30 eV variation in the incident electron energy. Based upon a predicted count rate of ~ 100 counts/minute, it is expected that a typical scan should take about five hours to complete. During a scan, the collected data is stored in a memory file in the host PC's main memory space. Once the scan is completed, the memory file is saved to disk.

The software needed to drive the (e,2e) data collection is essentially complete, although work to incorporate the ion channel is on-going.

The entire coincidence spectrometer was fabricated in the mechanical workshop at Flinders University. The experiment is controlled by electronics developed and constructed in the Flinders University electronics workshop. Figure 1a shows the external vacuum chambers and control electronics, while Fig. 1b shows the important internal elements of the spectrometer.

PRELIMINARY RESULTS

The electron gun has been optimised and characterised giving an incident electron beam of 200 μA current at 800 eV with a diameter of 2.5 mm. The gun produces very stable electron beams over long periods of time and therefore fulfils our original performance criteria. An energy resolution of ~ 0.9 eV is obtained for the scattered and ejected electron analysers using a mean analysing potential of 60 eV. This can be seen in Fig. 2 which shows an elastic scattering peak being moved across the detector as the incident beam energy is varied. This figure indicates a simultaneous energy detection of ~ 10 eV could be achieved, in agreement with theory. It also indicates the excellent linearity of the energy across the plate.

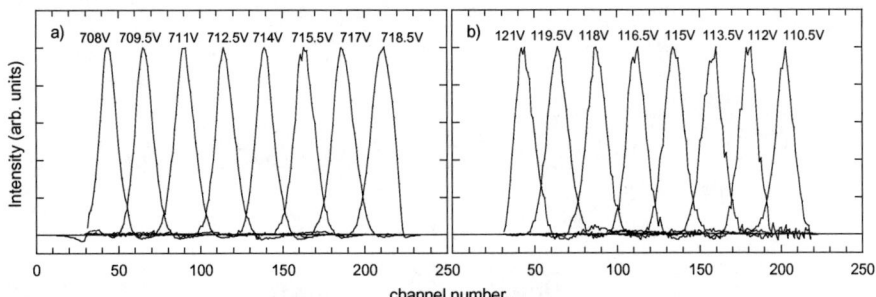

FIGURE 2. Elastic scattering peak being moved across the detector in the a) scattered electron analyser and b) ejected electron analyser. Data has been corrected for any inelastic scattering signal. The voltages shown are the kinetic energy of the incident electron beam.

SIMION MODELLING

CPO-3D [8] modelling of the cylindrical lens stacks currently installed suggested that they would be successful in providing ϕ resolved information over ± 5° and ± 4° for the scattered and ejected analysers, respectively. This corresponds to a target electron momentum range of 0.4 a.u. being sampled at a fixed geometry. However, SIMION [9] simulations of these optics give rather different results. A transmitted ϕ range of ± 4° and ± 2° for the scattered and ejected analyser, respectively, is obtained. This ϕ detection range corresponds to a target electron momentum range of ± 0.26 a.u. at a fixed geometry. SIMION simulations of the electron trajectory through the lens stack and hemispherical deflector give an arrival position spread for the ϕ range very close to the spatial resolution of the 2-D detectors, however, a much greater spread is achieved with CPO-3D. A theoretical detector image of the ejected electron optics in this configuration is given in Fig. 3a. Experimental measurements confirm the predictions of SIMION in that the ϕ spread is too narrow to obtain simultaneous momentum information with this lens stack. In these experiments the transmitted electrons were restricted by a calibration aperture consisting of a vertical series of 0.25 mm diameter holes at 1.5° intervals. Alternative sets of optics were therefore developed, based upon the slit lens design of Vos *et. al.* [10], to significantly improve the ϕ range and resolution. SIMION simulations of these lenses give a ϕ range of ± 6° in both analysers, corresponding to a target electron momentum range of ± 0.52 a.u. being sampled at a fixed geometry. A theoretical detector image of the ejected electron optics in this configuration is given in Fig. 3b, with the improved performance over the original design being manifest. These new lenses are currently being fabricated from titanium.

SUMMARY AND FUTURE WORK

All of the components for this apparatus, except the slit lenses, have been fabricated and are ready to be utilised. As mentioned above, the cylindrical lens stacks are installed and operational.

Preliminary testing of the apparatus with a well understood, simple, atomic target

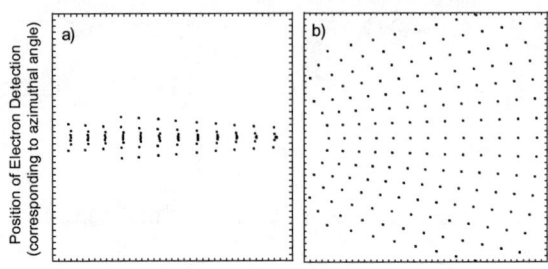

FIGURE 3. Theoretical detector images of the two alternative sets of optics for the ejected electrons modelled in SIMION. Axes represent the 40 mm active area of the detectors. a) shows the image predicted when using the cylindrical lens stacks while b) represents the predicted image for the slit lenses. In each case the electrons exit the interaction region in 1 eV kinetic energy intervals and 1° ϕ intervals.

gas will be carried out while the slit lens stacks are being fabricated. However, once they are available and characterised a pulsed (e,2e) experiment will be carried out, taking advantage of the supersonic expansion capabilities of the apparatus. This provides a means to measure conformationally pure (e,2e) spectra, as the molecules in a supersonic expansion are sufficiently cold to limit their isomers to the lowest energy conformer.

Implementation of the ToFMS, and preliminary characterisation and optimisation of the cluster beam and detection timing will also shortly be performed, initially using a laser as the ionisation source. Finally, these two aspects of the experiment will be combined to achieve the triple coincidence (e,2e+ion) experiment.

ACKNOWLEDGMENTS

This work was supported by the Australian Research Council. DSS acknowledges the Elaine Martin and Flinders University Travel Scholarships. KLN and DSS were supported by the Ferry Scholarship Trust.

REFERENCES

1. E Weigold and IE McCarthy, *Electron Momentum Spectroscopy*, Kluwer Academic/Plenum Press, New York (1999).
2. H Mackenzie-Ross, MJ Brunger, F Wang, W Adcock, N Trout, IE McCarthy and DA Winkler, *J. Electron Spectrosc. and Relat. Phenom.* **123**, 389 (2002).
3. PJ Storer, PhD thesis, Flinders University (unpublished) and private communication.
4. MJ Brunger and W Adcock, *J. Chem. Soc., Perkin Trans.* **2**, (2002)
5. Quantar Technology Inc.: www.quantar.com
6. WC Wiley and IH McLaren, *Rev. Sci. Instrum.* **26** (1955).
7. U Bosel, R Weinkauf and WE Schlang, *Int. J. Mass Spectrom. and Ion Processes* **112** (1992).
8. FH Read and N Bowring, CPO-3D, Manchester University.
9. DA Dahl, SIMION 3-D version 6.0 Users Manual, Princeton Electronic Systems, Inc. Princeton (1995).
10. M Vos, VA Sashin, C Bowles, AS Kheifets and E Weigold, *J. Phys. Chem. of Solids* **65** (2004).

Super-elastic Scattering from Ca and Rb in a Magnetic Angle Changing Spectrometer

Andrew Murray[†], Martyn Hussey, William MacGillivray[*] & George King

*University of Manchester, Manchester, UK *University of Southern Queensland, Queensland, Australia †Email: Andrew.Murray@manchester.ac.uk*

Abstract. Super-elastic scattering measurements at 20eV incident energy are presented for calcium and rubidium, contrasting differences between alkali and alkali-earth targets. The laser excitation of calcium at 423nm is detailed, indicating the effects of optical pumping on the super-elastic scattering process. A new type of spectrometer is described which exploits a controlled magnetic field in the interaction region, allowing super-elastically scattered electrons to be detected over all geometries. The effects of the magnetic field on the laser-atom interaction are considered for calcium.
Keywords: Super-elastic scattering, MAC spectrometer, Alkali-earth target.
PACS: 34.80.Dp, 33.80.Be

INTRODUCTION

Super-elastic scattering is a powerful technique allowing the excitation of atoms by electron impact to be detailed. This technique can be considered as the 'time inverse' of conventional electron-photon coincidence measurements, the photon detector being replaced by a polarized laser source, and the electron gun and analyser being swapped in position and energy (thereby reversing the direction of the incident and scattered electron momentum vectors) [1-3].

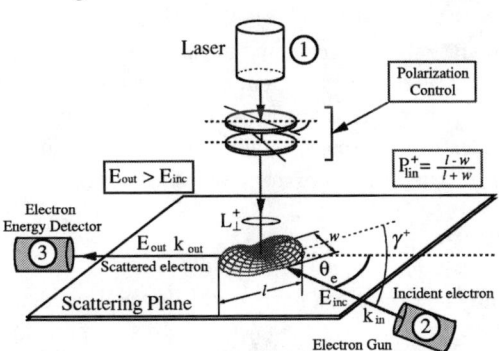

FIG 1. The superelastic scattering geometry, showing the position of components. For details, see text.

Figure 1 shows the configuration adopted. A laser beam of well defined wavelength, polarization and direction excites an atom, which is then de-excited by electron impact. The outgoing electrons leave the interaction region with higher energy that the incident beam (super-elastic scattering), and are detected by an energy-selecting analyser. The rate of super-elastically scattered electrons is determined as a function of scattering angle θ_e and polarization of the laser beam using the so-called pseudo-Stokes parameters [4-6]. These parameters are related to conventional coincidence Stokes parameters through optical pumping parameters that describe the laser-atom interaction [4]. For calcium, the relationship between the pseudo-Stokes parameters and conventional Stokes and Atomic Collision Parameters (ACP's) is simple [6], whereas for rubidium this needs to be determined both from experiment and through modeling [5].

Since only the *rate* of electrons is measured, information on the excitation process is obtained many times faster than using coincidence methods. This allows the interaction to be determined over a wide range of scattering angles θ_e, up to the limit imposed by the spectrometer. These limits are set by the physical size of the electron gun and analyser, preventing results being obtained at $\theta_e > 125°$ for most spectrometers.

To eliminate these restrictions a new Magnetic Angle Changing (MAC) spectrometer has been designed to allow the full scattering geometry to be explored, with $0° \leq \theta_e \leq 180°$. This spectrometer uses a controlled magnetic field in the interaction region so that the incident and scattered electrons are steered to new trajectories, while eliminating magnetic fields within the gun and analyser [7]. Figure 2 shows the trajectories for 20eV elastic and ~23eV super-elastically scattered electrons (as produced from calcium excitation), for $\theta_e = 0°$ and 180°. As can be seen, the MAC spectrometer steers the trajectories to angles which allow all electrons to be detected.

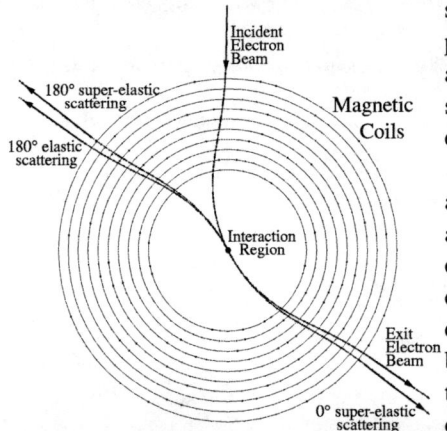

FIG 2. Electron trajectories in the new MAC spectrometer for 20eV incident energy.

In this paper results for Ca and Rb obtained at 20eV using two spectrometers of limited angular range ($\theta_e \leq 125°$) are reported, contrasting results between alkali and alkali-earth targets [5,6]. Differences are clearly evident, which may be due to correlations between the valence electrons in the alkali-earth target compared to the alkali target. Results from different theoretical models are also shown, no model reproducing the data over all scattering geometries at this energy [5,8].

New superelastic studies from calcium and rubidium are about to start using the MAC spectrometer. A complication arises since the laser-atom interaction occurs in a magnetic field. Determination of the optical pumping parameters is hence more complex, and these effects are currently being explored theoretically and experimentally. A brief description of the status of this work is given, together with a description of the new spectrometer.

RESULTS FOR Rb & Ca AT 20eV.

Super-elastic scattering data have recently been obtained for both calcium and rubidium targets at 20eV equivalent impact energy [5,6]. Pseudo-Stokes parameters and optical pumping parameters were measured, allowing the natural frame Atomic Collision Parameters (ACP's) $P_{lin}^+, \gamma, L_\perp^+$ to be ascertained [1,2]. The results of this analysis are shown in figure 3, where the ACP's are compared to theoretical calculations using Relativistic Distorted Wave Approximations (RDWA) [5,8], a Distorted Wave Born Approximation (DWBA) [5] and a Convergent Close Coupling approximation (CCC) [5].

The RDWA model of Srivastava et al [8] is shown for calcium. This model does not predict the deep minimum in P_{lin}^+ observed, but does reproduce the charge cloud angle γ accurately. The positions of the minima in $L_\perp^+$ are predicted, however the magnitude is not correct. By contrast, at an incident energy of 35eV (not shown here), the model reproduces the experimental data well, indicating that this model is accurate at higher energies.

The data for rubidium are very different from calcium. No deep minimum in P_{lin}^+ is seen in Rb, and the charge cloud angle γ reverses direction at $\theta_e \sim 45°$. $L_\perp^+$ indicates the charge

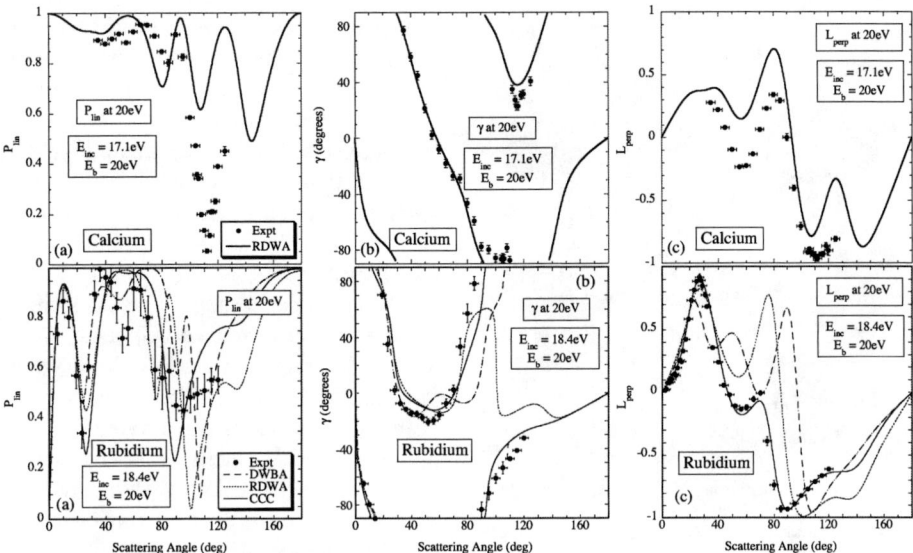

FIG 3. ACP's for Ca & Rb at 20eV compared to theory. For a description, see text.

cloud rapidly becomes almost fully oriented at low scattering angles, before reversing direction to be almost fully oriented in the opposite direction at $\theta_e \sim 85°$. The CCC model reproduces $L_\perp^+$ superbly, and predicts γ well, in contrast to other models. No theory reproduces P_{lin}^+ accurately for this target.

One of the key motivations of the present studies is to provide data at low to intermediate energies, allowing cross-over between theories accurate in different regimes. It is hoped that this will lead to refinement and unification of different models over all energies.

THE SUPER-ELASTIC MAC SPECTROMETER.

Figure 4 shows the new MAC spectrometer which is being developed. The oven is heated using thermo-coax which surrounds a crucible filled either with calcium or rubidium inside five heat shields constructed of 310 stainless steel. The oven emits through a molybdenum nozzle, and is collimated in the front chamber by a skimmer and heated aperture. This produces an atomic beam of narrow angular divergence and low Doppler profile orthogonal to the beam direction. The atomic beam is collected on a liquid nitrogen cooled trap located opposite the oven aperture (not shown).

The interaction region is located centrally between the MAC coils constructed from two pairs of PTFE coated copper wire wound around a copper former. The coils are fully shielded by a thin 310 grade stainless steel cylinder coated with colloidal graphite to reduce patch fields in the interaction region. The coil current is supplied using shielded twisted pair feed-wires, to eliminate fields from these wires.

The unselected-energy electron gun uses two 3-element lenses to produce a 1mm diameter electron beam of zero beam angle and 2° pencil angle in the interaction region. The energy of the electrons can be varied from ~5eV to 100eV, allowing experiments to be conducted from the low to intermediate energy regimes. The electron analyser uses a single 3-element

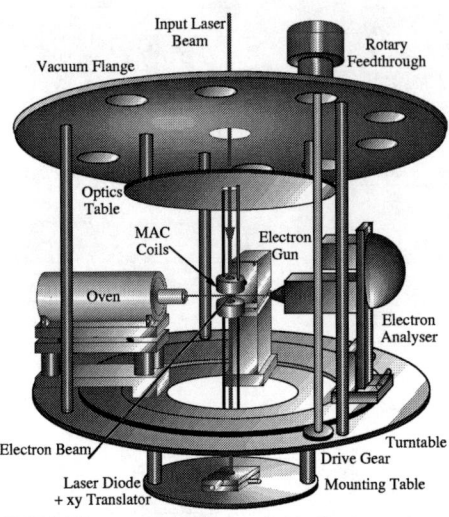

zoom lens to focus superelastically scattered electrons from the interaction region onto the entrance of a hemispherical energy selector, before passing these electrons to a channel electron multiplier for amplification and detection. The oven and electron gun are fixed in space, whereas the analyser can rotate around the interaction region on a rotary turntable.

The laser beam produced by a Coherent MBR-110 laser combined with a Coherent MBD-200 frequency doubler enters the vacuum chamber through the top flange after being polarized using a Glan-Taylor polarizer and $\lambda/4$ plate. The laser beam is accurately positioned through the interaction region orthogonal to the scattering plane, by following a tracer beam from a visible laser diode located on an xy-translator stage below the turntables. Fluorescence from the interaction is measured using a photomultiplier tube located outside the vacuum chamber (not shown).

FIG 4. The new MAC super-elastic Spectrometer, showing the position of various components.

All components of the spectrometer have now been constructed, and it is expected that the MAC spectrometer will begin operation within the next 3 months.

LASER-ATOM INTERACTIONS IN THE B-FIELD.

As noted above, the laser-atom interaction in the MAC spectrometer will occur in a magnetic field **B**, and so it is important to understand this process to derive the optical pumping parameters. The field required for deflection as shown in figure 2 is ~10G, and so the energy level perturbation of the atomic levels will be linear with the field. Zeeman splitting of the substates

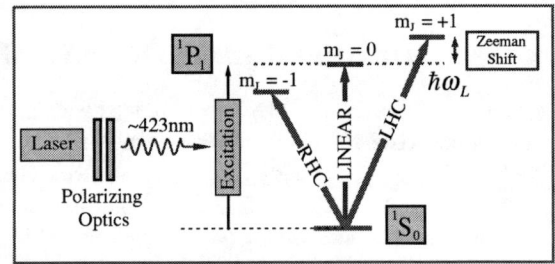

FIG 5. Perturbation of the calcium energy levels due to the B-field in the MAC spectrometer.

occurs, the energy splitting being given by the Larmor frequency ω_L such that $\Delta E = \hbar \omega_L = g_J \mu_B |\mathbf{B}|$. For calcium, the ground state is the 4^1S_0 state whereas the excited state is the 4^1P_1 state. Under the influence of the **B**-field, the energy levels therefore perturb as shown in figure 5, where the quantization axis is chosen along the laser beam for $\Delta m_J = \pm 1$ sub-states, and along the direction of the polarization vector for linearly polarized excitation.

Since the experimental geometry has the direction of the laser beam parallel to the **B**-field, the laser-atom interaction is relatively easily derived for $\sigma^\pm$ circularly polarized light. To derive the associated pseudo-Stokes parameter, the $\sigma^\pm$ laser radiation must be tuned and polarized for each $m_J = \pm 1$ sub-state, the rate of super-elastically scattered electrons being determined for each excitation. The pseudo-Stokes parameter is then derived from the scattering rates. This is the same as for conventional super-elastic experiments, except the laser wavelength also must be varied for each measurement.

By contrast, for linearly polarized π-excitation the **B**-field is perpendicular to the radiation polarization vector, and so to ensure a common quantization axis for analysis it is necessary to decompose the π-light into $\sigma^{\pm}$ components. The consequence of this decomposition is that the linear polarised light *simultaneously and coherently* excites both $m_J = \pm 1$ sub-states, the coherence being controlled by the magnitude of the **B**-field, the lifetime of the excited state and intensity of the laser beam. This is related to the Hanle effect, which describes the interaction for broadband, low power radiation [9].

To study these effects, experiments have been performed to measure the fluorescence from laser excitation of calcium in a **B**-field, the fluorescence being measured along the laser beam direction as shown in figure 6. These experiments allow the excited state to be characterised for a quantization axis as given above. For circular excitation, only one state is excited for each $\sigma^{\pm}$ polarization as expected, and the fluorescence is fully circularly polarized as measured using the λ/4 plate and linear polarizer positioned in front of the photodiode detector. For linear π-excitation, two peaks are seen as a function of laser de-tuning as shown in figure 7. The magnitude and position

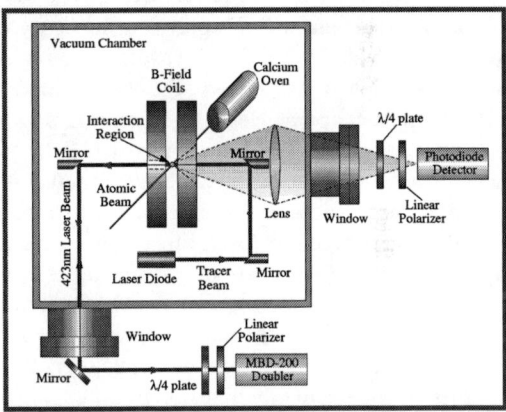

FIG 6. Experimental setup to measure fluorescence along the laser beam. The visible laser diode provides a tracer beam for accurate alignment of the main laser beam from the MBD-200 doubler at 423nm to the interaction region.

of these peaks varies as the polarization angle of the linear analyser in front of the detector is changed, this effect depending on the intensity of the laser beam and the magnitude of the magnetic field.

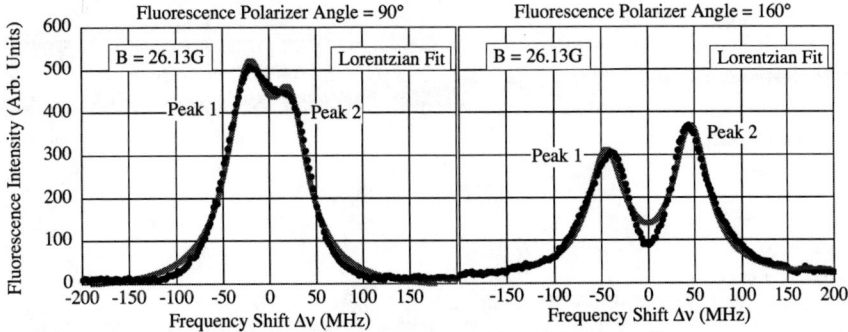

FIG 7. Example spectrum obtained from linear excitation of calcium with |**B**| = 26.13G as a function of laser detuning and analyzer polarizer angle. Both intensity and position of the peaks are seen to vary.

A fully quantum mechanical description of the laser-atom interaction in the **B**-field is currently being developed to understand these results. It is essential that this interaction is well characterised for a simple atom like calcium, before proceeding to rubidium with its rich hyperfine structure. Modeling these interactions for linear laser excitation will be challenging, as each hyperfine sub-state shifts in energy, influencing the coupling of the laser field to the atom in a complex way. This will in-turn influence the optical pumping parameters which allow the ACP's to be derived from the measured pseudo-Stokes parameters, considerably enhancing the difficulties expected in analysis of the experimental results.

CONCLUSIONS & SUMMARY.

Experimental results have been presented for super-elastic scattering of electrons from calcium and rubidium at 20eV equivalent incident energy, showing that significant differences occur between alkali and alkali-earth targets. Theoretical models describing the ACP's for both targets show mixed success, the charge cloud angle being most accurately determined, and the alignment parameter being predicted poorly. The CCC model predicts the angular momentum parameter superbly for rubidium, but has yet to be tested for calcium. The RDWA model works well for calcium at 35eV, but fails to predict the results at 20eV.

A new spectrometer has been described which will allow the ACP's to be determined over all scattering geometries from 5eV through to 100eV, providing the most stringent test of theory to date. This spectrometer is being assembled, and is expected to begin operation soon. The laser-atom interaction in the MAC spectrometer will be in a magnetic field, and so this interaction has been studied by observing fluorescence from laser excited calcium. These results are to be compared to a model being developed to describe this interaction, which will then be applied to rubidium. Optical pumping parameters which relate the super-elastic scattering measurements to the ACP's will then be determined from the model, and from additional experiments in the MAC spectrometer.

The results from these new experiments will provide new data for scattering models which attempt to describe the electron impact process. Only by providing accurate data over a wide range of energies can these models be honed into a unified theory. It is the purpose of this experimental programme to provide a wide body of data to help achieve this aim.

ACKNOWLEDGMENTS

The EPSRC, UK is acknowledged for funding this research programme, and for providing a postdoctoral fellowship for one of us (MH). The technicians in the engineering workshops of the Schuster Laboratory are also thanked for building the components of the new MAC super-elastic spectrometer.

REFERENCES

1. Andersen N, Gallagher J W & Hertel I V 1988 *Phys. Rep.* **165** 1
2. Andersen N & Bartschat K 1996 *Adv. At. Mol. Phys.* **36** 1
3. Hertel I V & Stoll W 1977 *Adv. At. Mol. Phys.* **13** 113
4. Farrell P M, MacGillivray W R & Standage M C 1991 *Phys Rev A* **44** 1828
5. Hall B V, Shen Y, Murray A J, Standage M C, MacGillivray W R & Bray I 2004 *J Phys B* **37** 1113
6. Murray A J & Cvejanovic D 2003 *J Phys B* **36** 4889
7. Linert I, King G C & Zubek M 2004 *J. Elect. Spec. Rel. Phenom.* **134** 1
8. Sravistava R, Zuo T, McEachran R P & Stauffer A D 1992 *J Phys B* **25** 3709
9. Corney A 1977 *"Atomic and Laser Spectroscopy"*, Clarendon Press – Oxford.

Spin Up-Down Asymmetry in the Excitation of Kr 5p' $[\frac{3}{2}]_2$ by Polarized Electrons

D H Yu[*,†], D Cvejanović[*], J F Williams[*], L Pravica[*], R Srivastava[**], A. Stauffer[‡], P A Hayes[*,§] and S Napier[*]

[*]*The University of Western Australia, Perth, Australia*
[†]*Present address: Australian Nuclear Science and Technology Organization, Sydney, Australia*
[**]*Indian Institute of Technology - Roorkee, Roorkee, India*
[‡]*York University, Toronto, Canada*
[§]*Present address: QR Sciences Ltd., Cannington Western Australia, Australia*

Abstract. The spin up - down asymmetry in the excitation of the 5p' $[\frac{3}{2}]_2$ state of krypton by spin polarized electrons is discussed. The experiment was done using the electron - photon coincidence method. The experimental data are compared with theoretical predictions using the Relativistic Distorted Wave (RDW) model.

Keywords: Spin, Asymmetry, Polarized electrons, Krypton
PACS: 34.80 DP

INTRODUCTION

In the field of electron scattering from free single atoms, spin-dependent interactions are weaker than, and usually masked by, the Coulomb interaction. Nevertheless, the use of incident spin-polarized electrons [1] enables the exploration of spin-dependent effects which can arise from the spin-orbit interaction of the incident (continuum) electron in the atomic field, spin-orbit interaction of the atomic electrons and exchange effects. Not all of the effects can be revealed and disentangled in a single experiment so the added complexity of using spin-polarized electrons requires more than the normal experimental expertize and theoretical interpretation.

Systematic studies using incident spin-polarized electrons have indicated classic spin effects. For sodium (with atomic number Z=11), when the fine structure of the target is resolved, a spin up-down asymmetry arises from electron exchange. For neon (with atomic number Z=10) integral polarizations have shown Stokes parameters dependent on the angular momentum couplings [2, 3]. For heavy atoms, such as mercury (Z=80), the spin-orbit interaction of the continuum electron is strong enough to cause spin asymmetry larger than that arising from exchange [4]. Here we report studies of krypton atoms (Z=36), for which spin effects are expected to be intermediate between light (quasi one electron alkali atoms) and heavy (mercury and possibly xenon with Z=54) atoms.

A relatively small number of previous experimental studies of excitation of krypton by polarized electrons have been reported. These include measurements of integral Stokes parameters for the $4p^5 5p[\frac{5}{2}]_3$, $4p^5 5p'[\frac{3}{2}]_2$ and $4p^5 5p'[\frac{3}{2}]_1$ states [5] and cascade effects in excitation of these same states [6]; integral Stokes parameters for the $4p^5 5p'[\frac{5}{2}]_{j=2,3}$

states [7]; and integral Stokes parameters for KrII $5p\,^4P^0_{3/2}$ transition, i.e. a simultaneous ionization and excitation [8]. The only differential measurements of left-right asymmetries for excitation of the $4p^55s\,^3P_1$, $^3P_2$ and $^1P_1$ were performed by Dümmler et al [4]. They showed both electron exchange and spin-orbit coupling within the target atom were the dominant mechanisms for producing a scattering asymmetry and were able to predict relationships between the sign and magnitude of the asymmetries between the different J states. But for xenon, where there is strong violation of LS-coupling and the continuum spin-orbit interaction is not negligible, the difference between observation and theory were not readily explained just by exchange and spin-orbit coupling within the target atom.

Our approach is to extend their work by measuring the asymmetry A from the coincidence rate corresponding to spin up and down incident electrons. Then asymmetry is defined as

$$A = \frac{I^\uparrow - I^\downarrow}{I^\uparrow + I^\downarrow} \tag{1}$$

where I stands for the measured intensity of true coincidences detected for the observed transition and specific scattering conditions, while arrows indicate the incident spin orientation in respect to the normal to the scattering plane, the z axis in the natural or y axis in the collision frame [9]. This is a first step with the important addition of coincidence detection of the radiated photon and scattered electron to identify the excited state.

The work of Dümmler et al [4] separated the fine-structure states using good electron energy resolution. For situations where the fine structure can not be resolved in energy loss, but can be in decay photon wavelength, and this is the case for the $4p^55p$ states of krypton considered here, we use the method of electron-photon coincidences to observe the separated J state. This approach is advantageous since cascade effects are minimised and asymmetry measurements can be made over a wider energy range. We present the first experimental and theoretical determinations of the spin up-down asymmetry for the $5p'\,[\frac{3}{2}]_2$ state of krypton. These results extend the previous data for the $4p^55s$ to the $4p^55p$ levels.

EXPERIMENT

A crossed electron-atom beams apparatus [10] used for the measurement of integrated Stokes parameters was modified by the addition of an electron energy analyzer. A schematic diagram showing essential components and scattering geometry is shown in figure 1. The momentum vectors of the incident and scattered electrons define the scattering plane (X-Z plane) and photons are detected perpendicular to the scattering plane, along the Y axis, as illustrated in figure 1.

The incident polarized electron beam, with a polarization of 75% and energy width FWHM≈300 meV, is produced by photo-emission from a strained GaAs crystal. This source of polarized electrons and the preparation of the crystal was described in detail by Hayes et al [11]. Control of the spin orientation was done using a Liquid Crystal Variable Retarder (LCVR) as a quarter-wave plate [12] to produce the desired helicity

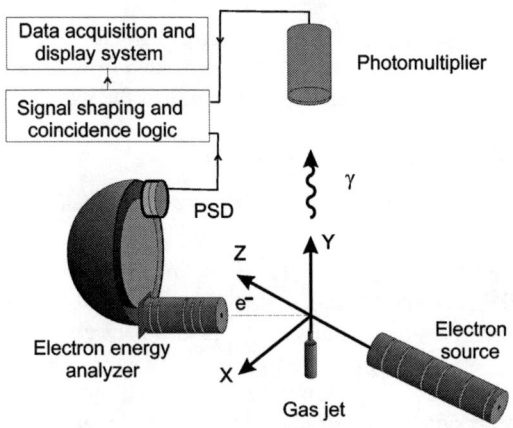

FIGURE 1. Experimental arrangement.

of the incident laser light. The spin polarization was measured using an optical neon polarimeter [7]. For each defined scattering angle and incident electron energy, the spin of the incident electrons was reversed after ten seconds. A PC-based data acquisition system, made in our lab, controlled the LCVR and recorded the data.

The electron analyzer consists of cylindrical input lenses which focus scattered electrons onto the entrance of a 180° hemispherical selector. Electrons dispersed according to their energy in the output plane of the 180° selector are detected using a Position Sensitive Detector (PSD) consisting of a chevron pair of 25 mm multichannel plates followed by a resistive anode. The design, characteristics and performance are the same as described previously [13, 14]. The analyzer can be rotated in the scattering plane from -80 to + 120°, with respect to the incident electron beam direction.

Spin up-down asymmetries can be obtained directly from the energy loss spectrum if the fine-structure components are resolved. With the present apparatus we chose to separate the $4p^5 5p$ states in krypton using the electron-photon coincidence method with good resolution of the photons. The decay photons λ=826.3 nm from the excited Kr $5p'[\frac{3}{2}]_2$ state, were selected by wavelength with a narrow band interference filter and detected by a photomultiplier in the direction perpendicular to the scattering plane. The electrons scattered with an energy-loss corresponding to excitation of the $5p'[\frac{3}{2}]_2$ state were detected in coincidence with the decay photons. In this way excitation of the $5p'[\frac{3}{2}]_2$ state was uniquely identified irrespective of the energy resolution of the electron spectrometer. The fast positive signal capacitively decoupled from the second channel plate was used to start a time-to-amplitude converter (TAC), while pulse height analysis and coincidence counting within a pre-selected energy range was performed by the data acquisition system. The overall time resolution of this setup for the present measurements was around 5 ns. The spin up-down asymmetry was determined from the TAC spectra after subtracting the background of accidental coincidences and taking into account the electron spin polarization. That background was primarily determined by the relatively broad electron energy resolution and consequently the electrons detected from states other than those identified in the photon channel.

The apparatus asymmetry was negligible as determined from observations of scattered electron intensities for scattering angles on either side of the incident beam direction and for changed spin directions.

THEORETICAL METHOD

The calculations were done using the Relativistic Distorted-Wave (RDW) method. We have previously used this method to calculate the excitation of the $4p^55p$ states of Kr by spin-polarized electrons [15] and details of the method are given there and in the cited references.

Since Kr is a relatively heavy atom, it is best described in the $j-j$ coupling scheme. In this scheme we represent the outer shell of the ground states by the simple single-configuration $4\bar{p}^24p^4$ where $\bar{p}$ represents a p-orbital with total angular momentum $j = 1/2$ while p has $j = 3/2$. The excited states with total angular momentum $J = 2$ are linear combinations of the configurations $4\bar{p}^24p^35\bar{p}$, $4\bar{p}^24p^35p$ and $4\bar{p}4p^45p$ all coupled to a $J = 2$ state. More details of these wave functions are given in [16].

We have used the MCDF program of Grant et al [17] to calculate Dirac-Fock wave functions based on these configurations. These produce separate wave functions and energies for the individual fine-structure states which are important for a heavy atom like krypton. The formulas for $I^\uparrow$ and $I^\downarrow$ are taken from [18] assuming the incident electrons are completely spin-polarized perpendicular to the scattering plane. From these we can calculate the asymmetry parameter A as given above.

RESULTS AND DISCUSSION

The experimental and theoretical spin up-down asymmetries from excitation by spin polarized electrons of the $5p'[\frac{3}{2}]_2$ state of krypton for two incident energies, 26 eV and 60 eV, are shown in figure 2 and 3 respectively. Due to differences in magnitudes of asymmetries and angular ranges, the figures are reproduced to give a better insight into angular and energy behavior of the experimental data and theoretical predictions. Experimental data are obtained in an angular range from 10 to 80°. At 26 eV, both theory and experiment indicate a change of sign of the asymmetry around a scattering angle of 50°. The general shape of the angular behavior, a maximum with positive asymmetry followed by a minimum with negative asymmetry, is seen in both experimental and theoretical data. However the angular positions are different. Similarly, the maximum values, reaching only 15% around a scattering angle of 35° in the experimental data at 26 eV, are over-estimated by theory. The present RDW theory predicts strong angular behaviour and larger asymmetries, approximately +0.28 around 80° and -0.35 around 140°.

At electron impact energy of 60 eV the observations and calculations are very different. Smaller values of the asymmetry have been measured for electron scattering angles less than 50°. In contrast, theory indicates asymmetries even larger than at 26 eV.

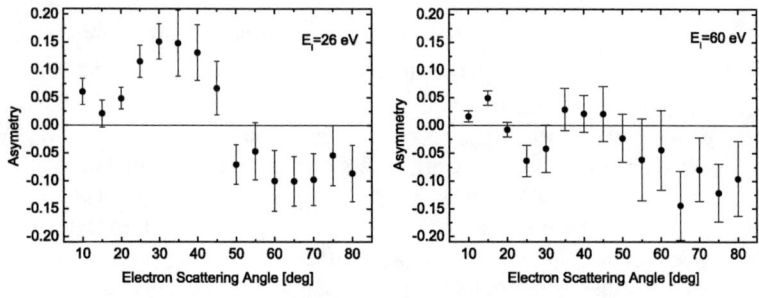

FIGURE 2. Spin up-down asymmetry at 26 eV and 60 eV : • present experiment.

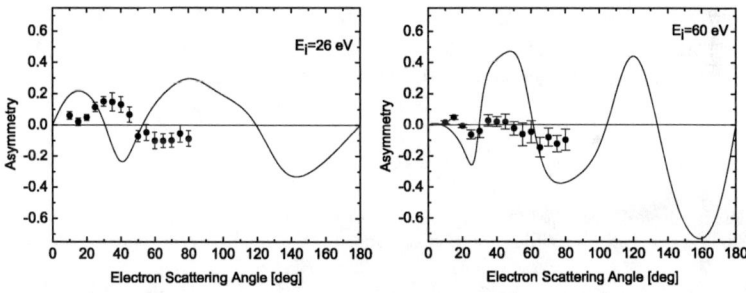

FIGURE 3. Spin up-down asymmetry at 26 eV and 60 eV: • present experiment; ——— present RDW theory

In contrast to excitation, the data for elastic scattering show good agreement between experimental and theoretical asymmetries [19]. These experimental results were compared with calculations obtained by the solution of the Dirac-Fock equations. Calculations were performed with inclusion of polarization and dynamic distortion potentials and with the addition of an absorption potential to model inelastic processes. Inclusion of an absorption potential seemed important at energies where the inelastic scattering cross sections are largest.

The energy behavior observed for the $5p'[\frac{3}{2}]_2$ state is interesting to compare with the situation reported by Dümler et al [4] for different J-components within the $4p^55s$ manifold in krypton. These authors report a decrease of asymmetry with increasing incident electron energy from 15 to 25 eV for the two of the J=1 states, $4p^55s[\frac{3}{2}]_1^0$ and $4p^55s'[\frac{1}{2}]_1^0$ which they represent as mixed singlet triplet components of LS coupled $^1P_1$ and $^3P_1$ states. The observed decrease in magnitude of the asymmetries is then explained by the energy dependence of the cross section for individual components. This model was in agreement with observations for the J=2 state, i.e. $^3P_2$, which should show no significant variation with energy if exchange scattering is the dominant mechanism.

The value of the present study emerges from the comparison of measurement with theory. Our measurements and modeling of a difficult scattering process indicate a lack of agreement. Further experimentally work should include other states from the same fine structure multiplet. Observations at lower energies may enable the exchange contributions to be identified more readily. Improvements to data collection efficiency will experimentally strengthen these observations.

ACKNOWLEDGMENTS

This research was supported by the Australian Research Council and the University of Western Australia, the Natural Science and Engineering Research Council of Canada and the Council of Scientific and Industrial Research of India.

REFERENCES

1. J. Kessler, editor, *Polarized Electrons*, Springer-Verlag, Berlin, 1985.
2. D. H. Yu, P. A. Hayes, J. F. Williams, and J. E. Furst, *J. Phys. B: At. Mol. Opt. Phys.*, **30**, 1799–1812 (1997).
3. D. H. Yu, P. A. Hayes, and J. F. Williams, *J. Phys. B: At. Mol. Opt. Phys.*, **32**, 1181–1191 (1999).
4. M. Dümmler, G. F. Hanne, and J. Kessler, *J. Phys. B: At. Mol. Opt. Phys.*, **28**, 2985–3001 (1995).
5. D. H. Yu, P. Hayes, J. F. Williams, V. Zeman, and K. Bartschat, *J. Phys. B: At. Mol. Opt. Phys.*, **33**, 1881–1894 (2000).
6. D. H. Yu, J. F. Williams, X. J. Chen, P. Hayes, and K. Bartschat, *Phys. Rev. A*, **67**, 032707 (2001).
7. J. E. Furst, W. M. K. P. Wijayaratna, D. H. Madison, and T. J. Gay, *Phys. Rev. A*, **47**, 3775–3787 (1993).
8. P. A. Hayes, D. H. Yu, and J. F. Williams, *J. Phys. B: At. Mol. Opt. Phys.*, **31**, L193–L200 (1998).
9. N. Andersen, J. W. Gallagher, and I. V. Hertel, *Phys. Rep.*, **165**, 1–188 (1988).
10. P. A. Hayes, D. H. Yu, J. E. Furst, M. Donath, and J. F. Williams, *J. Phys. B: At. Mol. Opt. Phys.*, **29**, 3989–4000 (1996).
11. P. A. Hayes, D. H. Yu, and J. F. Williams, *Rev. Sci. Instrum.*, **68**, 1708–1713 (1997).
12. J. E. Furst, D. H. Yu, P. A. Hayes, C. M. D'Souza, and J. F. Williams, *Rev. Sci. Instrum.*, **67**, 3813–3817 (1996).
13. D. K. Waterhouse, and J. F. Williams, *Rev. Sci. Instrum.*, **68**, 3363–3370 (1997).
14. P. A. Hayes, M. A. Bennett, J. Flexman, and J. F. Williams, *Rev. Sci. Instrum.*, **59**, 2445–2452 (1988).
15. S. Kaur, R. Srivastava, R. P. McEachran, and A. D. Stauffer, *J. Phys. B: At. Mol. Phys.*, **32**, 4331–4359 (1999).
16. S. Kaur, R. Srivastava, R. P. McEachran, and A. D. Stauffer, *J. Phys. B: At. Mol. Phys.*, **31**, 4833–4852 (1998).
17. I. P. Grant, B. J. McKenzie, P. H. Norrington, D. F. Mayers, and N. C. Pyper, *Comput. Phys. Commun.*, **21**, 207–231 (1980).
18. K. Bartschat, K. Blum, G. F. Hanne, and J. Kessler, *J. Phys. B: At. Mol. Phys.*, **14**, 3761–3776 (1981).
19. M. R. Went, R. P. McEachran, B. Lohman, and W. R. MacGillivray, *J. Phys. B: At. Mol. Opt. Phys.*, **35**, 4885–4897 (2002).

Universal Scaling of Resonances in Vector Correlation Photoionization Parameters

A.N. Grum-Grzhimailo* and M. Meyer[†]

Institute of Nuclear Physics, Moscow State University, Moscow 119992, Russia
[†]*LIXAM, UMR 8624, Centre Universitaire Paris-Sud, Bâtiment 350, 91405 Orsay Cedex, France*

Abstract. Two universal quantities are introduced for the Fano-like energy dependence of the vector correlation photoionization parameters in the region of an isolated autoionizing resonance: the width and the energy shift. This result can be used to connect data from different types of experiments and theoretical calculations.

Keywords: Photoionization, Autoionizing states, Resonances
PACS: 32.80.Dz, 32.80.Hd, 32.80.Fb

INTRODUCTION

The statement that the Fano-like profiles of the polarization and correlation parameters ('vector correlation parameters') in the region of an isolated autoionizing resonance possess a similar width and a similar energy shift with respect to the energy of the autoionizing state has been proved recently [1]. This result is a significant extension of previously considered particular cases: the angular anisotropy coefficient and the integral polarization of photoelectrons in ionization below the second ionization threshold [2], as well as the angular distribution of the secondary radiation in the dielectronic recombination [3]. Introducing this presentation we attract the attention of the community to this not widely known regularity, which appears to be useful at the modern stage of development of experimental technique. After presenting general formulas and a discussion we turn to a verification of the above statement by means of fluorescence polarimetry applied to the investigation of atomic photoionization by differently polarized beams of synchrotron radiation [1, 4, 5]

GENERAL FORMULAS

Consider photoionization of an atom into a particular ion state αJ, where J is the total angular momentum of the residual ion and α is the set of other quantum numbers specifying the ion state. In the region of an isolated autoionizing state the integral photoionization cross section on the state αJ is given by

$$\sigma_{\alpha J}(\varepsilon) = \sigma_0 \left[1 + \frac{2C_1 \varepsilon + C_2}{1 + \varepsilon^2} \right], \tag{1}$$

where σ_0 is the cross section without the autoionizing resonance and C_1, C_2 are the profile parameters introduced by Starace [6]. We consider the αJ-dependent values of

σ_0, C_1 and C_2 as constants in the vicinity of the resonance. The energy dependence of the cross section is therefore represented only by the variable $\varepsilon = (E - E_r)/(\Gamma/2)$, where E_r and Γ are the energy and the natural width of the autoionizing state, respectively. The parameter C_1 characterizes the asymmetry of the resonance in the cross section (1). The parameter C_2 determines the relative integral yield of the resonance:

$$\frac{1}{\sigma_0} \int_{-\infty}^{+\infty} (\sigma_{\alpha J}(\varepsilon) - \sigma_0) \, d\varepsilon = \pi C_2. \tag{2}$$

Assume that a physical parameter T is given by the general formula

$$T = \sum_{\substack{\ell \ell' j j' \\ J_t J'_t}} t_{\ell j J_t, \ell' j' J'_t} D_{\ell j J_t} D^*_{\ell' j' J'_t} \bigg/ \sum_{\ell j J_t} \left| D_{\ell j J_t} \right|^2, \tag{3}$$

where $D_{\ell j J_t}$ is the photoionization amplitude corresponding to the orbital (ℓ) and total (j) angular momentum of the photoelectron, respectively. The latter couples with J giving the total angular momentum of the channel J_t. The general form (3) is characteristic for the majority of the vector correlation parameters, which are independent of the angle of the photoemission (e.g. the anisotropy coefficients in the angular distribution of photoelectrons, the photoelectron spin polarization parameters and many others, including the time-reverse process of radiative electronic recombination [7]). The normalizing denominator of equation (3) is, up to a constant factor, the integral cross section (1). We will call parameters, which are described by equation (3), 'T-like'. It has been shown recently [1] that such T-like parameters can be transformed, in the region of an isolated autoionizing state, into the Fano-like form

$$T = \sigma_b^T + \sigma_a^T \frac{(q^T + \tilde{\varepsilon})^2}{1 + \tilde{\varepsilon}^2}, \tag{4}$$

where the energy independent parameters σ_b^T, σ_a^T and q^T are specific quantities for each individual T. The scaled energy deviation in equation (4) is given by

$$\tilde{\varepsilon} = (E - \tilde{E}_r)/(\tilde{\Gamma}/2) \tag{5}$$

with the scaled width

$$\tilde{\Gamma} = \left[1 + C_2 - C_1^2\right]^{\frac{1}{2}} \Gamma \equiv \chi \Gamma \tag{6}$$

and the shifted resonance energy

$$\tilde{E}_r = E_r - \frac{C_1}{2} \Gamma \equiv E_r + \Delta. \tag{7}$$

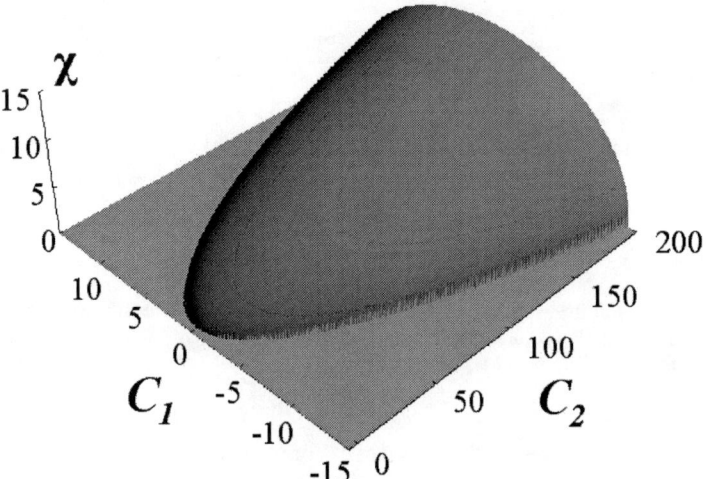

FIGURE 1. Scaling factor χ as function of the parameters C_1 and C_2. The region with $\chi = 0$ corresponds to forbidden ($\sigma_{\alpha J} < 0$) values of C_1 and C_2. The parabolic boundary $C_1^2 = 1 + C_2$ between the regions with $\sigma_{\alpha J} > 0$ and $\sigma_{\alpha J} < 0$ corresponds to a single photoionization channel and, therefore, to the zero cross section (1) in the minimum ('window') of the resonance profile.

DISCUSSION

Equations (5)-(7) are general for all T-like parameters and indicate that the Fano profiles for these parameters, being potentially very different, possess the same width $\tilde{\Gamma}$ and the same shift $\tilde{\Delta}$. Figure 1 shows the scaling factor χ as function of the parameters C_1 and C_2 (the shift $\tilde{\Delta}$ is a trivial function of C_1). The largest scaling factors (i.e. the largest 'broadening') of the resonance in the T-like parameters correspond to symmetric resonances with large yield in the cross sections. Note that the scaling factor χ vanishes in case of a single photoionization channel, when $C_1^2 = 1 + C_2$. The vanishing scaling factor corresponds to the absence of a resonance structure in the T-like parameters.

For autoionizing states decaying only to a single ion state αJ (e.g. for the states below the second ionization threshold), the integral cross section (1) coincides with the photoabsorption cross section. The parameters C_1 and C_2 are then directly related to the profile index q and the correlation index ρ in the Fano formula for the photoabsorption: $C_1 = q\rho^2, C_2 = (q^2 - 1)\rho^2$. In this case the scaling factor and the shift take the form

$$\chi = \sqrt{(1-\rho^2)(1+\rho^2 q^2)}, \tag{8}$$
$$\tilde{\Delta} = -q\rho^2 \tilde{\Gamma}/2. \tag{9}$$

To date there are only a few examples in the literature, where the above features of resonances in the T-like parameters can be traced. In order to observe the scaling and the shift effects, the T-like parameters have to be measured across the resonance. The largest effects occur for strong resonances in the cross section with weak 'background' from direct ionization, but scanning across the resonance in this case is difficult, because

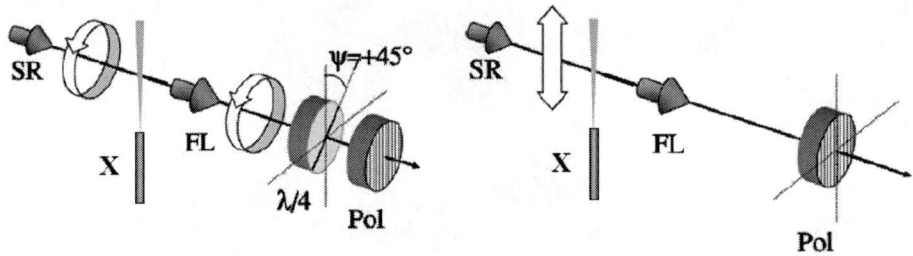

FIGURE 2. Scheme of the experimental set-up. Left: Excitation with circularly polarized synchrotron radiation (SR); circular polarization of the fluorescence (FL) is measured; Right: Excitation with linearly polarized SR; linear polarization of FL is measured.

of low statistics on the wings of the resonance. For resonances with strong intensities for direct ionization the effects are normally less pronounced.

The effect of broadening was observed for the angular anisotropy parameter β in the angular distribution of photoelectrons [8]. The data of [9] indicate similar broadening for the four different T-like parameters in resonance photoionization of thallium (see their figure 1). More examples are provided by theoretical calculations. Thus, the shift in the resonance position of calculated β values with respect to the resonance in the cross section was discussed for the giant autoionizing resonance in manganese [10, 11]. The broadening effect for β is clearly seen in figure 3 of [12], where the triply-excited resonances in Li were measured and calculated. The broadening of the resonance structure in the spin-polarization of thallium was discussed in [13]. Furthermore, the equivalents of expressions (8) and (9) were presented for β and the integral spin polarization of photoelectrons in theoretical studies of the resonance photoionization of sodium [2]. Similar expressions were found for the anisotropy coefficient in the angular distribution of secondary fluorescence in the dielectronic recombination on the excited states of helium [3].

With the advent of new high-brilliance VUV photon sources and increasing energy resolution of the electron detectors it is now becoming possible to obtain experimental results of high quality across the strong resonances and therefore to study the scaling effect experimentally in more detail. To confirm explicitly the universal scaling, an experiment has been performed [1], briefly described in the next section.

EXPERIMENT

Photoionization experiments, combined with the dispersed fluorescence polarimetry method, were performed on Xe atoms at the CIPO (Circular Polarization) beamline of the ELETTRA synchrotron radiation facility in Trieste (Italy). The scheme of the set-up is shown in figure 2. We measured the polarization of the fluorescence line 484.43 nm $5p^4(^3P_2)6p[1]_{3/2} \to 5p^4(^3P_2)6s[2]_{5/2}$ in the Xe II spectrum when the photon energy of the SR was scanned across the Xe** $4d_{5/2}^{-1}6p\,(J=1)$ resonance centered at 65.1 eV.

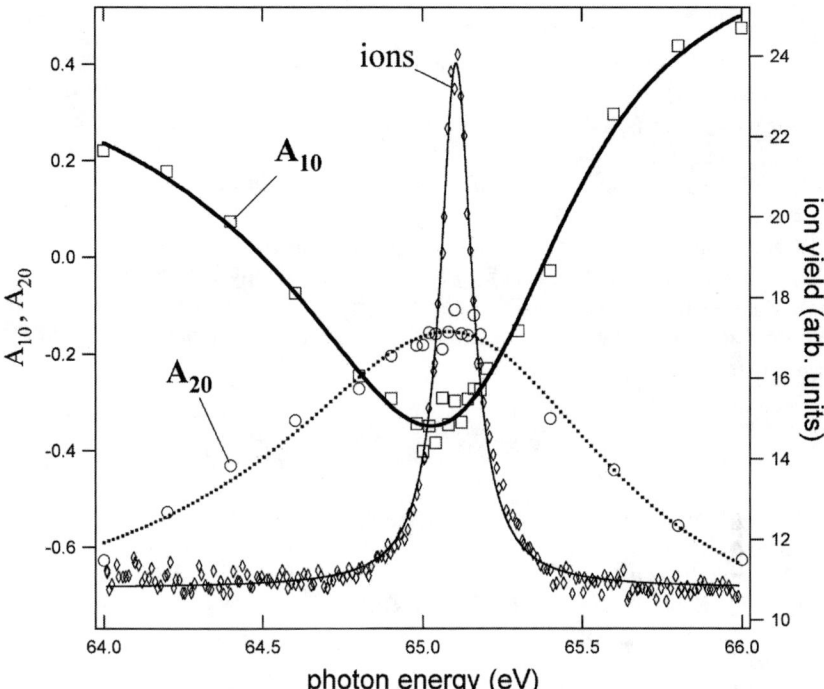

FIGURE 3. The Xe $4d_{5/2}^{-1}6p\,(J=1)$ resonance in the total ion yield, as well as in the alignment and the orientation parameters of the photoion in the $5p^4(^3P_2)6p[1]_{3/2}$ state. The energy dependence of the orientation and alignment parameters is fitted by equation (4).

Both circularly and linearly polarized synchrotron radiation was used and the degree of circular and linear polarization of the fluorescence was measured, respectively. From these measurements the orientation, $\mathcal{A}_{10}$, and the alignment, $\mathcal{A}_{20}$, parameters of the photoion Xe II $5p^4(^3P_2)6p[1]_{3/2}$ state were deduced, using the procedure described in detail in [14]. The orientation and alignment of photoions are the T-like parameters, as described above by equation (3). Note that the radiative cascades on the particular Xe II state under the consideration are weak [14] and hardly disturb its polarization.

The deduced data for $\mathcal{A}_{10}$ and $\mathcal{A}_{20}$ are displayed in figure 3 together with the Xe II ion yield. The position and the width of the resonance in the ion yield is in accordance with earlier measurements [15, 16]. Although the resonance profiles for $\mathcal{A}_{10}$ and $\mathcal{A}_{20}$ are very different, the widths of the resonances are equal within the experimental uncertainty: 1070±75 meV and 1010±140 meV, respectively. The corresponding scaling factor takes the value $\chi \sim 10$. The shift of the resonances, although being small (52±23 meV and 45±40 meV), is the same for $\mathcal{A}_{10}$ and $\mathcal{A}_{20}$ within the error bars. The obtained data directly confirm the scaling effect.

The extracted values for the Starace parameters $C_1 = -0.98 \pm 0.43$ and $C_2 = 101 \pm 14$ correspond to a large yield (2) and a weak asymmetry of the resonance in the integral photoionization/excitation cross section on the level Xe II $5p^4(^3P_2)6p[1]_{3/2}$. Note that the profile of the resonance in the integral cross section has been established without its measurement. The large uncertainty in the parameter C_1 is mainly caused by the large error bars in the shifts of the resonances. Independent observations of other T-like parameters can increase the accuracy.

The established relations between resonances in different observable quantities provide a way of cross comparison of data, both experimental and theoretical, in terms of reduced widths and shifts. Furthermore, the profile parameters of a resonance in the integral cross section can be established by measurements of any T-like vector-correlation parameter across this resonance. The latter possibility is particularly useful for strong, symmetric resonances in the cross section, when direct measurements of C_1 and C_2 (or q and ρ^2) parameters are difficult.

ACKNOWLEDGMENTS

It is a great pleasure to acknowledge the invaluable assistance of P. O'Keeffe and S. Aloïse during the course of the experiments and of S. Fritzsche in numerical calculations. ANG gratefully acknowledges support under grant 04-02-17236 from the Russian Foundation for Basic Research. The experimental work at the ELETTRA synchrotron radiation facility was supported by the European Community under contract no. HPRI-CT-1999-00033.

REFERENCES

1. A.N. Grum-Grzhimailo, S. Fritzsche, P. O'Keeffe, and M. Meyer, *J. Phys. B*, **38**, 2545 (2005).
2. A.N. Grum-Grzhimailo and B. Zhadamba, *Vestnik MGU, Ser. 3 Fiz. Astron.*, **28**, 19 (1987).
3. A.N. Grum-Grzhimailo, S. Danzan, O. Lhagva, and S.I. Strakhova, *Z. Phys. D*, **18**, 147 (1991).
4. P. O'Keeffe, S. Aloïse, M. Meyer, and A.N. Grum-Grzhimailo, *Phys. Rev. Lett.*, **90**, 023002 (2003).
5. P. O'Keeffe, S. Aloïse, S. Fritzsche, B. Lohmann, U. Kleiman, M. Meyer, and A.N. Grum-Grzhimailo, *Phys. Rev. A*, **70**, 012705 (2004).
6. A.F. Starace, *Phys. Rev. A*, **16**, 231 (1977).
7. V.V. Balashov, A.N. Grum-Grzhimailo, and N.M. Kabachnik, *Correlation and Polarization Phenomena in Atomic Collisions: A Practical Theory Course*, Kluwer Academic / Plenum Publishers, New York, 2000.
8. S. Southworth, U. Becker, C.M. Truesdale, P.H. Kobrin, D.W. Lindle, S. Owaki, D.A. and Shirley, *Phys. Rev. A*, **28**, 261 (1983).
9. M. Müller, N. Böwering, A. Svensson, and U. Heinzmann, *J. Phys. B*, **23**, 2267S (1990).
10. J.P. Connerade and V.K. Dolmatov, *J. Phys. B*, **29**, L831 (1996).
11. J.P. Connerade and V.K. Dolmatov, *J. Phys. B*, **30**, L181 (1997).
12. S. Diehl, D. Cubaynes, H.L. Zhou, L. VoKy, F.J. Wuilleumier, E.T. Kennedy, J.M. Bizau, S.T. Manson, T.J. Morgan, C. Blancard, N. Berrah, and J. Bozek, *Phys. Rev. Lett.*, **84**, 1677 (2000).
13. N.A. Cherepkov, *Opt. Spectrosc.*, **49**, 582 (1980).
14. M. Meyer, A. Marquette, A.N. Grum-Grzhimailo, U. Kleiman, and B. Lohmann, *Phys. Rev. A*, **64**, 022703 (2001).
15. G.C. King, M. Tronc, F.H. Read, and R.C. Bradford, *J. Phys. B*, **10**, 2479 (1977).
16. S. Masiu, E. Shigemasa, A. Yagishita, I.A. Sellin, *J. Phys. B*, **28**, 4529 (1995).

Correlation spectroscopy of condensed matter systems

F.O. Schumann, J. Kirschner, K. A. Kouzakov, and J. Berakdar

Max-Planck Institut für Mikrostrukturphysik, Weinberg 2, 06120 Halle, Germany

Abstract. We present experimental and theoretical evidence for the potential of the two-particle spectroscopy to explore the electron-electron interaction in condensed matter. The experiment consists of a single electron impinging onto a clean surface. Two electrons are then emitted simultaneously and their momentum vectors are resolved. The measured energy and angular pair correlations within the pair carry direct information akin to the electron-electron interaction in the sample. We also point out that the presence of the Fermi sea leads to a damping, and a suppression of the range of the electron-electron interaction.

Keywords: electronic correlation, electron spectroscopy
PACS: 79.20.Kz, 68.49.Jk

INTRODUCTION

Electronic correlations result in a variety of physical phenomena such as superconductivity and magnetism. Yet, the overwhelming majority of methods for the study of the nature of the electron-electron interaction in matter are based on an effective single-particle picture where electronic correlations are manifested as subsidiary structures in the measured quantities [1, 2]. Here we explore the potential of the two-particle spectroscopy for the study of the electron-electron interaction in matter, specifically a primary single electron incident on a surface results in the emission of (time) correlated electron pairs. We employ a novel time-of-flight coincidence set-up consisting of a small central collector surrounded by a resistive anode.

Here we report two key observations. If the electrons' energies E_1 and E_2 are tuned such that the pair emission from the top of the valence band is possible, a zone of reduced intensity with a diameter of 1.6 Å^{-1} is visible in the coincidence signal whenever the escaping electrons momenta are comparable in magnitudes and directions. This correlation and exchange induced hole disappears if the sample electron originates from below the top of the valence band which indicates the sensitivity of the xc-hole to inelastic, phase-breaking scattering processes. We comment on these findings from a theoretical point of view.

THEORETICAL FORMULATION

Electronic correlation in an N particle system is described conventionally by the reduced two-particle density matrix $\gamma_2(x_1,x_2,x_1',x_2')$. γ_2 is expressible in terms of the N-particle wave function Ψ as

$$\gamma_2(x_1,x_2,x_1',x_2') = N(N-1)\int \Psi(x_1,x_2,x_3,\cdots,x_N)\Psi^*(x_1',x_2',x_3,\cdots,x_N)\,dx_3\cdots dx_N. \quad (1)$$

Here x_j, $j=1\cdots N$ label the spin and the position coordinates. For fermions eq.(1) shows that $\gamma_2(x_1,x_2,x_1',x_2') = -\gamma_2(x_2,x_1,x_1',x_2')$. The two-particle density derives from γ_2 as $\rho_2(x_1,x_2) = \gamma_2(x_1,x_2,x_1,x_2)$. For fermions ρ_2 vanishes if $x_2 = x_1 = x$, i.e. $\rho_2(x,x) = 0$. For independent particles $\rho_2(x_1,x_2)$ is related to the single particle density $\rho(x)$ via $\rho_2(x_1,x_2) = \rho(x_1)\frac{N-1}{N}\rho(x_2)$. Thus, for overlapping fermions the antisymmetry of Ψ implies a correlation among the particles that results in the existence of the (Fermi) hole in the two-particle density for $x_1 = x_2$ and which we will expose in some details below. The electrostatic Coulomb repulsion between the electrons also contribute to the hole which is quantified conventionally by the xc hole [1, 3] $h_{xc}(x_1,x_2) = \frac{\rho_2(x_1,x_2)}{\rho(x_1)} - \rho(x_2)$. The pair-correlation function g, as defined in the literature, is given by $g(x_1,x_2) = h_{xc}(x_1,x_2)/\rho(x_2) + 1 = \rho_2(x_1,x_2)/[\rho(x_1)\rho(x_2)]$.

Role of exchange

To illustrate the role of exchange let us consider the two-particle probability density $g(\mathbf{r}_1,\mathbf{r}_2)$ (also called two-particle correlation function) associated with a two-particle plane wave $\Psi_{\mathbf{k}_1,\mathbf{k}_2} = \exp[i(\mathbf{k}_1\cdot\mathbf{r}_1 + \mathbf{k}_2\cdot\mathbf{r}_2)]$. The electrons' wave vectors are denoted by $\mathbf{k}_1, \mathbf{k}_2$. Assuming the spin and spatial degrees of freedom to be decoupled we find that g is uniform for the singlet channel. For the triplet channel however it reads

$$g(\mathbf{r}_1,\mathbf{r}_2) = \Psi^*_{\mathbf{k}_1,\mathbf{k}_2}\Psi_{\mathbf{k}_1,\mathbf{k}_2} = 1 - \cos[(\mathbf{k}_1-\mathbf{k}_2)\cdot(\mathbf{r}_1-\mathbf{r}_2)]. \quad (2)$$

Assuming that the two electrons are immersed in a Fermi sea and if there are N distinct $\mathbf{k}$ states occupied, we find for the ground-state average of g that

$$\langle g(\mathbf{r}_1,\mathbf{r}_2)\rangle = \frac{1}{N^2}\sum_{ij}(1-e^{i(\mathbf{k}_i-\mathbf{k}_j)\cdot\mathbf{R}}), \quad (3)$$

$$= 1 - F^2(k_F R), \quad (4)$$

where $\mathbf{R} = \mathbf{r}_2 - \mathbf{r}_1$, k_F is the Fermi wave vector, and the function F is given by

$$F(k_F R) = 3\frac{\sin(k_F R) - (k_F R)\cos(k_F R)}{(k_F R)^3}. \quad (5)$$

From these equations we see that the result of exchange is the appearance in *the triplet channel* of a hole (the Fermi hole) at $R=0$ with an undamped oscillations away from $R=0$. Averaging over the ground-state Fermi sea leads to a damping of the oscillations and hence to a finite range of the hole. The period of the oscillations are directly determined by the Fermi wave vector, as in the case of Friedel oscillations. In the singlet channel there is no trace of exchange observable in g, however in a more realistic description the Coulomb interaction will contribute additionally to the hole (correlation hole). In a spin-resolved experiment these two contribution can be investigated (but not disentangled).

Pair emission and the two-particle density

The measured coincidence cross sections in the present experiment can be related to ρ_2. To show this we remark that the probability P_{if} for the reaction under consideration is given as $P_{if} = S_{if}S_{if}^*$. Here the S matrix elements are given by $S_{if} = \langle \Psi_{E_f} | \Psi_{E_i} \rangle$ and Ψ_{E_i} (Ψ_{E_f}) is the normalized wave function describing the system in the initial (final) state with the appropriate boundary conditions. The initial state with energy E_i consists of the incident electron interacting with an electron in the valence band in the presences of all other particles in the system (over which we will average eventually). The two vacuum electrons have the energy $E_f = E_1 + E_2$.

Assuming the surrounding medium is not affected while the incident and the valence band electron are interacting and during the emission of the two electrons (i.e. within a frozen-core picture) we find that $\Psi_{E_i} \approx \psi_{E_i}(x_1, x_2) \chi(x_3, \cdots, x_N)$. ψ_{E_i} is the electron pair wave function in the initial state. The function χ describes the surrounding medium. The reduced density matrix (1) takes on the form $\gamma_2(x_1, x_2, x_1', x_2') \approx 2\psi(x_1, x_2) \psi^*(x_1', x_2')$. For a further progress we assume the emitted electron pair state ψ_{E_f} to be described by plane waves. Under these conditions the measured, spin (σ_j) unresolved probability reads $P_{if} \propto \sum_{\sigma_1, \sigma_2, \sigma_1', \sigma_2'} \tilde{\psi}_{E_i}(\sigma_1 \mathbf{k}_1, \sigma_2 \mathbf{k}_2) \tilde{\psi}_{E_i}^*(\sigma_1' \mathbf{k}_1, \sigma_2' \mathbf{k}_2)$, where $\tilde{\psi}_{E_i}$ is the double Fourier transform of ψ_{E_i}.

From the above relations we conclude that the present experiment measures the spin-averaged diagonal elements of the reduced density matrix in momentum space which is the spin-averaged, momentum-space two-particle probability density ρ_2, a quantity that we discussed above.

EXPERIMENTAL DETAILS

The experiments were conducted under UHV conditions featuring a novel time-of-flight spectrometer depicted in fig.1. The sample was a LiF(100) single crystal which was cleaned and annealed. During the measurements the sample was kept at a temperature of ~ 400 K, this avoids the charging up of the sample. Primary electrons delivered from a pulsed electron gun hit the sample with an angle of ~ 80 degree with respect to the surface normal. Ejected electrons can move towards the spectrometer, where a pair of hemispherical grids ensure a field free region between the sample and a multi-channel plate. The resulting electron avalanches hit two detectors. A central collector accepts electrons only within a solid angle of ~ 0.1 sr, this detected electrons we refer to as "e_1". A resistive anode is the second detector which allows for a spatial resolution of the impact position. Electrons registered within a solid angle of ~ 1 sr are termed in what follows as "e_2".

Using fast timing signals and an electronic coincidence set-up allow for the determination of the flight times, which then can be converted into the energies E_1, E_2 when considering the flight path of ~ 58 mm. The impact position on the resistive anode can now be converted into momentum space. In this way we map out the energy and momentum dependence of the electron pair correlation. The time resolution of the set-up is better than 1 ns as determined by the width of the elastic peak in the time-of-flight spectrum. Thus, the energy resolution depends on the kinetic energy, which for the results

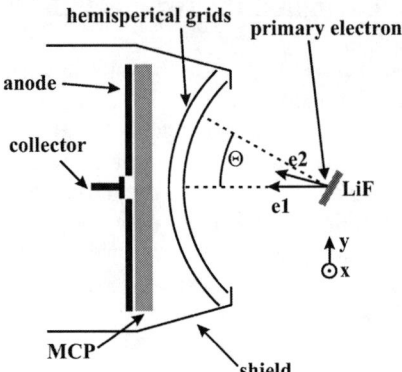

FIGURE 1. Electron pair detection technique. Two electrons with momenta k_1, k_2 and energies E_1 and E_2 are detected in coincidence by a resistive anode and central collector. The polar angle Θ denotes the angle between the surface normal and the central axis of the spectrometer.

presented here is 0.5 eV. The accuracy of the momentum resolution of the coincidence events is primarily determined by the acceptance angle of the central collector and the specific momenta under consideration, a typical value is 0.1 Å^{-1}, respectively.

EXPERIMENTAL RESULTS

In fig.2 we plot the 2D energy distribution of the coincidence electron pairs upon excitation with 32.7 eV electrons. We observe the onset of pair emission when the sum energy E_1+E_2 equals ~19 eV, which is indicated by the dashed diagonal line. This position can be easily understood when considering the bandstructure of LiF. The energy required to excite an electron from the top of the valence band to the vacuum level is ~ 14 eV, hence the maximum sum energy of the scattered electron and the valence band electron is expected to be 32.7-14=18.7 eV in agreement with the experiment.

More insight can be obtained if we take advantage of the lateral resolution of the set-up. In a first step we select only those coincidences for which the energies E_1 and E_2 are fixed. In other words we focus on a small region in the 2D energy distribution, as indicated in fig.2. In order to obtain sufficient statistics we actually select an energy window of ~ 1.6 eV around the respective energies. This has been indicated by the square boxes in fig.2 labelled a)-c). We can now proceed and plot the coincidence intensity as a function of the in-plane momentum $k_\parallel$ of the electron "e_2". We have selected three different regimes within the 2D energy distribution highlighted in fig.2 by the black squares a)-c). In the case a) and b) we are right at the onset of pair emission, in other words we move along the dashed diagonal line. Case c) describes the situation if emission below the highest occupied level is possible. In fig.3 we display the resulting momentum distributions. We would like to point out that all momentum plots display a zero intensity at a position where the central collector is positioned. The position and size of this "blind spot" depends on the polar angle Θ and the momentum of the

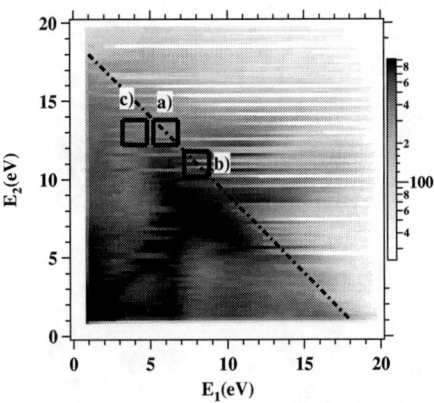

FIGURE 2. The 2D energy distribution for coincident pairs is shown for which the primary energy was 32.7 eV and the polar angle $\Theta=0°$. The energies E_1 and E_2 refer to the electron e_1 and e_2 electron, respectively, as displayed ion fig.1. The dashed diagonal line indicates the onset of the pair emission. The square boxes (width of 1.6 eV) indicate narrow regions which were chosen to map the coincidence intensity in momentum space, see fig.3.

electron "e_2". For the plots shown in fig.3 this "blind spot" is centered at $k_\parallel=0$ and has diameter of ~ 0.3 Å^{-1}. In fig.3 a) the energies are $E_1=6$ eV and $E_2=13$ eV (box a) in fig.2), respectively. We clearly observe that the region $k_\parallel=0$ (outside the "blind spot") is surrounded by a region of diminished intensity. The intensity increases for larger $k_\parallel$ values and reaches a maximum for $k_\parallel=0.8$Å^{-1}. It should be stressed however that these pictures do not conclusively provide information on the size of the exchange and correlation hole because the fall-off of the intensity at large momenta is experimentally inevitable due to the finite size of the detectors.

Fig. (3c) (in which $E_1=4$ eV and $E_1=13$ eV) demonstrates a new aspect of the correlated electron-pair emission: Here we observe that the ring of enhanced intensity is essentially filled up. Energetically the sum energy E_1+E_2 has been reduced from 19 eV to 17 eV. This energy difference allows for the excitation of other modes of the sample and opens thus the channel of inelastic scattering processes that lead to a decoherence of the escaping electrons' wave [4]. The influence of these phase-breaking processes on the correlation within the electron pair is illustrated in fig.3 a) and b).

We have also performed experiments where the polar angle was varied. The results are displayed in fig.3 for excitation with 30.7 eV electrons. The energies are selected to be at the onset of pair emission ($E_1=6$ eV, $E_2=11$ eV). For a polar angle of 0° the momentum distribution is equivalent to fig.3 a) and b). Changing the polar angle to 10° and then to 20° moves the ring of enhanced intensity with it such that it surrounds the "fixed" electron. This shows that the intensity enhancement is associated with the fixed electron.

We may summarize our observations obtained also with different primary energies as follows: (i) if we select the energies E_1 and E_2 such that the sum energy E_1+E_2 has the largest possible value for pair emission, the 2D momentum plots display an ring of

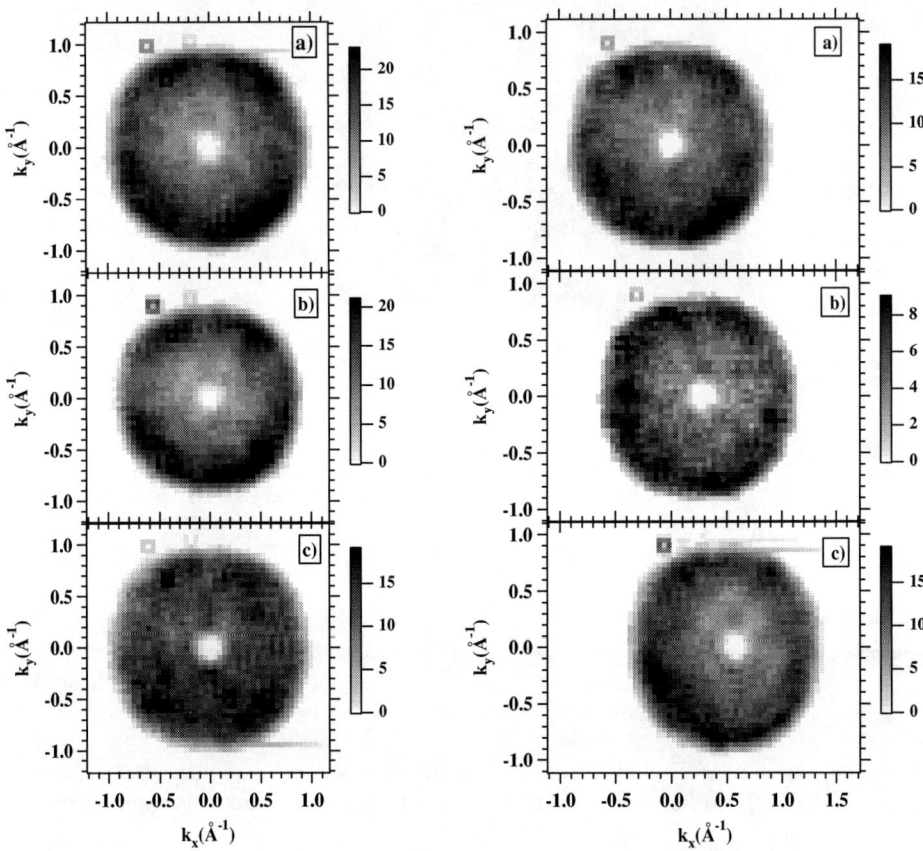

FIGURE 3. **Left panel** shows the 2D $k_\parallel$ distribution of the coincidence intensity ($\Theta=0°$) for different regions in the 2D energy plane, which refer to the boxes a)-c) of fig.2. **Right panel** shows 2D $k_\parallel$ plots obtained upon excitation with 30.7 eV primary electrons. The energies are $E_1=6$ eV and $E_2=11$ eV, respectively. Different polar angles Θ (see fig.1) have been used which are a) 0°, b) 10°, c) 20°.

enhanced intensity which is centered around the "fixed" electron. (ii) if the sum energy is below the maximum value a more or less uniform momentum distribution is the result.

REFERENCES

1. P. Fulde,*Electron Correlations in Molecules and Solids*, Springer Series in Solid State Sciences, Vol. **100** (Springer, Berlin, 1993).
2. W. Schattke, M. A. van Hove,(Eds.) *Solid-State Photoemission and Related Methods: Theory and Experiment* (Wiley-VCH, Weinheim, 2003).
3. N. Fominykh, J. Berakdar, J. Henk, and P. Bruno, Phys. Rev. Lett. **89**, 086402 (2002).
4. F.O. Schumann, J. Kirschner, and J. Berakdar, Phys. Rev. Lett. **95**, 117601 (2005).

A Hybrid DWBA–R-Matrix Approach for Charged-Particle Impact Ionization of Atoms

Klaus Bartschat

Department of Physics and Astronomy, Drake University, Des Moines, IA 50311, USA
e-mail: klaus.bartschat@drake.edu

Abstract. We have developed a hybrid method to treat electron-impact ionization of complex atoms and ions. The essential idea is to describe the interaction between a fast projectile and the target perturbatively, up to second order, while the initial bound state and the ejected-electron–residual-ion interaction are handled via a convergent R-matrix with pseudo-states (close-coupling) expansion. Results for ionization of $Ar(3s^23p^6)$, resulting in either $Ar^+(3s^23p^5)$ or $Ar^+(3s3p^6)$, and for simultaneous ionization–excitation of $He(1s^2)$ leading to $He^+(2p)$ by electron impact are presented.

Keywords: electron impact ionization, ionization-excitation, second-order DWBA, close-coupling
PACS: PACS 34.80.Dp

INTRODUCTION

In recent years, enormous progress has been made regarding the theoretical and computational treatment of electron-impact ionization processes. In particular, non-perturbative approaches, such as exterior complex scaling (ECS) [1], time-dependent close-coupling (TDCC) [2, 3], convergent close-coupling (CCC) [4], or R-matrix with pseudo-states (RMPS) [5–7], are now able to predict the total (TCS), single-differential (SDCS), double-differential (DDCS) and triple-differential (TDCS) cross section very accurately for the atomic hydrogen target. Similar success has been achieved, particularly by the CCC method, for ionization of helium — provided the ion is left in the $He^+(1s)$ ground state, i.e., one of the two target electrons is essentially a spectator [8, 9]. The very same methods, and others, such as the hyperspherical R-matrix with semiclassical outgoing waves (HRM-SOW) approach [10], have achieved most impressive results for the closely related double-photoionization problem of helium.

The theoretical situation, however, does not look as good for more complex targets, such as the heavy noble gases Ne–Xe, or more complicated processes, such as simultaneous ionization-excitation or double-ionization of helium by charged-particle impact. While the ECS [11] and TDCC [12] methods have been extended to the radial coordinates of three electrons, they have not yet been applied to the calculation of fully-differential cross sections involving three active electrons. These, as well as all of the other methods mentioned above, require major computational resources and further algorithm work to handle open-shell residual ions of even limited complexity.

Examples of such ions and processes include $Ar^+(3p)$ or $Ar^+(3s)$, as well as $He^+(2p)$ generated in the highly correlated ionization-excitation process starting from $He(1s^2)$. Especially for the heavy noble gases, non-perturbative methods based on the distorted-wave approach have been used for many years. Much work has been done by Madison and co-workers [13–16], who are now able to account for the proper asymptotic three-body Coulomb boundary condition in the final state [17]. However, these methods often suffer from the approximations made in the treatment of the target structure and the ejected-electron–residual-ion interaction. Especially in the case of asymmetric energy sharing, where one of the outgoing electrons often has an energy of just a few eV, the neglect of channel coupling and sometimes also of exchange effects, or the approximate treatment of exchange via a local potential [18], can become problematic. Finally, for highly correlated processes such as simultaneous ionization-excitation or double ionization by charged-particle impact, accounting for second-order effects in the projectile–target interaction is often very important.

In order to address these problems, we have developed a hybrid method to treat electron-impact ionization of atoms and ions. The essential idea is to describe the interaction between a fast projectile and the target perturbatively, up to second order, while the initial bound state and the ejected-electron–residual-ion interaction are handled via a convergent R-matrix (close-coupling) expansion [19, 20]. A code for calculating total and single-differential cross sections, using a first-order distorted-wave approach for the projectile, was published in 1993 [21]. Its use for positron impact is straightforward [22].

Much of the recent development work focused on the calculation of angle-differential parameters, such as the TDCS and the angular distribution and polarization of light emitted from an excited ionic state [23]. For some kinematical situations studied experimentally, it was critical to extend the code and account for second-order effects [24], to allow for a plane-wave description of the projectile [25], and to ensure a convergent treatment of the close-coupling part of the model via an RMPS treatment. While the method does not account for the three-body asymptotic Coulomb boundary condition, the much improved description of the "near-target" regime yielded very satisfactory results for ionization of several collision systems, including He [25], Rb [26], and Ar [27].

THEORETICAL FORMULATION

The theoretical and computational methods for a first-order distorted-wave treatment of the process have been outlined before [19–21], as have the extensions to account for second-order effects in the projectile–target interaction [24] and the use of plane waves for high incident energies [25]. Here we will focus on a general discussion of the ingredients in a first-order theory and specify some details of the current model for electron-impact ionization of argon.

In our model, the first-order amplitude is given by [19]

$$f_{L_0 M_0 S_0 M_{S_0}, \mu_0 \to L_f M_f S_f M_{S_f}, \mu_1 \mu_2}(\boldsymbol{k}_0, \boldsymbol{k}_1, \boldsymbol{k}_2) =$$
$$-\frac{1}{(2\pi)^{5/2}} \langle \varphi_{\boldsymbol{k}_1 \mu_1}^{(-)}(x) \Psi_{L_f M_f S_f M_{S_f}}^{\boldsymbol{k}_2 \mu_2 (-)}(X) | V(x,X) | \Psi_{L_0 M_0 S_0 M_{S_0}}(X) \varphi_{\boldsymbol{k}_0 \mu_0}^{(+)}(x) \rangle. \quad (1)$$

Here X denotes a set of electronic spatial and spin coordinates in the $(N+1)$-electron atom, while $x = \{\boldsymbol{r}, \sigma\}$ represents the coordinates of the colliding electron. The projectile in the initial and final states is described by the functions $\varphi_{\boldsymbol{k}_0 \mu_0}^{(+)}(x)$ and $\varphi_{\boldsymbol{k}_1 \mu_1}^{(-)}(x)$, respectively, $\Psi_{L_0 M_0 S_0 M_{S_0}}(X)$ is the initial bound state of the target, $\Psi_{L_f M_f S_f M_{S_f}}^{;k_2 \mu_2 (-)}(X)$ the scattering state of the ejected-electron–residual-ion system, and $V(x,X)$ the Coulomb interaction between the projectile and the atomic electrons as well as the nucleus.

We use a non-relativistic model and also neglect exchange between the "fast" projectile and the target electrons. In a plane-wave treatment, which is often a reasonable approximation for the He target, the integral over the coordinates of the fast electron can be performed through the Bethe integral [25]. In a distorted-wave treatment for the projectile, which is critical for a target such as argon at the relatively low incident energies we will consider, we use a partial-wave description for $\varphi_{k_0 \mu_0}^{(+)}(x)$ and $\varphi_{k_1 \mu_1}^{(-)}(x)$. This simplifies the angular integration over the projectile coordinates in the second-order term. As noted by Reid et al. [24], however, we still need to make approximations to evaluate the second-order term. These include the use of only the pole term in a principal-value integral, effectively dropping the real part of the Green's function for propagating through the intermediate states, the choice of an average excitation energy for the intermediate state, and limiting the evaluation of integrals to within the R-matrix sphere.

Another important aspect is the treatment of the ejected-electron–residual-ion interaction, i.e., the function $\Psi_{L_f M_f S_f M_{S_f}}^{;k_2 \mu_2 (-)}$ in (1). In principle, coupling a large number of discrete and pseudo-states will ultimately lead to a converged result for the electron–residual-ion collision problem, as well as the initial bound state. As mentioned above, this is the basic idea behind the CCC [4] and RMPS [5–7] methods.

Typical simplifications in the treatment of this part of the problem include (i) limiting the number of coupled states, (ii) using a distorted wave for the ejected electron, with exact, approximate, or no exchange accounted for, or even just an unperturbed Coulomb wave. When checking the effect of including the three-body Coulomb boundary condition for electron-impact ionization of argon, Prideaux and Madison [15] and Prideaux et al. [16] effectively used a one-state model with local or no exchange for the ionization problem, together with single-configuration target wavefunctions. With increasing projectile and ejected-electron energies, some of these approximations become less critical.

Coming back to the target description, we note that the initial bound state for complex targets is often represented by single-configuration Hartree-Fock (HF), multi-configuration Hartree-Fock (MCHF), or frozen-core multi-configuration Hartree-Fock (FC-MCHF) descriptions. A variant of the latter method is achieved in the R-matrix approach by running the electron–ion collision problem with modified boundary conditions to obtain the bound states of the system. In the work described below for argon [28], we used the multi-configuration ionic target description given by Burke and Taylor [29] for the corresponding photoionization problem. In addition to the 2-state model, in which the $Ar^+(3s^2 3p^5)^2P^o$ and $Ar^+(3s 3p^6)^2S$ states are closely coupled, we performed 1-state and 5-state calculations with the orbitals and configuration expansions obtained by diagonalizing the ionic target hamiltonian. We also used single-configuration descriptions of the $Ar^+(3s^2 3p^5)^2P^o$ and $Ar^+(3s 3p^6)^2S$ states and calculated a 4s valence orbital to include the ionic states with configurations $Ar^+(3s 3p^5 4s)$, $Ar^+(3s^2 3p^4 4s)$, and $Ar^+(3s 3p^4 4s^2)$.

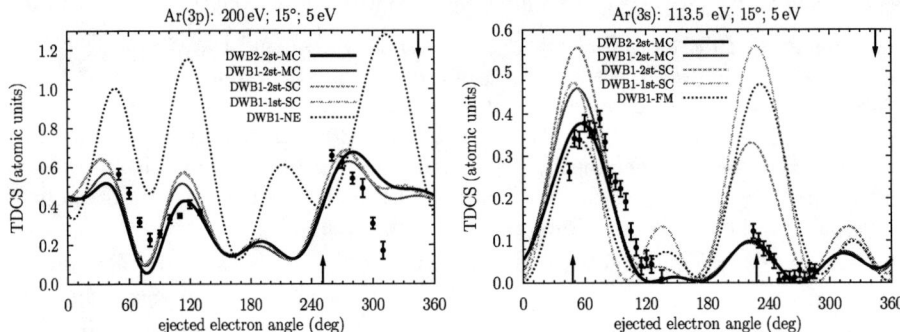

FIGURE 1. Left: Fully-differential cross sections for 200 eV co-planar electron-impact ionization of the 3p shell of argon. The arrow from the top axis indicates the scattering angle (15°) at which the fast electron is observed. This corresponds to an angle of 345° in the coordinate system used for the ejected electron, whose energy is given in the title. The arrows from the bottom axis show the directions of $\pm\hat{q}$, where q is the momentum transfer. The experimental data of Stevenson et al. [27] are visually normalized to the DWB2-2st-MC results in the binary region. Right: Similar results for 113.5 eV electron-impact ionization of the 3s shell of argon. The experimental data are from Haynes and Lohmann [30].

RESULTS

Figure 1 shows fully-differential cross sections for 200 eV and 113.5 eV co-planar electron-impact ionization of the 3p and 3s shells of argon. First-order (DWB1) or second-order (DWB2) descriptions for the fast projectile were combined with 1-, 2-, 5-, or 10-state (#-st in the legends of the panels) close-coupling expansions for the ejected-electron−residual-ion scattering process, Also, the effects of using single-configuration (SC) and many-configuration (MC) representations for the initial bound state and the final ionic states were studied. The results are compared with experimental data from Stevenson et al. [27] and from Haynes and Lohmann [30], and with distorted-wave results from Prideaux and Madison [15] and Prideaux et al. [16]. Exchange effects in the latter works were approximated by the local Furness-McCarthy [18] potential (DWB1-FM) for 3s ionization, while they were often neglected (DWB1-NE) for 3p ionization because of apparent problems with the local-potential approximation [13, 14].

The overall agreement between the predictions of our presumably best model, the DWB2-2st-MC approach, with the experimental data is very satisfactory. It should be noted, however, that the experimental results have been normalized by a visual fit to the DWB2-2st-MC results in the binary region. For 3p ionization, the results from all our models are very similar, indicating that both second-order and channel-coupling effects play a relatively small role. However, there is a substantial difference with the DWB1-NE results, indicating that a proper treatment of exchange effects is absolutely critical. It is also interesting to note that problems with our theory apparently emerge for angles around 300° and above. They may be due to the neglect of post-collision interaction (PCI) between the two outgoing electrons. Since the fast outgoing electron is observed at 345°, the PCI effect may become important in this angular range.

For the weak ionization channel, on the other hand, where the Ar^+ ion is left in the

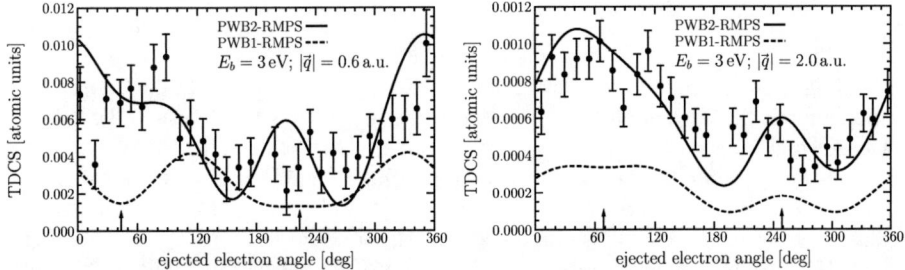

FIGURE 2. TDCS for ionization–excitation of He$(1s^2)$ to He$^+$(2p) by 500 eV electrons as function of the slow (3 eV) electron emission angle. Results obtained in first-order (PWB1) and second-order (PWB2) models combined with an RMPS treatment of the ejected-electron–residual-ion interaction are compared with the experimental data of Sakhelashvili et al. [31]. The arrows indicate the directions of $\pm\vec{q}$.

$(3s3p^6)^2$S state, we see a dramatic change in the model sensitivity of the results. Channel coupling and the target description strongly affect the predictions, while second-order effects remain comparatively small. Since the local exchange potential of Furness and McCarthy [18] works reasonably well for this case [13], one would expect the DWB1-FM results to lie somewhere in the vicinity of the DWB1-1st-SC results. This is indeed the case, although the one-electron orbitals used by Prideaux and Madison were slightly different from ours. We note very satisfactory agreement of the DWB2-2st-MC predictions with the experimental data for this slow-electron energy of 5 eV. Unfortunately, the growing height of the recoil peak with increasing energy [30] of that electron is not reproduced in the model [28].

An example for ionization-excitation of He$(1s^2)$ to He$^+$(2p) is shown in figure 2. The experimental data from Heidelberg [31] are relative but *cross-normalized* to each other using a *common factor* determined by a visual fit to the magnitude of the second-order RMPS results at $|q| = 2$ a.u. and $E_b = 3$ eV. At small momentum transfer, $|q| = 0.6$ a.u., only the second-order RMPS model reproduces the experimental data in a satisfactory way. Note the broken symmetry of the angular emission pattern with respect to the momentum transfer axis. Also, accounting for only the dipole term of the multipole expansion of the Coulomb interaction in the calculation of the second-order amplitude [32] is apparently not sufficient to reproduce the experimental data [31]. Finally, at large momentum transfer, the second-order and first-order results agree fairly well with each other and the experiment regarding the *shape* of the curves, but the first-order results would need to be multiplied by about 2.5(!) to obtain the same magnitude. This shows the importance of cross-normalizing the experimental datasets whenever possible.

CONCLUSIONS

We have presented a hybrid approach to treat charged-particle impact ionization of complex targets. The method is most suitable for the cases of asymmetric energy sharing between a fast projectile and a slow ejected electron. Channel-coupling and exchange effects in the interaction of the ejected electron and the residual ion can be treated to convergence, while the projectile−target interaction is limited to second-order processes. Also, highly sophisticated multi-configuration expansions can readily be used for the initial bound and the final ionic states. The biggest drawback of the current implementation is the neglect of the PCI effect, which we will try to address in the future.

More experiments are required to guide further theoretical efforts. While these datasets should preferably be absolute, relative results for different kinematics should definitely be *cross-normalized* to each other, in order to better assess the strengths and weaknesses of current theoretical approaches. The latest results from the Heidelberg group [31] for simultaneous ionization−excitation of helium are a showcase for how such cross normalization may clearly favor one approach over an other.

ACKNOWLEDGMENTS

This work would not have been possible without the contributions from many colleagues and the financial support provided by the United States National Science Foundation.

REFERENCES

1. T.N. Rescigno, M. Baertschy, W.A. Isaacs, and C.W. McCurdy, Science **286** (1999) 2474.
2. M.S. Pindzola and F. Robicheaux, Phys. Rev. A **54** (1996) 2142.
3. J.P. Colgan *et al.*, Phys. Rev. A **65** (2002) 042721.
4. I. Bray, D.V. Fursa, A. Kheifets, and A.T. Stelbovics, J. Phys. B **35** (2002) R117.
5. K. Bartschat, E.T. Hudson, M.P. Scott, P.G. Burke, and V.M. Burke, J. Phys. B **29** (1996) 115.
6. K. Bartschat and I. Bray, Phys. Rev. A **54** (1996) R1002.
7. K. Bartschat, Comp. Phys. Commun. **114** (1998) 168.
8. I. Bray, Phys. Rev. Lett. **89** (2002) 273201.
9. A.T. Stelbovics,I. Bray, D.V. Fursa, and K. Bartschat, Phys. Rev. A **71** (2005) 052716.
10. L. Malegat, P. Selles, and A.K. Kazansky, Phys. Rev. Lett. **85** (2000) 4450.
11. C.W. McCurdy, D.A. Horner, and T.N. Rescigno, Phys. Rev. A **65** (2002) 042714.
12. M.S. Pindzola *et al.*, Phys. Rev. A **70** (2004) 032705.
13. D.A. Biava *et al.*, J. Phys. B **35** (2002) 293.
14. D.A. Biava, K. Bartschat, H.P. Saha, and D.H. Madison, J. Phys. B **35** (2002) 5121.
15. A. Prideaux and D.H. Madison, Phys. Rev. A **67** (2003) 052710.
16. A. Prideaux, D.H. Madison, and K. Bartschat, Phys. Rev. A **71** (2005) 032702.
17. M. Brauner, J.S. Briggs, and H. Klar, J. Phys. B **22** (1989) 2265.
18. J.B. Furness and I.E. McCarthy, J. Phys. B **6** (1973) 2280.
19. K. Bartschat and P.G. Burke, J. Phys. B **20** (1987) 3191 and **21** (1988) 2969.
20. K. Bartschat and P.G. Burke, J. Phys. B **21** (1988) 2969.
21. K. Bartschat, Comp. Phys. Commun. **75** (1993) 219.
22. K. Bartschat, Phys. Rev. A **71** (2005) 032718.
23. K. Bartschat and A.N. Grum-Grzhimailo, J. Phys. B **35** (2002) 5035.
24. R.H.G. Reid, K. Bartschat, and A. Raeker, J. Phys. B **31** (1998) 563; *corr.* J. Phys. B **33** (2000) 5261.
25. Y. Fang and K. Bartschat, J. Phys. B **34** (2001) L19.
26. M.A. Haynes, B. Lohmann, I. Bray, and K. Bartschat, Phys. Rev. A **71** (2004) 044704.
27. M. Stevenson *et al*, J. Phys. B **38** (2005) 433.
28. K. Bartschat and O.K. Vorov, Phys. Rev. A **72** (2005) 022728.
29. P.G. Burke and K.T. Taylor, J. Phys. B **8** (1975) 2620.
30. M.A. Haynes and B. Lohmann, J. Phys. B **33** (2000) 4711.
31. G. Sakhelashvili *et al.*, Phys. Rev. Lett. **95** (2005) 033201.
32. A.S. Kheifets, Phys. Rev. A **69** (2004) 032712.

Conference Programme

(Note: The sessions marked E, P, J refer to the (e,2e), or to the Polarization or to the Joint sessions, respectively.)

Thursday, 28 July 2005

8:45 **Joint Session J1 : Opening**

Chair : J. Williams

9:00   J. Briggs, University of Freiburg (Germany)

Few – particle fragmentation : (e,2e), PDI and beyond

9:30   U. Heinzmann, University of Bielefeld (Germany)

Phase and time-resolved photoelectron and Auger electron spectroscopy

10:00   J. Ullrich, Max-Planck-Institute for Nuclear Physics (Germany)

Three-body quantum dynamics of helium single ionization by heavy ion impact

10:30 **Coffee break**

11:00 **Session E1 : (e,2e) and (e,3e)**

Chair : M. Schulz

11:00   L.U. Ancarani, Université de Metz (France)

The use of correlated wavefunctions in describing the double ionization of helium

11:30   M. Foster, University of Missouri-Rolla (USA)

How well do we understand single ionization of atoms by charged particle impact?

12:00   M. Dürr, Max-Planck-Institute for Nuclear Physics (Germany)

Fragmentation of helium in collisions with slow electrons : three- and four-body dynamics

11:00 **Session P1 : Correlation in excitation and ionization**

Chair : A. Temkin

11:00   N.L. Manakov, Voronezh State University (Russia)

Nondipole effects in double photoionization of helium

11:30   J.G. Childers, California State University (USA)

Low energy electron scattering from atomic hydrogen

12:00   S. Otranto, University of Missouri-Rolla (USA)

Kinetic correlation in photo-double-ionization processes: the He-isoelectronic sequence

12:30 **Lunch**

13:30 **Poster Session**

15:00 Joint Session J2
Chair : L. Malegat
15:00   A. Stelbovics, Murdoch University (Australia)
Theory of electron impact ionization of atoms
15:30   H. Schmidt-Böcking, J.W. Goethe Universität (Germany)
Multi-coincidence studies of photo and Auger electrons from fixed-in-space molecules using the COLTRIMS technique

16:00 Coffee break

16:30 Session E2 : Electron and photon interactions with molecules
Chair : A. Huetz
16:30   O. A. Fojón, Instituto de Física, Rosario (Argentina)
Interference effects in single ionization of H2 molecules by photon and electron impact
17:00   M.S. Schöffler, J.W. Goethe Universität (Germany)
Transfer ionization in fast proton He collisions. Revealing the non-S^2 contributions in the ground state
17:30   J. Colgan, Los Alamos national Laboratory (USA)
Double photo ionization of He and H2

16:30 Session P2 : Spin-dependent processes I
Chair : G. F. Hanne
16:30   J.D. Bozek, Lawrence Berkeley National Laboratory (USA)
Spin-resolved photoelectron spectroscopy of rare gas clusters
17:00   J. Lower, Australian National University (Australia)
Ionization of atoms with spin polarized electrons
17:30   F. Jüttemann, Universität Münster (Germany)
Study of coherence in electron-impact excitation of mercury

18:00 End

Friday, 29 July 2005

9:00 Joint Session J3

Chair : K. Bartschat

9:00   D.H. Madison, University of Missouri-Rolla (USA)
Recent developments in the theoretical calculation of electron-impact ionization of atoms and molecules

9:30   A. Crowe, University of Newcastle-upon-Tyne (United Kingdom)
New results on excitation, ionization and ionization-excitation by electron impact

10:00 C.W. McCurdy, University of California, Davis (USA)
Non perturbative calculation of excitation autoionization of helium using the ECS method: sharp resonance features in the SDCS

10:30 Coffee break

11:00 Session E3 : (e,3-1e) processes

Chair : P. Bolognesi

11:00 F. Catoire, Université de Paris-Sud, Orsay (France)
Double ionization of argon through direct double ionzation and Auger decay

11:30 N. Watanabe, Institute of Molecular Science (Japan)
(e,2e) and (e,3-1e) studies on double processes of He at large momentum transfer

12:00 K. Kouzakov, Moscow StateUniversity (Russia)
(e,3-1e) reactions at large momentum transfer

11:00 Session P3 : Spin-dependent processes II

Chair : A. Stauffer

11:00 T.J. Gay, University of Nebraska, Lincoln (USA)
Polarized molecular fluorescence from polarized-electron impact

11:30 B. Langer, Max-Born-Institut (Germany)
Spin-resolved studies in Ar, Kr and Xe: intrinsic and dynamic parameters

12:00 W.E. Guinea, Griffith University (Australia)
Spin-resolved collisions of electrons with rubidium atoms: a search for relativistic effects

12:30 Lunch

13:30 Poster Session

15:00 Joint Session J4

Chair : J. Berakdar

15:00   R.C. Bilodeau, Western Michigan University (USA)
Multi-Auger decay in negative ion photodetachment

15:30   F. Penent, Université Pierre et Marie Curie (France)
Multiple direct and sequential Auger effect in the rare gases

16:00 Coffee break

16:30 Session E4 : Photo double ionization

Chair : D. Dowek

16:30   F. Martín, Universidad Autonoma de Madrid (Spain)
First principles calculations of the double photoionization of atoms and molecules using B-splines and Exterior Complex Scaling

17:00   M. Gisselbrecht, Université Paris-Sud, Orsay (France)
Photo double ionization of fixed in space H_2

17:30   P. Bolognesi, IMIP, Monterotondo (Italy)
Complete photoionization experiments by photoelectron - Auger electron coincidence measurements

16:30 Session P4 : Correlation in molecules

Chair : T. Gay

16:30   G. Pruemper, Tohoku University (Japan)
Doppler energy shifts and intramolecular backscattering of Auger electrons emitted from molecules under ultrafast dissociation

17:00   B. Joulakian, Université Paul Verlaine (France)
Correlation effects in electron-diatomic molecule inelastic collisions

17:30   T. Odagiri, Tokyo Institute of Technology (Japan)
(γ, 2γ) studies on multiply excited states of H_2 and N_2 in the vacuum ultraviolet range

18:00 End

Saturday, 30 July 2005

9:00 Session E5 : Interactions with solids and clusters

Chair : D. Fursa

9:00 K. Mitsuke, Institute for Molecular Science, Okazaki (Japan)
   Photoionization and photodissociation dynamics of fullerenes and endohedral metallofullerenes

9:30 C. Bowles, Australian National University (Australia)
   Direct measurement of spectral momentum densities of single crystals using high energy EMS spectroscopy

10:00 K.L. Nixon, Flinders University (Australia)
   Intermolecular Interaction probed via Electron Momentum Spectroscopy of van der Waals Molecules

9:00 Session P5 : Excitation and autoionization

Chair : R. Srivastava

9:00 A.J. Murray, University of Manchester (United Kingdom)
   Low energy super-elastic electron scattering using a new magnetic angle changing spectrometer

9:30 D. Cvejanovic, University of Western Australia (Australia)
   Spin up-down asymmetry in the excitation of Kr $5P'[3/2]_2$ by polarized electrons

10:00 A.N. Grum-Grzhimailo, Moscow State University (Russia)
   Universal scaling of resonances in vector correlation photoionization parameters

10:30 Coffee break

11:00 Joint Session J5

Chair : D. Madison

11:00 J. Berakdar, Max-Planck Institute of Microstructure Physics (Germany)
   Correlation spectroscopy of condensed matter systems

11:30 K. Bartschat, Drake University (USA)
   A hybrid DWBA- R-matrix approach for charged particle impact ionization of atoms and ions

12:00 Concluding remarks, general discussion and close of the conference

Posters

P.1 CHANNEL-COUPLING AND SECOND-ORDER EFFECTS IN ELECTRON IMPACT IONIZATION OF AR(3S) AND AR(3P), K. Bartschat and O. Vorov

P.2 A TIME-DEPENDENT STUDY OF THE ANGULAR DISTRIBUTION OF PHOTOELECTRONS IN THE REGION OF LASER-INDUCED CONTINUUM STRUCTURES, A. N. Grum-Grzhimailo, K. Bartschat and S. I. Strakhova

P.3 HOW DOES SINGLE IONIZATION WITH EXCITATION MERGE INTO DOUBLE IONIZATION, C. Bouri, P. Selles, L. Malegat, A. K. Kazansky, J.M. Teuler and M. Kwato Njock

P.4 EXCITATION OF HE + PARABOLIC STATES WITH LARGE N BY PHOTON IMPACT ON HELIUM, C. Bouri, L. Malegat, P. Selles, A.K. Kazansky and M.G. Kwato Njock

P.5 SPIN RESOLVED (e-2e) STUDY OF THE ARGON 2P STATES, S. Bellm, J. Lower, C. Whelan and B. Lohmann

P.6 SEARCH FOR INTERFERENCE EFFECTS IN THE ELECTRON IMPACT IONIZATION OF H2, D. Milne-Brownlie and B. Lohmann

P.7 VALIDITY OF APPROXIMATING THE SECOND ORDER AMPLITUDE AS A PRODUCT OF FIRST ORDER AMPLITUDES, Z. Chen and D H Madison

P.8 INTERFERENCE EFFECTS IN THE PHOTOIONIZATION OF THE HYDROGEN MOLECULAR ION, R. Della Picca, P. D. Fainstein, M. L. Martiarena and A. Dubois

P.9 PHOTOEMISSION IN THE MOLECULAR FRAME FOR INNER-SHELL IONIZATION OF LINEAR MOLECULES INDUCED BY CIRCULARLY POLARIZED LIGHT, W. B. Li, A. Haouas, J. C. Houver, R. Guillemin, L.Journel, M. Lebech, R. R. Lucchese, M. Simon and D. Dowek

P.10 FRAGMENT EMISSION FOLLOWING MULTIPLE IONIZATION IN 20 eV - 200 eV e^-+H_2O COLLISIONS, F Frémont, C Leclercq, A Hajaji, A.Naja, J. Soret, R. Lelièvre and J-Y Chesnel

P.11 ELECTRON-IMPACT IONIZATION OF CALCIUM FOR EQUAL-ENERGY SHARING KINEMATICS, D. V. Fursa, I. Bray, A. T. Stelbovics

P.12 THE TRIPLE DIFFERENTIAL CROSS SECTION AND INTERFERENCE EFFECTS OF ELECTRON IMPACT IONIZATION OF MOLECULES, J. Gao, D. H. Madison and J. L. Peacher

P.13 PROJECTILE INFLUENCE ON THE ANGULAR DISTRIBUTIONS IN (E,3E) PROCESSES, G. Gasaneo, S. Otranto, K. V. Rodriguez and R. H. Pratt

P.14 STURMIAN WAVE FUNCTIONS FOR COULOMB SCREENED POTENTIALS, L. Frapiccin, V. Y. Gonzalez, J. M. Randazzo, F. D.Colavecchia and G. Gasaneo

P.15 ACCURATE HYLLERAAS-LIKE FUNCTIONS FOR THE HE ATOM WITH CORRECT CUSP CONDITIONS, K. V Rodríguez and G. Gasaneo

P.16 MULTI PARTICLE EFFECTS IN THE COULOMB CONTINUUM, J. R. Gotz, M. Walter and J. Briggs

P.17 SPIN-EXCHANGE COLLISIONS OF POLARIZED ELECTRONS WITH OPEN-SHELL-TARGETS, I. Holtkötter, G. F. Hanne

P.18 CUSP CONDITIONS AND CONVERGENCES IN CLOSE-COUPLING WAVEFUNCTIONS FOR INNER-SHELL ELECTRONS, D. M. Mitnik and J. E. Miraglia

P.19 IONIZATION STUDIES OF THE ALKALI & ALKALI EARTH METALS: Na, Mg, K & Ca (e,2e) STUDIES FROM NEAR THRESHOLD TO THE INTERMEDIATE ENERGY REGIME, A. J. Murray

P.20 (e,2e) IONIZATION STUDIES OF H_2 IN THE INTERMEDIATE ENERGY REGIME, A. J. Murray

P.21 LOW ENERGY (e,2e) IONIZATION MEASUREMENTS FROM THE $3\sigma_g$ AND $1\pi_u$ ORBITALS OF N_2 AND THE $1\pi_g$ AND $4\sigma_g$ ORBITALS OF CO_2, M. J. Hussey and A. J. Murray

P.22 MULTIPHOTON IONIZATION OF HYDROGEN MOLECULE BY ULTRASHORT LASER PULSES, A. Palacios, H. Bachau and F. Martín

P.23 RESONANT EFFECTS IN THE COULOMB EXPLOSION OF H_2^+ BY ULTRASHORT LASER PULSES, A. Palacios, H. Bachau and F. Martín

P.24 ACCURATE CALCULATION OF TRIPLE DIFFERENTIAL CROSS SECTIONS FOR DOUBLE PHOTOIONIZATION OF THE HYDROGEN MOLECULE, W. Vanroose, F. Martín, T. N. Rescigno and C. W. McCurdy

P.25 $O(m\alpha/\omega)$ CORRECTIONS TO THE SHAKE-OFF (SHAKE-UP) CORRELATION EFFECTS IN DOUBLE (SINGLE) IONIZATION OF H^- AND Li^+ BY ABSORPTION OF A PHOTON, T. Suric, R. H. Pratt

P.26 IONIZATION WITH EXCHANGE IN Ps & ATOMS COLLISION, H.Ray

P.27 EVIDENCE FOR COHERENT TWO-CENTER PHOTOELECTRON SCATTERING IN DIATOMIC MOLECULES: CO vs. N_2, B. Zimmerman, M. Braune, O. Geiner, R. Hentges, S. Korica, G. Prümper, A. Reinköster, J. Viefhaus, B. Langer, R. Dörner, V. McKoy and U. Becker

P.28 COINCIDENCE STUDY OF DOUBLE ELECTRON EMISSION ASSOCIATED WITH K-SHELL PHOTOIONIZATION OF C_{60}, A. Reinköster, M. Braune, S. Korica, D. Rolles, J. Viefhaus, B. Langer and U.Becker

P.29 PERTURBATIVE ANALYSIS OF DOUBLE PHOTOIONIZATION MECHANISMS, A. Y. Istomin, N. L. Manakov, and A. F. Starace

P.30 ALIGNMENT AND ORIENTATION IN ELECTRON SCATTERING FROM EXCITED BARIUM ATOMS, R. Srivastava and A. D. Stauffer

P.31 ELECTRON-IMPACT IONIZATION MECHANISM NEAR THRESHOLD FOR HYDROGEN, J. F. Williams, P. L Bartlett and A. T Stelbovics

P.32 PHOTO-DOUBLE IONIZATION OF MOLECULAR BROMINE STUDIED BY THRESHOLD PHOTOELECTRONS COINCIDENCE SPECTROSCOPY, A. J. Yencha, S. P. Lee and G. C. King

P.33 INVESTIGATIONS OF ELECTRON DICHROISM SIGNALS FROM TARGETS OF CHIRAL MOLECULE, A. H. Zimnol, T. Meyer, V. Hamelbeck and G. F. Hanne

P.34 (e,2e) AND (e,3-1e) STUDIES ON THE DOUBLE PROCESSES OF He AT LARGE MOMENTUM TRANSFER : THE SECOND BORN CALCULATION, P. S. Vinitsky, Y. V. Popov, K. A. Kousakov, N. Watanabe and M. Takahashi

P.35 AN (e,2e) INVESTIGATION OF He AND Ar AT CLOSE TO MINIMUM MOMENTUM TRANSFER AND LARGE ENERGY KINEMATICS, F. Catoire, M Casagrande, A. Lahmam-Bennani and C. dal Capello

List of Participants

Guillermo Acuña
Edificio IAFE
Ciudad Universitaria
Buenos Aires 1430
Argentina
acuna@iafe.uba.ar

Lorenzo Ugo Ancarani
LPMC, Institut de Physique
Université de Metz
1 bd Arago
Metz 57078
France
ancarani@sciences.univ-metz.fr

Yohko Awaya
General Education
Musashino Art University
1-736 Ogawa, Kodaira
Tokyo 187-8505
Japan
awaya@musabi.ac.jp

Klaus Bartschat
Physics & Astronomy
Drake University
2507 University Avenue
Des Moines, IA 50311
USA
klaus.bartschat@drake.edu

Susan Bellm
Atomic and Molecular Physics Laboratories
Australian National University
Canberra ACT 0200
Australia
susan.bellm@anu.edu.au

Jamal Berakdar
MPI of Microstructure Physics
Weinberg 2
Halle 6120
Germany
jber@mpi-halle.de

Rene Bilodeau
Physics
Western Michigan University and Lawrence Berkeley National Laboratory—ALS
1 Cyclotron Rd.
Berkeley 94720-8235
USA
rcbilodeau@lbl.gov

Gisela A. Bocan
Atomic Collisions
IAFE-CONICET
University of Buenos Aires
Buenos Aires 1428
Argentina
gbocan@iafe.uba.ar

Paola Bolognesi
Istituto di metodologie Inorganiche e dei Plasmi
Consiglio Nazionale delle Ricerche
Area della Ricerca di Roma 1 - Via Salaria Km 29.3
Monterotondo Scalo (Roma) 10
Italy
paola.bolognesi@imip.cnr.it

Cameron Bowles
Atomic And Molecular Physics Laboratories
Australian National University
Canberra 0200
Australia
cameron.bowles@anu.edu.au

John Bozek
Advanced Light Source
Lawrence Berkeley National Laboroatory
MS 6R2100, 1 Cyclotron Road
Berkeley 94720
USA
jdbozek@lbl.gov

John Briggs
Theoretical Quantum Dynamics
Univ. of Freiburg
H.Herder Str. 3
Freiburg 79104
Germany
john.briggs@physik.uni-freiburg.de

Fabrice Catoire
LCAM, Bât. 351, Université Paris-Sud XI
Orsay 91405
France
catoire@lcam.u-psud.fr

Celsus Bouri
LIXAM,
Bât. 350, Université Paris-Sud XI
Orsay 91405
France
Celsus.Bouri@lixam.u-psud.fr

Zhangjin Chen
Department of Physics
University of Missouri-Rolla
Rolla 65401
USA
zhangjin@umr.edu

Greg Childers
Dept. of Physics
California State University Fullerton
800 N. State College Blvd
Fullerton, CA 92834
USA
gchilders@fullerton.edu

Flavio Colavecchia
Div. Colisiones Atómicas
Centro Atómico Bariloche
Av. Bustillo 9500
S. C. de Bariloche 8400
Argentina
flavioc@cab.cnea.gov.ar

James Colgan
Theoretical Division
Los Alamos National Laboratory
T4, MS B283
Los Alamos NM 87545
U.S.
jcolgan@lanl.gov

Albert Crowe
Physics
University of Newcastle
Herschel Building
Newcastle upon Tyne NE1 7RU
UK
albert.crowe@ncl.ac.uk

Danica Cvejanovic
Physics
University of Western Australia
35 Stirling Highway
Crawley, WA 6009
Australia
danica.cvejanovic@uwa.edu.au

Bruno deHarak
Physics and Astronomy
University of Kentucky
3713 Forest Green Dr.
Lexington 40517
USA
badeha2@uky.edu

Renata Della Picca
Colisiones Atomicas
CNEA-Conicet-Inst. Balseiro
Av. Bustillo km 9,5
Bariloche 8400
Argentina
renata@cab.cnea.gov.ar

Danielle Dowek
LCAM, Bât 351, Université Paris-Sud
Orsay 91405
France
dowek@lcam.u-psud.fr

Martin Dürr
Experimental Few-Particle Quantum Dynamics
Max Planck Institute for Nuclear Physics
Saupfercheckweg 1
Heidelberg 69117
Germany
martin.duerr@mpi-hd.mpg.de

Tamer El-Kafrawy
Physics Department
Faculty of Science, Ain Shams
Physics Department, Faculty of Science, Ain Shams
Cairo
Egypt
tamer_elkafrawy@hotmail.com

Marisa Faraggi
Instituto de Astronomía y Física del Espacio.(CONICET-UBA)
Casilla de Correo 67, Sucursal 28
(1428) Buenos Aires,
Argentina
faraggi@iafe.uba.ar

Ray Flannery
Physics
Georgia Institute of Technology
837 State Street
ATLANTA 30332-0430
USA
ray.flannery@physics.gatech.edu

Omar Fojon
Fisica Instituto de Fisica Rosario (CONICET-UNR)
Pellegrini
Rosario 2000
Argentina
ofojon@fceia.unr.edu.ar

Matthew Foster
Physics
University of Missouri - Rolla
111 Gene Drive
Rolla, MO 65401
USA
foster@umr.edu

Ana Laura Frapiccini
Física
Universidad Nacional del Sur
Av. Alem 1253
Bahía Blanca 8000
Argentina
afrapic@uns.edu.ar

François Frémont
CIRIL Ensicaen
6, bd du Mal Juin
Caen 14050
France
francois.fremont@ensicaen.fr

Dmitry Fursa
Physics and Energy Studies,
Murdoch University
Perth 6150
Australia
d.fursa@murdoch.edu.au

Junfang Gao
Department of Physics
University of Missouri-Rolla
Rolla 65401
USA
jgzm6@umr.edu

Andrea García
IAFE
Chacabuco 64
Buenos Aires 1653
Argentina
agarcia.@iafe.uba.ar

Carlos Roberto Garibotti
Division Colisiones Atómicas
CAB and CONICET
Centro Atomico
Bariloche 8400
Argentina
gari@cab.cnea.gov.ar

Gustavo Gasaneo
Departamento de Física
Universidad Nacional del Sur
Ave Alem 1253
Bahía Blanca 8000
Argentina
ggasaneo@criba.edu.ar

Timothy Gay
Physics
Behlen Lab, University of Nebraska
Lincoln, Nebraska 68588-0111
USA
tgay1@unl.edu

Mathieu Gisselbrecht
LIXAM - CNRS
Bât. 350, Université Paris-Sud XI
Orsay 91405
France
mathieu.gisselbrecht@lixam.u-psud.fr

Valeria Yanina González
Universidad Nacional del Sur
Alem 1253
Bahía Blanca 8000
Argentina
vgonzal@uns.edu.ar

María Silvia Gravielle
IAFE
Buenos Aires 1428
Argentina
msilvia@iafe.uba.ar

Kim Green
Physics
Herschel Building, University of Newcastle
Newcastle Upon Tyne NE1 7RU
United Kingdom
kim.green@ncl.ac.uk

Alexei Grum-Grzhimailo
Institute of Nuclear Physics
Moscow State University
Vorobyevy Gory
Moscow 119992
Russia
algrgr1492@yahoo.com

William Guinea
Centre for Quantum Dynamics
School of Science, Griffith University
Nathan 4111
Australia
W.Guinea@griffith.edu.au

Olivier Guyétand
LIXAM-CNRS
Bât. 350, Université Paris-Sud XI
Orsay 91405
France
olivier.guyetand@lixam.u-psud.fr

Jens Götz
Physikalisches Institut
Albert-Ludwigs-Universität
Hermann-Herder-Str. 3
Freiburg 79104
Germany
goetzj@uni-freiburg.de

Volker Hamelbeck
Physikalisches Institut
University of Muenster
Wilhelm-Klemm-Str. 10
Muenster 48149
Germany
hamelbe@uni-muenster.de

G. Friedrich Hanne
Physics
University of Muenster
Wilhelm-Klemm-Str. 10
Muenster D-48149
Germany
hanne@uni-muenster.de

Ulrich Heinzmann
Physics
University Bielefeld
Bielefeld D-33501
Germany
uheinzm@physik.uni-bielefeld.de

Ingo Holtkoetter
Physikalisches Institut
WWU Münster
Wilhelm-Klemm-Str.10
Münster 48149
Germany
holtkoi@nwz.uni-muenster.de

Alain Huetz
LIXAM-CNRS
Bât. 350, Université Paris-Sud XI
Orsay 91405
France
alain.huetz@lixam.u-psud.fr

Amy Huntington
Physics
Herschel Building, University of Newcastle
Newcastle Upon Tyne NE1 7RU
United Kingdom
amy.huntington@ncl.ac.uk

Martyn Hussey
School of Physics and Astronomy
The University of Manchester
Schuster Laboratory
Manchester M13 9PL
UK
martyn.hussey@manchester.ac.uk

John Adamsom Oyeyemi
Electronics Department
John Royal Limited
Sw8/1370, Liberty Road, Oke-Ado, Ibadan,
Oyo-State
Ibadan 234
Nigeria
johnroyals2000@yahoo.co.uk

Joulakian Bhogos
University Paul Verlaine-Metz
I.P.E.M. L.P.M.C. 1, blv Arago
Metz 57078
France
joulak@univ-metz.fr

Frank Jüttemann
Physikalisches Institut
Universität Münster
Wilhelm-Klemm-Str. 10
Münster 48149
Germany
juttema@uni-muenster.de

Konstantin Kouzakov
Nuclear Physics and Quantum Theory of
Collisions
Moscow State University
Vorob'evy gory
Moscow 119992Russia
kouzakov@srd.sinp.msu.ru

Eldar Kurtaliev
Physics
Samarkand State University
University blvd.,15
Samarkand 703004
Uzbekistan
kurtaliev@rambler.ru

Azzedine Lahmam-Bennani
LCAM,
Bât. 351, Université Paris-Sud
Orsay cedex F-91405
France
azzedine.l-bennani@lcam.u-psud.fr

Burkhard Langer
A1 Max-Born-Institut
Max-Born-Straße 2A
Berlin 12489
Germany
langer@gpta.de

Gary Leighton
Physics
University of Newcastle
Herschel building
Newcastle-Upon-Tyne NE1 7RU
England
G.J.Leighton@ncl.ac.uk

Jon Levin
Physics
U. Tennessee
401 Nielsen Physics Building
Knoxville, Tennessee 37996
USA
jlevin@utk.edu

Birgit Lohmann
Centre for Quantum Dynamics
School of Science, Griffith University
Kessels Road
Nathan 4111
Australia
B.Lohmann@griffith.edu.au

Julian Lower
Atomic and Molecular Physics Laboratories
Australian National University
Canberra 0200
Australia
Julian.Lower@anu.edu.au

Don Madison
Physics
University of Missouri
400 Lariat Lane
Rolla, Missouri 65401
USA
madison@umr.edu

Laurence Malegat
LIXAM,
Bât. 350, Universite Paris-Sud
ORSAY 91405
FRANCE
Laurence.Malegat@lixam.u-psud.fr

Nikolai Manakov
Physics Department
Voronezh State University
University Square, 1
Voronezh 394006
Russia
manakov@phys.vsu.ru

Fernando Martin
Quimica. C-9.
Universidad Autonoma de Madrid
Facultad de Ciencias
Madrid 28049
Spain
fernando.martin@uam.es

Nicholas Martin
Dept. Physics and Astronomy, University of Kentucky
Lexington KY 40506-0
USA
nmartin@uky.edu

Michio Matsuzawa
Dept.Appl.Phys.&Chem.
Univ.Electr-Communications
Katakura-cho 1074-94
Hachiohji 192-0914
JAPAN
michio@nna.so-net.ne.jp

Bill McCurdy
Chemistry and Applied Science
University of California, Davis
One Cyclotron Rd.
Berkeley, California 94720
U.S.A.
cwmccurdy@lbl.gov

Danielle Milne-Brownlie
Centre for Quantum Dynamics
School of Science, Griffith University
Nathan 4111
Australia
D.Milne-Brownlie@griffith.edu.au

Jorge Miraglia
Fisica
Edificio IAFE, Ciudad Universitaria
Capital Federal C1428EGA
Argentina
miraglia@iafe.uba.ar

Dario Mitnik
Departamento de Fisica
IAFE / Universidad de Buenos Aires
CC 67, Suc. 28
Buenos Aires C1428EGA
Argentina
dmitnik@df.uba.ar

Koichiro Mitsuke
Vacuum UV Photo-science
Institute for Molecular Science
Myodaiji
Okazaki 444-8585
Japan
mitsuke@ims.ac.jp

Claudia Montanari
Departamento de Fisica
IAFE / Universidad de Buenos Aires
CC 67, Suc. 28
Buenos Aires C1428EGA
Argentina
mclaudia@iafe.uba.ar

Andrew Murray
Physics & Astronomy
University of Manchester
Schuster Laboratory, Brunswick St.
Manchester M13 9PL
United Kingdom
Andrew.Murray@manchester.ac.uk

Stuart Napier
Physics
University of Western Australia
35 Stirling Hwy
Crawley
Perth 6009
Australia
san@physics.uwa.edu.au

Kate Nixon
School of Chemistry, Physics and Earth Sciences
Flinders University
Adelaide 5001
Australia
Kate.Nixon@flinders.edu.au

Takeshi Odagiri
Chemistry
Tokyo Institute of Technology
O-okayama 2-12-1, Meguro-ku
Tokyo 152-8551
Japan
joe@chem.titech.ac.jp

Sebastian Otranto
1870 Miner Circle
Rolla 65401
USA
otrantos@umr.edu

Alicia Palacios
Quimica
Universidad Autonoma De Madrid
Cantoblanco, Ctra. Colmenar, Km. 15
Madrid 28049
Spain
alicia.palacios@uam.es

Francis Penent
CNRS and UPMC
LCP-MR, 11 rue P. et M. Curie
Paris 75005
France
penent@ccr.jussieu.fr

Richard Pratt
Physics
U. of Pittsburgh
Pittsburgh, PA 15260
USA
rpratt@pitt.edu

Georg Pruemper
Institute of Multidisciplinary Research
Tohoku University
Sendai-shi, Katahira 2-1-1-, Aoba-ku
Sendai 980-857
Japan
pruemper@tagen.tohoku.ac.jp

Juan Martín Randazzo
Departamento de Física
Universidad Nacional del Sur
Av. Alem 1253
Bahía Blanca 8000
Argentina

Hasi Ray
Physics
IITR
Roorkee 247667
India
hasi_ray@yahoo.com

Axel Reinköster
Fritz-Haber-Institut der MPG
Faradayweg 4-6
Berlin 14195
Germany
reinkoc@fhi-berlin.mpg.de

Roberto Rivarola
Instituto de Física Rosario (UNR-CONICET)
Av. Pellegrini 250
Rosario 2000
Argentina
rivarola@fceia.unr.edu.ar

Karina Rodriguez
Fisica
Universidad nacional del Sur
Alem 1253
Bahía Blanca 8000
Argentine
krodri@criba.edu.ar

Maysara Salakhitdinova
Physics
Samarkand State University
University blvd.,15
Samarkand 703004
Uzbekistan
smaysara@yandex.ru

Horst Schmidt-Boecking
Institute of Nuclear Physics
University of Frankfurt
Frankfurt 60438
Germany
schmidtb@atom.uni-frankfurt.de

Michael Schulz
Physics
University of Missouri-Rolla
10931 Greenlefe Dr.
Rolla 65401
USA
schulz@umr.edu

Markus Schöffler
Institut für Kernphysik
J. W. Goethe Universität, Frankfurt
Max-von-Laue-Straße 1
Frankfurt 60438
Germany
schoeffler@atom.uni-frankfurt.de

Patricia Selles
LIXAM-CNRS
Bât. 350, Université Paris-Sud XI
Orsay 91405
France
patricia.selles@lixam.u-psud.fr

Omer Sise
Afyon Kocatepe University, Department of Physics
Afyon 3000
Turkey
omersise@aku.edu.tr

Daniel Slaughter
SoCPES
Flinders University
Adelaide 5001
AUSTRALIA
daniel.slaughter@flinders.edu.au

Winthrop Smith
Physics
University of Connecticut
156 Hillyndale Road
Storrs 6268
United States
winthrop.smith@uconn.edu

Emma Sokell
Experimental Physics
University College Dublin
Belfield
Dublin D4
Ireland
emma.sokell@ucd.ie

Rajesh Srivastava
I. I. T. Roorkee
Dept. of Physics
Roorkee 247667
India
rajsrfph@iitr.ernet.in

Anthony F. Starace
Physics and Astronomy
University of Nebraska
116 Brace Laboratory
Lincoln NE 68588-0111
U.S.A.
astarace1@unl.edu

Al Stauffer
Physics & Astronomy
York University
4700 Keele Street
Toronto M3J 1P3
Canada
stauffer@yorku.ca

Andris Stelbovics
Murdoch University
South Street
Murdoch WA 6150
Australia
a.stelbovics@murdoch.edu.au

Swaraj Tayal
Department of Physics, Clark Atlanta University
Atlanta, Georgia
USA
stayal@cau.edu

Aaron Temkin
NASA, Goddard
Greenbelt,MD
USA
Aaron.Temkin-1@nasa.gov

Marco Tomaselli
AP
GSI
Planckstr.1
Darmstadt 64291
Germany
m.tomaselli@gsi.de

Joachim Ullrich
Max-Planck-Institut fuer Kernphysik
Heidelberg 69029
Germany
j.ullrich@mpi-hd.mpg.de

Pavel Vinitsky
Moscow State University, Physical Department
Vorob'ovy Gory
Moscow 119992
Russia
vinitsky@srd.sinp.msu.ru

Noboru Watanabe
Tohoku University
Katahira 2-1-1 Aoba-ku
Sendai 980-8577
Japan
noboru@tagen.tohoku.ac.jp

Jim Williams
Physics
University Of Western Australia
5 Stirling Hy
Perth 6009
Australia
jfw@physics.uwa.edu.au

Andrew J. Yencha
Chemistry
State University of New York at Albany
1400 Washington Ave.
Albany 12222
U.S.A.
yencha@albany.edu

Andre H. Zimnol
Physikalisches Institut
WWU Muenster
Wilhelm-Klemm-Str. 10
Muenster 48149
Germany
zimnol@uni-muenster.de

AUTHOR INDEX

A

Ackerman, G. D., 120
Aguilar, A., 120
Ancarani, L. U., 1
Andric, L., 126
Aoto, T., 126
Außendorf, G., 66
Avaldi, L., 138

B

Bartlett, P. L., 36
Bartschat, K., 203
Bellm, S., 60
Berakdar, J., 197
Berrah, N., 120
Bilodeau, R. C., 120
Bolognesi, P., 138
Bowles, C., 167
Bozek, J. D., 54, 108, 120
Bray, I., 36
Brunger, M. J., 173

C

Carravetta, V., 144
Catoire, F., 90
Childers, J. G., 24
Chuluunbaatar, O., 96
Colgan, J., 48
Coreno, M., 138
Crowe, A., 78
Cvejanović, D., 185

D

Dal Cappello, C., 1, 90
Daniell, M. L., 114
De Fanis, A., 138
Dimopoulou, C., 12
Dorn, A., 12
Duguet, A., 90
Dumitriu, I., 120

Dunn, A., 173
Dürr, M., 12

E

Eland, J. H. D., 126
Ellis, M., 173
Euripides, P., 173

F

Feifel, R., 126
Feyer, V., 138
Fojón, O. A., 42
Foster, M., 7
Fursa, D. V., 36
Furst, J. E., 108

G

Gallup, G. A., 108
Gao, J., 72
Garibotti, C. R., 30
Gay, T. J., 108
Gibson, N. D., 120
Gilbert, B., 173
Green, A. S., 108
Grum-Grzhimailo, A. N., 191
Guinea, W. E., 114

H

Hanne, G. F., 66, 114
Hasan, A., 7
Hayes, P. A., 185
Hewitt, G., 173
Hikosaka, Y., 126
Horner, D. A., 84, 132
Hoshino, M., 144
Hussey, M., 179

I

Istomin, A. Y., 18
Ito, K., 126

J

Joulakian, B., 150
Jüttemann, F., 66

K

Kadyrov, A. S., 36
Katayanagi, H., 161
Khajuria, Y., 96
Khakoo, M. A., 24
Kheifets, A. S., 167
Kilcoyne, A. L. D., 54, 108
King, G., 179
Kirschner, J., 197
Kitajima, M., 144
Kou, J., 161
Kouzakov, K. A., 96, 102, 197
Kubozono, Y., 161

L

Lablanquie, P., 126
Lahmam-Bennani, A., 90
Lahmidi, N., 150
Lawrance, W. D., 173
Liu, X. J., 144
Lohmann, B., 60, 114
Lower, J., 60

M

MacGillivray, W. R., 114, 179
Machacek, J. R., 108
Madison, D. H., 7, 60, 72
Manakov, N.L., 18
Martín, F., 132
Maseberg, J. W., 108
McCurdy, C. W., 84, 132
McLaughlin, K. W., 108
Meremianin, A. V., 18

Meyer, M., 191
Mitsuke, K., 161
Montagnese, T., 1
Mori, T., 161
Muramatsu, Y., 144
Murray, A., 179

N

Najjari, B., 12
Napier, S., 185
Nixon, K. L., 173
Northeast, R., 173

O

Odagiri, T., 156
Otranto, S., 30

P

Palaudoux, J., 126
Panajotovic, R., 60
Peacher, J. L., 7, 72
Penent, F., 126
Pindzola, M. S., 48
Popov, Y. V., 96, 102
Pravica, L., 185
Prideaux, A., 60
Prosperi, T., 138
Prümper, G., 144

R

Rescigno, T. N., 84, 132
Rivarola, R. D., 42
Robicheaux, F., 48
Rosenberry, M. A., 108
Rude, B. S., 54

S

Schulz, M., 7
Schumann, F. O., 197
Serov, V., 150

Slaughter, D. S., 173
Srivastava, R., 185
Starace, A. F., 18
Stauffer, A., 185
Stegen, Z., 60
Stelbovics, A. T., 36
Stevenson, M. A., 114
Stia, C. R., 42

T

Takahashi, M., 96, 102
Tamenori, Y., 144
Tanaka, H., 144
Turchini, S., 138
Turri, G., 120

U

Udagawa, Y., 96
Ueda, K., 144
Ullrich, J., 12

V

Vanroose, W., 132
Vinitsky, P. S., 96, 102
Vos, M., 167

W

Walter, C. W., 120
Watanabe, N., 96, 102
Weigold, E., 60
Went, M. R., 114, 167
Whelan, C. T., 60
Williams, J. F., 185

Y

Yu, D. H., 185

Z

Zema, N., 138